Lecture Notes in Computer Science 5029

Commenced Publication in 1973
Founding and Former Series Editors:
Gerhard Goos, Juris Hartmanis, and Jan van Leeuwen

Paolo Ferragina   Gad M. Landau (Eds.)

Combinatorial Pattern Matching

19th Annual Symposium, CPM 2008
Pisa, Italy, June 18-20, 2008
Proceedings

 Springer

Volume Editors

Paolo Ferragina
University of Pisa, Dipartimento di Informatica
Largo B. Pontecorvo 3, 56127 Pisa, Italy
E-mail: ferragina@di.unipi.it

Gad M. Landau
University of Haifa, Department of Computer Science
Mount Carmel, Haifa 31905, Israel
E-mail: landau@cs.haifa.ac.il

Library of Congress Control Number: 2008928843

CR Subject Classification (1998): F.2.2, I.5.4, I.5.0, H.3.3, J.3, E.4, G.2.1, E.1

LNCS Sublibrary: SL 1 – Theoretical Computer Science and General Issues

ISSN	0302-9743
ISBN	3-540-69066-2 Springer Berlin Heidelberg New York
ISBN	978-3-540-69066-5 Springer Berlin Heidelberg New York

Springer is a part of Springer Science+Business Media

springer.com

© Springer-Verlag Berlin Heidelberg 2008

Typesetting: Camera-ready by author, data conversion by Scientific Publishing Services, Chennai, India
Printed on acid-free paper   SPIN: 12281893   06/3180   5 4 3 2 1 0

Preface

The papers contained in this volume were presented at the 19th Annual Symposium on Combinatorial Pattern Matching (CPM 2008) held at the University of Pisa, Italy, June 18–20, 2008.

All the papers presented at the conference are original research contributions on computational pattern matching and analysis. They were selected from 78 submissions. Each submission was reviewed by at least three reviewers. The committee decided to accept 25 papers. The programme also includes three invited talks by Daniel M. Gusfield from the University of California, Davis, USA, J. Ian Munro from the University of Waterloo, Canada, and Prabhakar Raghavan from Yahoo! Research, USA.

The objective of the annual CPM meetings is to provide an international forum for research in combinatorial pattern matching and related applications. It addresses issues of searching and matching strings and more complicated patterns such as trees, regular expressions, graphs, point sets, and arrays. The goal is to derive non-trivial combinatorial properties of such structures and to exploit these properties in order to either achieve superior performance for the corresponding computational problems or pinpoint conditions under which searches cannot be performed efficiently. The meeting also deals with problems in computational biology, data compression, data mining, coding, information retrieval, natural language processing and pattern recognition.

The Annual Symposium on Combinatorial Pattern Matching started in 1990, and has since taken place every year. Previous CPM meetings were held in Paris, London (UK), Tucson, Padova, Asilomar, Helsinki, Laguna Beach, Aarhus, Piscataway, Warwick, Montreal, Jerusalem, Fukuoka, Morelia, Istanbul, Jeju Island, Barcelona and London (Canada). Starting from the third meeting, proceedings have been published in the LNCS series, volumes 644, 684, 807, 937, 1075, 1264, 1448, 1645, 1848, 2089, 2373, 2676, 3109, 3537, 4009, and 4580. Selected papers from the first meeting appeared in volume 92 of *Theoretical Computer Science*, from the 11th meeting in volume 2 of *Journal of Discrete Algorithms*, from the 12th meeting in volume 146 of *Discrete Applied Mathematics*, from the 14th meeting in volume 3 of *Journal of Discrete Algorithms*, from the 15th meeting in volume 368 of *Theoretical Computer Science* and from the 16th meeting in volume 5 of *Journal of Discrete Algorithms*.

The whole submission and review process was carried out with the help of the EasyChair system. The conference was sponsored by the University of Pisa and by Yahoo! Research. Special thanks are due to the members of the Program Committee who worked very hard to ensure the timely review of all the submitted manuscripts, and participated in a stimulating discussion that allowed all of us to choose the best papers to be presented at the conference.

April 2008

Paolo Ferragina
Gad M. Landau

Organization

Program Committee

Srinivasa Aluru	Iowa State University
Lars Arge	University of Aarhus
Ricardo Baeza-Yates	Yahoo! Research
Edgar Chavez	Universidad Michoacana
Ken Church	Microsoft Research
Raphael Clifford	University of Bristol
Paolo Ferragina (Co-chair)	University of Pisa
Roberto Grossi	University of Pisa
Tao Jiang	University of California, Riverside
Dong K. Kim	Hanyang University
Gregory Kucherov	CNRS, Lille
Gad M. Landau (Co-chair)	Haifa University and Polytechnic University
Stefano Lonardi	University of California, Riverside
Bin Ma	University of Western Ontario
Veli Makinen	University of Helsinki
Yossi Matias	Google, Tel Aviv
Alistair Moffat	University of Melbourne
Ian Munro	University of Waterloo
Gene Myers	Howard Hughes Medical Institute
Wojciech Plandowski	Uniwersytet Warszawski
Rajeev Raman	University of Leicester
Kunihiko Sadakane	Kyushu University
Peter Sanders	Karlsruhe University
Marinella Sciortino	University of Palermo
Dafna Sheinwald	IBM, Haifa
Stephane Vialette	University of Marne-la-Vallee
Frances Yao	City University of Hong Kong
Kaizhong Zhang	University of Western Ontario

Steering Committee

Alberto Apostolico	University of Padova and Georgia Tech
Maxime Crochemore	King's College
Zvi Galil	Tel Aviv University

Organizing Committee

Anna Bernasconi	Roberto Grossi
Paolo Ferragina	Danny Hermelin

Fabrizio Luccio
Igor Nitto
Linda Pagli
Nadia Pisanti

Giuseppe Prencipe
Romano Venturini
Oren Weimann

External Referees

Spyros Angelopoulos
Vo Anh
Frederique Bassino
Philip Bille
Marek Biskup
Guillaume Blin
Gerth Brodal
Marie-Pierre Bal
Arturo Carpi
Giuseppa Castiglione
Eric Chen
Shihyen Chen
Joselito Chua
Hagai Cohen
Shane Culpepper
Robert Dabrowski
Reza Dorri-Giv
Arash Farzan
Guillaume Fertin
Johannes Fischer
Gianni Franceschini
Kimmo Fredriksson
Travis Gagie
Mathieu Giraud
Massimiliano Goldwurm
Mark Greve
Franciszek Grzegorek
Sylvie Hamel
Elena Harris
Aram Harrow
Tzvika Hartman
Meng He
Danny Hermelin
Benjamin Jackson
Jesper Jansson
Anantharaman Kalyanaraman
Orgad Keller
Pang Ko

Roman Kolpakov
Robert Krauthgamer
Inbok Lee
Avivit Levy
Weiming Li
Xiaowen Liu
Jingping Liu
Giovanni Manzini
Jrme Monnot
Ashley Montanaro
Hiroyoshi Morita
Gabriel Moruz
Joong Chae Na
Pat Nicholson
Francois Nicolas
Igor Nitto
Igor Nor
Heejin Park
Giulio Pavesi
Christian Storm Pedersen
Gemma Piella
Natasa Przulj
Simon Puglisi
Mathieu Raffinot
Srinivasa Rao
Antonio Restivo
Romeo Rizzi
Wojciech Rytter
Benjamin Sach
Cenk Sahinalp
Abhinav Sarje
Srinivasa Rao Satti
Roded Sharan
Johannes Singler
Ranjan Sinha
Jouni Siren
Vladimir Vacic
Rossano Venturini

Niko Valimaki
Tomasz Wale
Lusheng Wang
Bob Wang
William Webber
Zhan Wu

Xiao Yang
Qiaofeng Yang
Jie Zheng
Jaroslaw Zola
Roelof van Zwol
Yonghui Wu

Sponsoring Institutions

Yahoo! Research
University of Pisa

Table of Contents

Invited Talks

Contributed Papers

ReCombinatorics: Combinatorial Algorithms for Studying the History of Recombination in Populations

Dan Gusfield

Department of Computer Science, University of California,
Davis, CA 95616, USA
gusfield@cs.ucdavis.edu

Abstract. The work discussed in this talk falls into the emerging area of Population Genomics. I will first introduce the area and then talk about specific problems and combinatorial algorithms involved in the inference of recombination from population data.

A phylogenetic network (or Ancestral Recombination Graph) is a generalization of a tree, allowing structural properties that are not tree-like. With the growth of genomic and population data (coming for example from the HAPMAP project) much of which does not fit ideal tree models, and the increasing appreciation of the genomic role of such phenomena as recombination (crossing-over and gene-conversion), recurrent and back mutation, horizontal gene transfer, and mobile genetic elements, there is greater need to understand the algorithmics and combinatorics of phylogenetic networks.

In this talk I will survey a range of our recent algorithmic, mathematical and practical results on phylogenetic networks with recombination and show applications of these results to several issues in Population Genomics.

Various parts of this work are joint work with Satish Eddhu, Chuck Langley, Dean Hickerson, Yun S. Song, Yufeng Wu, V. Bansal, V. Bafna and Z. Ding. All the papers and associated software can be accessed at http://wwwcsif.cs.ucdavis.edu/~gusfield/

References

1. Gusfield, D.: Optimal, efficient reconstruction of root-unknown phylogenetic networks with constrained recombination. J. Computer and Systems Sciences 70, 381–398 (2005)
2. Gusfield, D., Bansal, V.: A fundamental decomposition theory for phylogenetic networks and incompatible characters. In: Miyano, S., Mesirov, J., Kasif, S., Istrail, S., Pevzner, P.A., Waterman, M. (eds.) RECOMB 2005. LNCS (LNBI), vol. 3500, pp. 217–232. Springer, Heidelberg (2005)
3. Gusfield, D., Bansal, V., Bafna, V., Song, Y.S.: a decomposition theory for phylogenetic networks and incompatible characters. J. Computational Biology (December 2007)

P. Ferragina and G. Landau (Eds.): CPM 2008, LNCS 5029, pp. 1–2, 2008.

4. Gusfield, D., Eddhu, S., Langley, C.: Optimal efficient Reconstruction of phylogenetic networks with constrained recombination. Journal of Bioinformatics and Computational Biology 2(1), 173–213 (2004)
5. Gusfield, D., Eddhu, S., Langley, C.: The fine structure of galls in phylogenetic networks. Inf. J. on Computing, Special issue on Computational Biology 16(4), 459–469 (2004)
6. Gusfield, D., Hickerson, D., Eddhu, S.: A fundamental, efficiently computed lower bound on the number of recombinations needed in a phylogenetic history. Discrete Applied Math Special issue on Computational Biology (2007)
7. Song, Y., Gusfield, D., Ding, Z., Langley, C., Wu, Y.: Algorithms to distinguish the role of gene-conversion from single-crossover recombination in the derivation of SNP sequences in populations. In: Apostolico, A., Guerra, C., Istrail, S., Pevzner, P.A., Waterman, M. (eds.) RECOMB 2006. LNCS (LNBI), vol. 3909, pp. 231–245. Springer, Heidelberg (2006)
8. Song, Y., Wu, Y., Gusfield, D.: Efficient computation of close lower and upper bounds on the minimum number of needed recombinations in the evolution of biological sequences. In: Bioinformatics, Proceedings of the ISMB 2005 Conference, vol. 21, pp. 413–422 (2005)
9. Wu, Y.: Association mapping of complex diseases with ancestral recombination graphs: models and efficient algorithms. In: Speed, T., Huang, H. (eds.) RECOMB 2007. LNCS (LNBI), vol. 4453, pp. 488–502. Springer, Heidelberg (2007)
10. Wu, Y., Gusfield, D.: A new recombination lower bound and the minimum perfect phylogenetic forest problem. In: Proceedings of the 13th Annual International Conference on Combinatorics and Computing, pp. 16–26 (2007)
11. Wu, Y., Gusfield, D.: Improved algorithms for inferring the minimum mosaic of a set of recombinants. In: Ma, B., Zhang, K. (eds.) CPM 2007. LNCS, vol. 4580, pp. 150–161. Springer, Heidelberg (2007)
12. Wu, Y., Gusfield, D.: Efficient computation of minimum recombination with genotypes (not haplotypes). In: Proceedings of The Computational Systems Biology Conference (2006)

Lower Bounds for Succinct Data Structures

J. Ian Munro

Cheriton School of Computer Science, University of Waterloo,
Waterloo, Ontario N2L 3G1, Canada
imunro@uwaterloo.ca

Abstract. Indexing text files with methods such as suffix trees and suffix arrays permits extremely fast search for substrings. Unfortunately the space cost of these can dominate that of the raw data. For example, the naive implementation of a suffix tree on genetic information could take 80 times as much space as the raw data. Succinct data structures offer a technique by which the extra space of the indexing can be kept, at least in principle, to a "little oh" with respect to the raw data. This begs the question of how much extra space is necessary to support fast substring searches of other queries such as the rank/select problem or representing a permutation so that both the forward permutation and its inverse can be determined quickly. We survey some lower bounds on this type of problem, most notably the work of Demaine and López-Ortiz and of Golynski.

P. Ferragina and G. Landau (Eds.): CPM 2008, LNCS 5029, p. 3, 2008.
© Springer-Verlag Berlin Heidelberg 2008

The Changing Face of Web Search
(Abstract)

Prabhakar Raghavan

Yahoo! Research

Abstract. Web search has come to dominate our consciousness as a convenience we take for granted, as a medium for connecting advertisers and buyers, and as a fast-growing revenue source for the companies that provide this service. Following a brief overview of the state of the art and how we got there, this talk covers a spectrum of technical challenges arising in web search - ranging from spam detection to auction mechanisms.

P. Ferragina and G. Landau (Eds.): CPM 2008, LNCS 5029, p. 4, 2008.

Two-Dimensional Pattern Matching with Combined Scaling and Rotation[*]

Christian Hundt[1],[**] and Maciej Liśkiewicz[2],[***]

[1] Institut für Informatik, Universität Rostock, Germany
christian.hundt@uni-rostock.de
[2] Institut für Theoretische Informatik, Universität zu Lübeck, Germany
liskiewi@tcs.uni-luebeck.de

Abstract. The problem of two-dimensional pattern matching invariant under a given class of admissible transformations $\mathcal{F}$ is to find in text T matches of transformed versions $f(P)$ of the pattern P, for all f in $\mathcal{F}$. In this paper, pattern matching invariant under compositions of real scaling and rotation are investigated. We give a new discretization technique for this class of transformations and prove sharp lower and upper bounds on the number of different possibilities to transform a pattern in this way. Subsequently, we present the first efficient pattern matching algorithm invariant under compositions of scaling and rotation. The algorithm works in time $O(m^2 n^6)$ for patterns of size m^2 and texts of size n^2. Our method can also be applied to the image matching problem, the well known issue in the image processing research.

Keywords: combinatorial pattern matching, digital image matching, discrete rotations and scalings, discrete algorithms.

1 Introduction

The research in two-dimensional pattern matching (2D-PM, for short) in the combinatorial setting is strongly motivated by image retrieval with applications in such areas as optical character recognition, medical imaging, video compression, computer vision, searching aerial photographs, etc. Thus, in combinatorial pattern matching two-dimensional patterns P and texts T typically model digital images. Recently, many efforts have been made in the study of algorithms finding beside all matches of P in T also the locations of *transformed* versions $f(P)$ of P, given that f belongs to a specified class of admissible transformations $\mathcal{F}$. We call this kind of pattern matching $\mathcal{F}$-*invariant*, for short. Naturally, the combinatorial definitions of pattern transformations (like e.g. scaling or rotation of P) model real digital image transformations.

---

[*] Supported by DFG research grant RE 672/5-1.
[**] The work on this paper was done during the stay of the first author at the University of Lübeck.
[***] On leave from Instytut Informatyki, Uniwersytet Wrocławski, Poland.

P. Ferragina and G. Landau (Eds.): CPM 2008, LNCS 5029, pp. 5–17, 2008.

Motivated by the applications in image processing we assume that both P and T are distorted by noise, e.g., due to the digitalization process. In this case it is unlikely to find exact occurrences of P or $f(P)$ in T any more. An important extension of the (exact) $\mathcal{F}$-invariant 2D-PM involved with such a setting is called *robustly $\mathcal{F}$-invariant* pattern matching. Now, the task is to find a match M in text T and a transformation f such that the distance between M and $f(P)$ is minimized under a given distance measure (in the literature it is also called *approximate matching*).

Since robustly $\mathcal{F}$-invariant 2D-PM models matching in applications with digital images it exhibit a natural and close relation to the so called *image matching problem* (IM, for short) the well known issue in the image processing research. The problem consists in finding for two given digital images A and B an admissible transformation f in $\mathcal{F}$ that changes A closest to B.

In this paper we consider both matching problems (2D-PM and IM) for the class $\mathcal{F}_{sr}$ comprising all compositions of scaling and rotation. This basic subset of linear transformations applied to 2D-PM has a vast area of applications in various image processing settings of highly practical importance. In computer vision, e.g., one searches digital camera images for rotated and scaled versions of objects with known shape, like latin letters [21]. In video compression, algorithms for 2D-PM invariant under $\mathcal{F}_{sr}$ can be used to compress sequences of frames efficiently (see e.g. [27]). In medical imaging (see e.g. [9,25]) IM is applied to images of one object taken in different times, from different perspectives or using different medical image devices. In this area $\mathcal{F}_{sr}$ plays an important role because it simulates distortions which arise from small changes in viewing point and certain kinds of patient movement.

Despite the high practical relevance, to our knowledge, no efficient matching algorithms are known both for robustly $\mathcal{F}_{sr}$-invariant 2D-PM and for IM. For 2D-PM no efficient algorithms are known even for the non-robust (exact) case. On the other hand the special cases of the matching problems under solely scalings or solely rotations have been a subject of intensive study (see e.g. [8,5,3,4,6] for scalings, resp. [16,15,2,14,7] for rotations) yielding significant progress in the construction of efficient algorithms. However, the matching problems under *compositions* of scaling and rotation seem intrinsically harder. The straightforward approach of combining two algorithms, the first one for matching with scalings and the second for matching with rotations, does not work. This follows from the fact that neither the set of all patterns which are obtained by scaling all rotated patterns nor the set obtained by rotating all scaled patterns does coincides with the set of patterns obtained by combined scaling and rotation. In contrast to the continuous compositions of scaling and rotation, the compositions on patterns are neither commutative nor transitive.

In this paper we give the first efficient algorithm for robustly invariant 2D-PM under $\mathcal{F}_{sr}$. Since this is the most general matching problem the algorithm can also be applied for the simpler exact invariant 2D-PM as well as for IM under $\mathcal{F}_{sr}$. Basically, our algorithm solves the problem as follows: firstly it constructs the data structure for $\mathcal{D}(P, \mathcal{F}_{sr})$, the set of transformed patterns $f(P)$, for all

$f \in \mathcal{F}_{sr}$; secondly it computes the optimal image matching between P and all relevant subtexts M of T. To tackle IM under $\mathcal{F}_{sr}$ in turn the algorithm enumerates incrementally all elements in $\mathcal{D}(P, \mathcal{F}_{sr})$. The main achievement of this paper is the development of a new method for the discretization of all real compositions of scaling and rotation which allows to get an efficient data structure for the set $\mathcal{D}(P, \mathcal{F}_{sr})$. Moreover, sharp lower and upper bounds on the cardinality of this set are obtained by exploring structural properties of the discrete space.

Through the rest of this section we give a short overview on previous work and an informal discussion of our results and techniques. Section 2 contains formal definitions and preliminaries. In Section 3 we show a relation between $\mathcal{D}(P, \mathcal{F}_{sr})$ and a certain line arrangement. In Section 4 we provide the new matching algorithm and its analysis. Because of space limitations we omitted all proofs.

1.1 Previous Work

Recently, the studies in combinatorial pattern matching have been concentrated on 2D-PM invariant under solely scalings or solely rotations. After a series of improving results, the best known algorithm for 2D-PM with scalings is given by Amir and Chencinski [6]. It solves the problem in $O(mn^2)$ time, where the size of the text is n^2 and the size of the pattern is m^2. For 2D-PM with rotations, the best known algorithm is due to Amir et al. [7] solving the problem in time $O(m^2n^2)$. The above algorithms find all exact matches M of transformed P in T, i.e., M is identical to some scaled pattern $s(P)$ respectively to some rotated pattern $r(P)$. The best known algorithm for robustly invariant pattern matching with rotations is due to Frederikson et al. [15] and it works in time $O(m^3n^2)$.

Apart from algorithmic achievements, improved techniques for the analysis of 2D-PM have been developed in the last decade. Noticeably, the work [2] was the first pattern matching paper that heavily builds on the use of combinatorial geometry. With our paper we continue the research in this direction exploiting geometrical properties to design efficient and practical algorithms.

In contrast to combinatorial pattern matching the image processing community uses rather continuous analysis instead of combinatorial approaches when studying the IM problem. For an overview and discussion of selected issues in these field we refer to [10,9,21,25,24,2,18] and the references therein.

Another technique for IM is to use feature based approaches (see e.g. [1,23]). One extracts salient features (points, lines, regions etc. in the real plane) from images A and B and subsequently, one tries to find a transformation f which transforms the geometrical objects of A closest to those of B. But, this approach relies heavily on the quality of feature extraction and feature matching, two highly non-trivial tasks. Feature matching, e.g., remains difficult even for points (called also geometric point set matching), the well studied problem which consists in finding for two given point sets P and Q and some admissible space $\mathcal{F}$ a function $f \in \mathcal{F}$ that transforms P closest to Q (see [19] for a survey and [22,26] for some related problems). In fact, the known algorithms for this problem give only approximate solutions and particularly they do not guarantee to find the global optimum, even for such a simple class of transformations as compositions of rotation and translation [20]. Interestingly, in [20] Indyk et al. give a discretization technique to reduce the geometric point matching problem to a combinatorial pattern matching related to that one considered in this paper.

1.2   Our Contributions

In this paper we present a new discretization technique for the space of all transformed patterns under compositions of scaling and rotation which enables efficient incremental enumeration for the elements in $\mathcal{D}(P, \mathcal{F}_{\mathtt{sr}})$. The proposed method works in linear time, with respect to the cardinality $|\mathcal{D}(P, \mathcal{F}_{\mathtt{sr}})|$, and leads to linear time search to find for a given subtext of T the closest element in $\mathcal{D}(P, \mathcal{F}_{\mathtt{sr}})$. We obtain that in worst-case $|\mathcal{D}(P, \mathcal{F}_{\mathtt{sr}})|$ is in $\Omega\left(\frac{m^2 n^3}{\ln m}\right) \cap O(m^2 n^4)$ for patterns P of size m^2 and texts T of size[1] n^2. We use these results to provide a fast $O(m^2 n^6)$ time robustly invariant 2D-PM algorithm. The presented algorithm uses only integer arithmetic which means that no numerical problems occur due to the use of floating point arithmetic.

In our setting, each (real valued) transformation f in $\mathcal{F}_{\mathtt{sr}}$ is a composition of rotation with an angle ϕ and scaling with a factor s. The values ϕ, s are represented then by an appropriate point in the parameter space $\mathbb{R}^2$. To obtain our results, we introduce a discretization technique to partition the parameter space into a finite number of subspaces $\varphi_1, \varphi_2, \ldots, \varphi_t$ such that for any subspace φ_i and for all points $(p, q), (p', q') \in \varphi_i$ it is true that the transformation f corresponding to (p, q) gives the same pattern as the transformation f' corresponding to (p', q'). According to the practice in combinatorial geometry the subspaces $\varphi_1, \varphi_2, \ldots, \varphi_t$ are called faces.

To obtain the space partition we define a certain set of lines $\mathcal{H}$ which cut the parameter space $\mathbb{R}^2$ into faces $\varphi_1, \varphi_2, \ldots, \varphi_t$. Then, to enumerate the patterns in $\mathcal{D}(P, \mathcal{F}_{\mathtt{sr}})$ we simply search the faces of the line arrangement defined by $\mathcal{H}$. For each face visited during the search we choose one contained point and identify the corresponding transformation f. By the equivalence between the points in one face, we can compute with f the pattern $f(P)$, which is associated to all points of the face. Finally, we get all patterns in $\mathcal{D}(P, \mathcal{F}_{\mathtt{sr}})$ by this procedure. Moreover, we choose an enumeration order on $\mathcal{D}(P, \mathcal{F}_{\mathtt{sr}})$ implied by the geometrical incidence between the corresponding faces. This allows that patterns enumerated successively differ only in few pixels. Such a geometric approach is new both in combinatorial pattern matching and image matching[2].

---

[1] More precisely, in this paper we consider patterns of size $(2m + 1) \times (2m + 1)$ and texts of size $(2n + 1) \times (2n + 1)$.

[2] In [17] Hagedoorn uses a similar idea of creating an appropriate division of the parameter space of the underlying transformation class. However, he applies that technique for a *geometrical pattern matching* (called *Spacial Pattern Matching* in the thesis) and not the combinatorial setting considered in this paper.

In our algorithm we find the closest match between a pattern $P' \in \mathcal{D}(P, \mathcal{F}_{sr})$ with a subtext M of T by solving image matching parallel for all possible translations of the center of P' to a location (t_1, t_2) of T. Each IM in turn is solved by the use of the above described enumeration technique. Consequently, our algorithm is general enough for both robustly 2D-PM invariant under $\mathcal{F}_{sr}$ as well as IM under $\mathcal{F}_{sr}$. Furthermore, the approach is easily applied to other kinds of transformations, in particular to solely scalings or solely rotations.

2 Preliminaries

In this paper the pattern P as well as the text T are two-dimensional arrays of pixels, i.e, of unit squares covering a certain area of the real plane $\mathbb{R}^2$. In P the pixels are indexed over the set $\mathcal{M} = \{(i, j) \mid -m \leq i, j \leq m\}$ and in T over $\mathcal{N} = \{(i, j) \mid -n \leq i, j \leq n\}$. We call $\mathcal{M}$ the support of P and $\mathcal{N}$ the support of T. The pixel with index (i, j) has its geometric center point at coordinates (i, j). Each pixel (i, j) has a color $P\langle i, j \rangle$ ($T\langle i, j \rangle$ resp.) from a finite set $\Sigma = \{0, 1, \ldots, \sigma\}$ of color values. To simplify the dealing with borders we let $P\langle i, j \rangle = \bot$ if $(i, j) \notin \mathcal{M}$ and $T\langle i, j \rangle = \bot$ if $(i, j) \notin \mathcal{N}$, where $\bot$ is a special color marking the exterior of P and T. For a given pattern P, text T and pixel index $(t_1, t_2) \in \mathcal{N}$ the distortion $\Delta_{t_1, t_2}(P, T)$ between P and T at (t_1, t_2) is measured by $\sum \delta(P\langle i - t_1, j - t_2 \rangle, T\langle i, j \rangle)$ where $\delta(a, b)$ is a function charging mismatches, for example,

$$\delta(a, b) = \begin{cases} 0, & \text{if } a = \bot \text{ or } b = \bot \\ |a - b|, & \text{otherwise.} \end{cases}$$

Throughout this paper transformations are injective functions $f : \mathbb{R}^2 \to \mathbb{R}^2$. Applying a transformation $f : \mathbb{R}^2 \to \mathbb{R}^2$ to the pattern P we get the two-dimensional array of pixels $f(P)$ which has support $\mathcal{N}_2 = \{(i, j) \mid -2n \leq i, j \leq 2n\}$. Define for $g = f^{-1}$ the mapping $\gamma_g : \mathcal{N}_2 \to \mathbb{Z}^2$ which determines for any pixel (i, j) in $f(P)$ the corresponding pixel (i', j') in P. We define $\gamma_g(i, j) = [g(i, j)]$, where $[(x, y)] := ([x], [y])$ denotes rounding all components of a vector $(x, y) \in \mathbb{R}^2$. The color value of pixel (i, j) in $f(P)$ is defined as the color value of the pixel $(i', j') = \gamma_g(i, j)$ in P. Hence, we choose the pixel which geometrically contains the point $f^{-1}(i, j)$ in its square area (for an example see Fig. 1). With this setting we model *nearest-neighbor* interpolation, commonly used in the image processing.

For any pattern P and all sets $\mathcal{F}$ we define the set $\mathcal{D}(P, \mathcal{F}) = \{f(P) \mid f \in \mathcal{F}\}$. Then we call the following optimization problem the two-dimensional pattern matching robustly invariant under $\mathcal{F}$:

Problem 1. For given pattern P with support $\mathcal{M}$ and text T with support $\mathcal{N}$, find in the set $\mathcal{D}(P, \mathcal{F})$ a pattern P' and a pixel index $(t_1, t_2) \in \mathcal{N}$ minimizing the distortion $\Delta_{t_1, t_2}(P', T)$.

If we interpret the pattern P and the text T as digital images A and B, then the image matching problem under $\mathcal{F}$ can be defined as the following restricted version of robustly 2D-PM:

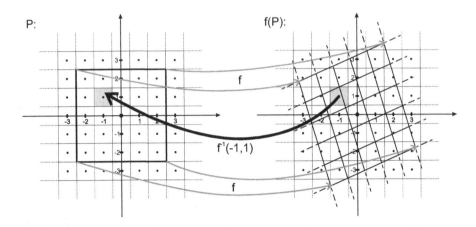

Fig. 1. Pattern P and the transformed $f(P)$. The color value $f(P)\langle-1,1\rangle$ is equal to the value $P\langle-1,1\rangle$ since $(-1,1)$ is the closest lattice point to $f^{-1}(-1,1)$.

Problem 2. For given images A and B with support $\mathcal{M}$ and $\mathcal{N}$ find in the set $\mathcal{D}(A,\mathcal{F})$ an image A' minimizing the distortion $\Delta_{0,0}(A',B)$.

For the analysis of complexity aspects we will apply the unit cost model for arithmetic operations. Therefore, we assume that mathematical basic operations can be done in constant time.

In this paper we are interested in $\mathcal{F}_{\text{sr}}$ the transformations combining scaling and rotation. Any transformation f in $\mathcal{F}_{\text{sr}}$ can be uniquely described by[3]

$$f(x,y) = \begin{pmatrix} s\cos\phi & s\sin\phi \\ -s\sin\phi & s\cos\phi \end{pmatrix} \cdot \begin{pmatrix} x \\ y \end{pmatrix}$$

for some $s,\phi \in \mathbb{R}$, with $s \neq 0$. The pattern $f(P)$ is defined by f but the pixel values of $f(P)$ can be computed by the inverse f^{-1}. Notice that according to our definition for any $f \in \mathcal{F}_{\text{sr}}$ it is true $f^{-1} \in \mathcal{F}_{\text{sr}}$. Hence, the considered class of transformations is closed under inversion.

All transformations in the set $\mathcal{F}_{\text{sr}}$ can be characterized by the two parameters $p = s\cos\phi$ and $q = s\sin\phi$. Hence, each such transformation f can be characterized by a vector $(p,q)^T$ in $\mathbb{R}^2$. Notice that $(0,0)^T$ is the only point in $\mathbb{R}^2$ which corresponds to a non-injective transformation and by this does not characterize a transformation in $\mathcal{F}_{\text{sr}}$. However, for convenience we will simply ignore this exception.

For our approach being connected to combinatorial geometry we need some further definitions: We denote by H a set of linear equations h of the form $h : c_1 p + c_2 q = c_3$. Let $(p,q) \in \mathbb{R}^2$. Then we define for each $h \in H$ the value $h(p,q) = c_1 p + c_2 q - c_3$. Each equation h describes a line $\ell = \{(p,q) \mid h(p,q) = 0\}$

---

[3] For the sake of completeness, the definition is very general. Particularly, it allows all real scaling factors $s \neq 0$. However, for many practical applications a reasonable restriction is to assume that $s \geq 1$ or that $1/c \leq s \leq c$, for a given constant $c > 1$. The matching algorithms presented in this paper are easily applicable for such cases.

in $\mathbb{R}^2$. Notice the difference: h is a algebraic expression whereas ℓ is a subspace of $\mathbb{R}^2$. Denote by $\mathcal{H}$ the set of all lines defined by the equations in H. Now define for all $h \in H$ the following additional subspaces of $\mathbb{R}^2$:

$$\ell^+ = \{(p, q) \mid h(p, q) > 0\}, \quad \text{and} \quad \ell^- = \{(p, q) \mid h(p, q) < 0\}.$$

For a finite set of equations $H = \{h_1, \ldots, h_t\}$ consider the following partition of $\mathbb{R}^2$ into subspaces:

$$\mathcal{A}(H) = \left\{ \varphi \subseteq \mathbb{R}^2 \;\middle|\; \varphi = \bigcap_{w=1}^{t} \ell_w^{s_w} \text{ for some } s_1, \ldots, s_t \in \{+, -, 0\} \right\},$$

where ℓ_w is the line corresponding to h_w and ℓ_w^0 denotes just ℓ_w. In literature the set $\mathcal{A}(H)$ is called the line arrangement given by the lines $\mathcal{H}$. See [11] or [13] for detailed information on line arrangements.

We call the elements of $\mathcal{A}(H)$ faces. A face is called a d-face if its dimension is d, for $d \in \{0, 1, 2\}$. Thus, a 0-face is a point, a 1-face is a line, half-line or line segment and 2-face is a convex region on the plane given by the intersection of a finite number of half-planes. A face φ' is a subface of another face φ if the dimension of φ' is one less than of φ and φ' is contained in the boundary of φ. We also say that φ and φ' are incident and that φ is a superface of φ'.

The incidence graph $\mathcal{I}(H)$ of $\mathcal{A}(H)$ contains a node $v(\varphi)$ for each face φ and $v(\varphi)$ and $v(\varphi')$ are connected by an edge if the faces φ and φ' are incident. The incidence graph is described in detail in [12] (see also [11]).

3   Exploring the Set $\mathcal{D}(P, \mathcal{F}_{sr})$

In this section we will present the structure of the set $\mathcal{D}(P, \mathcal{F}_{sr})$ and show how to estimate the worst case number of contained patterns. We now define the set $H_{sr,m,n}$ of equations. For all $(i, j) \in \mathcal{N}_2$ and $k \in \{-m - 1, \ldots, m + 1\}$ let

$$h_{ijk} : ip + jq + 0.5 - k = 0$$

be equations in $H_{sr,m,n}$. Each of the equations $h_{ijk} \in H_{sr,m,n}$ describes a line ℓ in $\mathbb{R}^2$ which partitions the parameter space into three parts ℓ^+, ℓ and ℓ^-. The transformed patterns $f(P)$ differ to each other with respect to the color value of pixel (i, j), depending on which of the three subspaces the point representing $g = f^{-1}$ is situated in. The following lemma states this relationship:

Lemma 1. *Let $h_{ijk} \in H_{sr,m,n}$ and let ℓ_x be the line described by h_{ijk}. Furthermore, let $f \in \mathcal{F}_{sr}$ be a transformation, $g = f^{-1}$ its inverse and r the point representing g. Consider $\gamma_g(i, j) = (i', j')$. Then $i' < k$ if $r \in \ell_x^-$ and $i' \geq k$ if $r \in \ell_x$ or $r \in \ell_x^+$. Analogously, if we consider $h_{j(-i)k} \in H_{sr,m,n}$, which describes the line ℓ_y, then $j' < k$ if $r \in \ell_y^-$ and $j' \geq k$ if $r \in \ell_y$ or $r \in \ell_y^+$.*

With the help of $H_{sr,m,n}$ we are now ready to provide the relation between the set $\mathcal{D}(P, \mathcal{F}_{sr})$ and the set $\mathcal{A}(H_{sr,m,n})$ of faces in $\mathbb{R}^2$.

Theorem 1. *For the pattern P there exists a surjective mapping*

$$\Gamma_{m,n} : \mathcal{A}(H_{sr,m,n}) \to \mathcal{D}(P, \mathcal{F}_{sr}).$$

By the theorem it suffices to estimate the number of faces in $\mathcal{A}(H_{sr,m,n})$ to get a bound on the cardinality of $\mathcal{D}(P, \mathcal{F}_{sr})$. Furthermore, the surjective mapping $\Gamma_{m,n}$ enables a simple method to enumerate the patterns in $\mathcal{D}(P, \mathcal{F}_{sr})$. One simply has to construct $\mathcal{A}(H_{sr,m,n})$, traverse its faces φ in an appropriate way and each time compute $\Gamma_{m,n}(\varphi)$ to obtain another transformed pattern of $\mathcal{D}(P, \mathcal{F}_{sr})$. Figure 2 shows the two-dimensional parameter space $\mathbb{R}^2$ partitioned by the lines $\mathcal{H}_{sr,2,2}$. The figure shows the set for patterns and texts of all in all 5^2 pixels. Although we have $m = n = 2$ the displayed structure seems already quite complex.

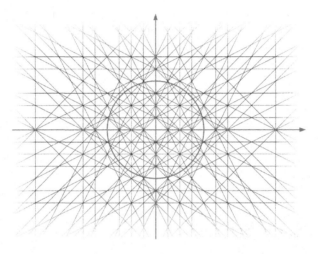

Fig. 2. The parameter space $\mathbb{R}^2$ partitioned by $\mathcal{H}_{sr,2,2}$ for patterns P and texts T of $5^2 = 25$ pixels. The points in $\mathbb{R}^2$ represents compositions of real scaling and rotation. For $p, p' \in \mathbb{R}^2$ representing f and f' the transformed patterns $f(P)$ and $f'(P)$ are equal if p and p' belong to the same face. The unit circle represents all compositions with scaling factor $s = 1$.

After we have exposed the geometrical structure behind $\mathcal{D}(P, \mathcal{F}_{sr})$ we want to obtain bounds on the cardinality of this set. An impression on this cardinality is given by $|\mathcal{A}(H_{sr,m,n})|$. Due to the surjective mapping $\Gamma_{m,n}$ we know that $\mathcal{D}(P, \mathcal{F}_{sr})$ cannot contain a larger number of patterns than the number of faces in $\mathcal{A}(H_{sr,m,n})$. However, since $\Gamma_{m,n}$ is not bijective it may happen for certain patterns and texts that $\mathcal{D}(P, \mathcal{F}_{sr})$ is significantly smaller than $\mathcal{A}(H_{sr,m,n})$. But it is still worth determining also a lower bound on $|\mathcal{A}(H_{sr,m,n})|$ because of two reasons: On the one hand, there may be patterns P for which the cardinality of $\mathcal{D}(P, \mathcal{F}_{sr})$ and $\mathcal{A}(H_{sr,m,n})$ are asymptotically equal and on the other hand it is the case that our algorithm always searches the whole set $\mathcal{A}(H_{sr,m,n})$ and thus, a lower bound on the size enables the estimate of a lower bound on the running time. In the next lemma we show narrow lower and upper bounds on $|\mathcal{A}(H_{sr,m,n})|$ using methods from combinatorial geometry:

Lemma 2. $|\mathcal{A}(H_{\mathtt{sr},m,n})| \in \Omega\left(\frac{m^2 n^3}{\ln m}\right) \cap O(m^2 n^4)$.

The first achievement of this section is a structural analysis of the set $\mathcal{D}(P, \mathcal{F}_{\mathtt{sr}})$ for given pattern P and text T. Furthermore, by the above lemma we have a good impression of the worst-case bounds on $|\mathcal{D}(P, \mathcal{F}_{\mathtt{sr}})|$.

4   The 2D-PM Algorithm

To solve the robust 2D-PM we perform a number of image matching tasks. In fact, Δ_{t_1,t_2} allows the translation of the center of patterns $P' \in \mathcal{D}(P, \mathcal{F}_{\mathtt{sr}})$ to any pixel of T by the specification of (t_1, t_2). In order to solve the IM problem under $\mathcal{F}_{\mathtt{sr}}$ for all patterns in $\mathcal{D}(P, \mathcal{F}_{\mathtt{sr}})$ and T we use the mapping $\Gamma_{m,n}$ introduced in Theorem 1 to enumerate $\mathcal{D}(P, \mathcal{F}_{\mathtt{sr}})$. To implement the mapping efficiently we perform a search of $\mathcal{A}(H_{\mathtt{sr},m,n})$ and choose for each encountered face one representative point (p, q) in $\mathbb{R}^2$, which can be encoded by rational numbers of length $O(\log(m \cdot n))$. The point represents an inverse transformation g, i.e., $g(x,y) = \left(\begin{smallmatrix} p & q \\ -q & p \end{smallmatrix}\right) \cdot \left(\begin{smallmatrix} x \\ y \end{smallmatrix}\right)$. Hence, for $f = g^{-1}$ the pattern $f(P)$ can be computed using the mapping γ_g. Finally, the distortion between $f(P)$ and T at center of pixel (t_1, t_2) can be computed easily by the procedure Δ_{t_1,t_2}. The property described in Theorem 1 guarantees that in this way all patterns in $\mathcal{D}(P, \mathcal{F}_{\mathtt{sr}})$ will be tested.

However, this straightforward enumeration approach of visiting systematically all faces $\varphi \in \mathcal{A}(H_{\mathtt{sr},m,n})$ and computing $f(P)$ for some f determined by the representative point of the corresponding face φ is not efficient. The time complexity of such a method is at least $|\mathcal{A}(H_{\mathtt{sr},m,n})|$ times $O(n^2)$, where the last term describes the cost of processing the pixels of $f(P)$. However, using our approach we can improve this complexity by computing $f(P)$ incrementally. It turns out that the incident faces correspond to very similar patterns. In fact, Lemma 1 and Theorem 1 imply the following:

Corollary 1. Let $\varphi, \varphi' \in \mathcal{A}(H_{\mathtt{sr},m,n})$ be two faces in $\mathbb{R}^2$ with φ the superface of φ'. Furthermore, let $(i, j) \in \mathcal{N}_2$. Then the two patterns $\Gamma_{m,n}(\varphi)$ and $\Gamma_{m,n}(\varphi')$ can differ at the pixel (i, j), only if there exists k in $\{-m-1, \ldots, m+1\}$ such that the line ℓ described by h_{ijk} or $h_{j(-i)k}$ contains as a subspace the face φ'.

By the above corollary it is profitable to enumerate the faces in the order implied by their incidence to guarantee minimal changes in $f(P)$ when going from one face to the next one. Using this property our algorithm enumerates $\mathcal{D}(P, \mathcal{F}_{\mathtt{sr}})$ in linear running time with respect to the cardinality of $\mathcal{A}(H_{\mathtt{sr},m,n})$. The incidence graph $\mathcal{I}(H_{\mathtt{sr},m,n})$ described in Section 2 is an appropriate data structure to describe the faces in $\mathcal{A}(H_{\mathtt{sr},m,n})$ and their incidences. See e.g. Edelsbrunner [11] for a detailed survey on line arrangements and the complexity to construct the graph $\mathcal{I}(H_{\mathtt{sr},m,n})$. Our algorithm performs the searching of $\mathcal{A}(H_{\mathtt{sr},m,n})$ by DFS-traversing the corresponding incidence graph $\mathcal{I}(H_{\mathtt{sr},m,n})$. This kind of traversal guarantees the enumeration due to the geometrical incidence.

4.1 Implementation

In our setting the following additional auxiliary information are stored for every node $v(\varphi)$: (1) a representative inverse transformation $g(\varphi)$ and (2) a set $Update(\varphi)$ of pixel coordinates.

We first define representative coordinates for each face in $\mathcal{A}(H_{\mathrm{sr},m,n})$. The coordinates of the representative point $p(\varphi)$ of a 0-face φ is just the vertex φ itself. Otherwise, if $\varphi_1, \varphi_2, \ldots, \varphi_t$ are the subfaces of φ then $p(\varphi) := \frac{1}{t} \sum_{w=1}^{t} p(\varphi_w)$. Now we define $g(\varphi)$ to be the transformation represented by the point $p(\varphi)$. Note that $g(\varphi)$ can be encoded by rational numbers of length $O(\log(m \cdot n))$ due to its relation to $p(\varphi)$. We will make an exception for the face φ_0 containing the point $(0,0)$. This face would have $p(\varphi_0) = (0,0)$, which is not allowed. Instead we can easily choose another point in φ_0 that can be encoded with $O(\log(m \cdot n))$ bits to get $p(\varphi_0)$ and $g(\varphi_0)$.

The sets $Update$ are defined as follows: (1) For all 2-faces φ we let

$$Update(\varphi) := \emptyset.$$

(2) If φ is a 1-face, i.e., a half-line or a line segment of a line $\ell \in \mathcal{H}_{\mathrm{sr},m,n}$, then

$$Update(\varphi) := \{(i,j) \mid \exists k \in \{-m-1, \ldots, m+1\} : h_{ijk} \text{ or } h_{j(-i)k} \text{ describes } \ell.\}.$$

(3) If $\varphi_1, \ldots, \varphi_t$ are the superfaces of a 0-face φ, then

$$Update(\varphi) := \bigcup_{w=1}^{t} Update(\varphi_w).$$

Our algorithm performs DFS on the graph $\mathcal{I}(H_{\mathrm{sr},m,n})$ and solves in parallel the IMs for all $(t_1, t_2) \in \mathcal{N}$. Visiting a node $v(\varphi)$ the algorithm stores for all $(t_1, t_2) \in \mathcal{N}$ the current distortion value Δ_{t_1,t_2} between $f(P)$ and T for $f = g^{-1}(\varphi)$. Next, when traversing from φ to an incident (sub or super) face φ' the algorithm has to compute incrementally from $f(P)$ the pattern $f'(P)$ for $f' = g^{-1}(\varphi')$. According to Corollary 1 it suffice to update only the pixel values, coordinates of which are elements in $Update(\varphi)$ or $Update(\varphi')$. For pattern P and text T we use the algorithm 2D-PatternMatching listed in Figure 3.

4.2 Analysis

As a conclusion of this section we give a bound on the running time of the robustly 2D-PM algorithm listed in Figure 3.

Theorem 2. *The 2D-PM robustly invariant under $\mathcal{F}_{\mathrm{sr}}$ can be done in time* $O(m^2 n^6)$.

Notice that the running time of the algorithm is proportional to $|\mathcal{A}(H_{\mathrm{sr},m,n})|$ times the number of parallel IM problems solved. Hence, the running time is bounded from below by $\Omega\left(\frac{m^2 n^5}{\ln m}\right)$.

Algorithm 2D-PatternMatching /* *2D-PM robustly invariant under* $\mathcal{F}_{sr}$ */
Input: Pattern P of support $\mathcal{M}$ and text T of support $\mathcal{N}$.
Output: Pattern $f(P)$ and pixel $(t_1, t_2) \in \mathcal{N}$, with $f = \arg\min_{f' \in \mathcal{F}_{sr}} \{\Delta_{t_1,t_2}(f'(P), T)\}$.

---

1.   **Procedure** SEARCH$(v(\varphi))$; /* *Depth first searching* */
2.   **begin**
3.   mark node $v(\varphi)$ as *seen*;
4.   **for all** *unseen* neighbors $v(\varphi')$ of $v(\varphi)$ **do begin**
5.   **for all** (i,j) in $Update(\varphi) \cup Update(\varphi')$ **do begin**
6.   **for all** (t_1,t_2) in $\mathcal{N}$ **do** $\Delta[t_1,t_2] = \Delta[t_1,t_2] - \delta(P'\langle i - t_1, j - t_2\rangle, T\langle i,j\rangle)$;
7. $P'\langle i,j\rangle = P\langle \gamma_{g(\varphi')}(i,j)\rangle$;
8.   **for all** (t_1,t_2) in $\mathcal{N}$ **do** $\Delta[t_1,t_2] = \Delta[t_1,t_2] + \delta(P'\langle i - t_1, j - t_2\rangle, T\langle i,j\rangle)$;
9.   **end**;
10.   **for all** (t_1,t_2) in $\mathcal{N}$ **do if** $(\Delta[t_1,t_2] < \Delta_{opt})$ **then begin**
11. $\Delta_{opt} = \Delta[t_1,t_2]$; $\varphi_{opt} = \varphi'$; $t_{opt} = (t_1,t_2)$;
12.   **end**;
13.   call SEARCH(φ');
14.   **for all** (i,j) in $Update(\varphi) \cup Update(\varphi')$ **do begin**
15.   **for all** (t_1,t_2) in $\mathcal{N}$ **do** $\Delta[t_1,t_2] = \Delta[t_1,t_2] - \delta(P'\langle i - t_1, j - t_2\rangle, T\langle i,j\rangle)$;
16. $P'\langle i,j\rangle = P\langle \gamma_{g(\varphi)}(i,j)\rangle$;
17.   **for all** (t_1,t_2) in $\mathcal{N}$ **do** $\Delta[t_1,t_2] = \Delta[t_1,t_2] + \delta(P'\langle i - t_1, j - t_2\rangle, T\langle i,j\rangle)$;
18.   **end**;
19.   **end**;
20.   **end**;

21.   **begin** /* *Main( ) for 2D-PM* */
22.   construct the incidence graph $\mathcal{I}(H_{sr,m,n})$;
23.   set all nodes in $\mathcal{I}(H_{sr,m,n})$ as unseen;
24.   let φ_{id} corresponds to the identity mapping;
25. $\varphi_{opt} = \varphi_{id}$; $\Delta_{opt} = \infty$; $P' = P$;
26.   **for all** (t_1,t_2) in $\mathcal{N}$ **do begin**
27. $\Delta[t_1,t_2] = \Delta_{t_1,t_2}(P,T)$;
28.   **if** $(\Delta[t_1,t_2] < \Delta_{opt})$ **then begin**
29. $\Delta_{opt} = \Delta[t_1,t_2]$; $t_{opt} = (t_1,t_2)$;
30.   **end**;
31.   **end**;
32.   call SEARCH$(v(\varphi_{id}))$;   /* *start incremental enumeration* */
33. $f := g^{-1}(\varphi_{opt})$; return $f(P)$ and t_{opt};
34.   **end**.

Fig. 3. The combinatorial pattern matching algorithm solving $(2n+1) \times (2n+1)$ image matching problems. The main procedure prepares only the DFS-search of $\mathcal{I}(H_{sr,m,n})$. The search itself is realized recursively by the SEARCH procedure. With each call one face φ becomes *seen*. Then the neighborhood of φ is processed by updating the pixels which have possibly changed and estimating all new distortions.

5   Conclusions and Future Work

In this work we have analyzed the two-dimensional pattern matching robustly invariant under transformations combining scaling and rotation as well as the image matching problem under the same class of transformations. We introduced a general polynomial time searching strategy which takes advantage of the set structure.

To provide precise bounds for the running time of the algorithm we examined the complexity of the set structure. As the main result we gave narrow bounds for combined scalings and rotations $\Omega\left(\frac{m^2 n^3}{\ln m}\right) \cap O(m^2 n^4)$. We conjecture that the lower bound for the structural complexity of the set of combined scalings and rotations is $\Omega(m^2 n^4)$ which to prove remains future work.

References

1. Bovik, A. (ed.): Handbook of Image and Video Processing. Academic Press, San Diego, California (2000)
2. Amir, A., Butman, A., Crochemore, M., Landau, G., Schaps, M.: Two-dimensional pattern matching with rotations. Theor. Comput. Sci. 314(1-2), 173–187 (2004)
3. Amir, A., Butman, A., Lewenstein, M.: Real scaled matching. Information Processing Letters 70(4), 185–190 (1999)
4. Amir, A., Butman, A., Lewenstein, M., Porat, E.: Real two-dimensional scaled matching. In: Dehne, F., Sack, J.-R., Smid, M. (eds.) WADS 2003. LNCS, vol. 2748, pp. 353–364. Springer, Heidelberg (2003)
5. Amir, A., Calinescu, G.: Alphabet independent and dictionary scaled matching. In: Hirschberg, D.S., Meyers, G. (eds.) CPM 1996. LNCS, vol. 1075, pp. 320–334. Springer, Heidelberg (1996)
6. Amir, A., Chencinski, E.: Faster two-dimensional scaled matching. In: Lewenstein, M., Valiente, G. (eds.) CPM 2006. LNCS, vol. 4009, pp. 200–210. Springer, Heidelberg (2006)
7. Amir, A., Kapah, O., Tsur, D.: Faster two-dimensional pattern matching with rotations. In: Sahinalp, S.C., Muthukrishnan, S.M., Dogrusoz, U. (eds.) CPM 2004. LNCS, vol. 3109, pp. 409–419. Springer, Heidelberg (2004)
8. Amir, A., Landau, G.M., Vishkin, U.: Efficient pattern matching with scaling. Journal of Algorithms 13(1), 2–32 (1992)
9. Brown, L.G.: A survey of image registration techniques. ACM Computing Surveys 24(4), 325–376 (1992)
10. Cox, I.J., Bloom, J.A., Miller, M.L.: Digital Watermarking, Principles and Practice. Morgan Kaufmann, San Francisco (2001)
11. Edelsbrunner, H.: Algorithms in Combinatorial Geometry. Springer, Berlin (1987)
12. Edelsbrunner, H., O'Rourke, J., Seidel, R.: Constructing arrangements of lines and hyperplanes with applications. SIAM J. Comput. 15, 341–363 (1986)
13. de Berg, M., van Kreveld, M., Overmars, M., Schwarzkopf, O.: Computational Geometry, Algorithms and Applications. Springer, Berlin (2000)
14. Fredriksson, K., Mäkinen, V., Navarro, G.: Rotation and lighting invariant template matching. In: Farach-Colton, M. (ed.) LATIN 2004. LNCS, vol. 2976, pp. 39–48. Springer, Heidelberg (2004)

15. Fredriksson, K., Navarro, G., Ukkonen, E.: Optimal exact and fast approximate two-dimensional pattern matching allowing rotations. In: Apostolico, A., Takeda, M. (eds.) CPM 2002. LNCS, vol. 2373, pp. 235–248. Springer, Heidelberg (2002)
16. Fredriksson, K., Ukkonen, E.: A rotation invariant filter for two-dimensional string matching. In: Farach-Colton, M. (ed.) CPM 1998. LNCS, vol. 1448, pp. 118–125. Springer, Heidelberg (1998)
17. Hagedoorn, M.: Pattern matching using similarity measures, PhD thesis, Univ. Utrecht, ISBN 90-393-2460-3 (2000)
18. Hundt, C., Liśkiewicz, M.: On the complexity of affine image matching. In: Thomas, W., Weil, P. (eds.) STACS 2007. LNCS, vol. 4393, pp. 284–295. Springer, Heidelberg (2007)
19. Indyk, P.: Algorithmic aspects of geometric embeddings. In: Proc. 42nd Annual Symposium on Foundations of Computer Science (FOCS 2001), pp. 10–33 (2001)
20. Indyk, P., Motwani, R., Venkatasubramanian, S.: Geometric matching under noise: Combinatorial bounds and algorithms. In: Proc. 10th ACM-SIAM Symposium on Discrete Algorithms (SODA 1999), pp. 354–360 (1999)
21. Kasturi, R., Jain, R.C.: Computer Vision: Principles. IEEE Computer Society Press, Los Alamitos (1991)
22. Kenyon, C., Rabani, Y., Sinclair, A.: Low distortion maps between point sets. In: Proc. 36th ACM Symposium on Theory of Computing (STOC 2004), pp. 272–280 (2004)
23. Kropatsch, W.G., Bischof, H. (eds.): Digital Image Analysis - Selected Techniques and Applications. Springer, Berlin (2001)
24. Landau, G.M., Vishkin, U.: Pattern matching in a digitized image. Algorithmica 12(3/4), 375–408 (1994)
25. Maintz, J.B.A., Viergever, M.A.: A survey of medical image registration. Medical Image Analysis 2(1), 1–36 (1998)
26. Papadimitriou, C., Safra, S.: The complexity of low-distortion embeddings between point sets. In: Proc. 16th ACM-SIAM Symposium on Discrete Algorithms (SODA 2005), pp. 112–118 (2005)
27. Shi, Y.Q., Sun, H.: Image and video Compression for multimedia engineering. CRC Press, Boca Raton (2000)

Searching for Gapped Palindromes

Roman Kolpakov[1] and Gregory Kucherov[2]

[1] Moscow University, 119899 Moscow, Russia
foroman@mail.ru
[2] LIFL/CNRS/INRIA, Parc scientifique de la Haute Borne, 40, Avenue Halley 59650
Villeneuve d'Ascq, France
Gregory.Kucherov@lifl.fr

Abstract. We study the problem of finding, in a given word, all *maximal* gapped palindromes verifying two types of constraints, that we call *long-armed* and *length-constrained* palindromes. For both classes, we propose algorithms that run in time $O(n + S)$, where S is the number of output palindromes. Both algorithms can be extended to compute biological gapped palindromes within the same time bound.

1 Introduction

A palindrome is a word that reads the same backward and forward. Palindromes have long drawn attention of computer science researchers. In word combinatorics, for example, studies have been made on palindromes occurring in Fibonacci words [Dro95], or in general Sturmian words [DP99, DLDL05]. More generally, a so-called *palindrome complexity* of words has been studied [ABCD03].

From an algorithmic perspective, identifying palindromic structures turned out to be an important test case for different algorithmic problems. For example, a number of works have been done on recognition of palindromic words on different types of Turing machines [Sli73, Gal78, Sli81, BBD+03]. Palindrome computation has also been an important problem for parallel models of computation [ABG94, BG95], as well as for distributed models such as systolic arrays [Col69, vdSSer].

Interestingly, a problem related to palindrome recognition was also considered in the seminal Knuth-Morris-Pratt paper presenting the well-known string matching algorithm [KMP77]. The relation between classical string matching and palindrome detection is not purely coincidental. Both the detection of a pattern occurrence and the detection of an even prefix palindrome (even palindrome occurring at the beginning of the input string) can be solved on the 2-way deterministic push-down automaton (2-DPDA), and therefore by Cook's theorem [Coo71], it can be solved by a linear algorithm on the usual RAM model.

Manacher [Man75] proposed a beautiful linear-time algorithm that computes the shortest prefix palindrome in the on-line fashion, i.e. in time proportional to its length. Actually, the algorithm is able to compute much more, namely to compute for each position of the word, the length of the biggest palindrome centered at this position. This gives the exhaustive representation of all palindromes present in the word.

P. Ferragina and G. Landau (Eds.): CPM 2008, LNCS 5029, pp. 18–30, 2008.
© Springer-Verlag Berlin Heidelberg 2008

Words with palindromic structure are important in DNA and RNA sequences, as they reflect the capacity of molecules to fold, i.e. to form double-stranded *stems*, which insures a stable state of those molecules with low free energy. However, in those applications, the reversal of palindromes should be combined with the *complementarity* relation on nucleotides, where c is complementary to g and a is complementary to t (or to u, in case of RNA). Moreover, biologically meaningful palindromes are *gapped*, i.e. contain a *spacer* between left and right copies. Those palindromes correspond, in particular, to *hairpin* structures of RNA molecules, but are also significant in DNA (see e.g. [WGC+04, LJDL07]). A linear-time algorithm for computing palindromes with *fixed* spacer length is presented in [Gus97]. A method for computing *approximate* biological palindromes has been proposed e.g. in [PB02].

Results. In this paper, we are concerned with gapped palindromes, i.e. subwords of the form vuv^T for some u, v, where v^T is v spelled in the reverse order. Occurrences of v and v^T are called respectively *left* and *right* arm of the palindrome. We propose algorithms for computing two natural classes of gapped palindromes. The first class, that we call *long-armed palindromes*, verifies the condition $|u| \leq |v|$, i.e. requires that the length of the palindrome arm is no less than the length of the spacer. The second class is called *length-constrained palindromes* and is specified by lower and upper length bounds on the spacer length $MinGap \leq |u| \leq MaxGap$, and a lower bound on the arm length $MinLen \leq |v|$, where $MinGap, MaxGap, MinLen$ are constants. Moreover, for both definitions, palindromes are additionally required to be *maximal*, i.e. their arms cannot be extended outward or inward preserving the palindromic structure. For both classes, our algorithms run in worst-case time $O(n + S)$, where n is the length of the input word and S is the number of output palindromes, for an alphabet of constant size. (For length-constrained palindromes, our algorithm is actually independent on the alphabet size.) We note that because of the variable spacer length, the above-mentioned algorithm from [Gus97] cannot be efficiently applied to our problems. Both algorithms can be modified to find biological long-armed and length-constrained palindromes within the same running time.

2   Basic Definitions

Let w^T denote the reversal of w. An even palindrome is a word of the form vv^T, where v is some word. An odd palindrome is a word vav^T, where v is a word, and a a letter of the alphabet. A *gapped palindrome* is a word of the form vuv^T for some words u, v such that $|u| \geq 2$. Occurrences of v and v^T are called respectively *left arm* and *right arm* of the palindrome.

In this paper, we will be interested in two classes of palindromes. A gapped palindrome vuv^T is *long-armed* if $|u| \leq |v|$. For pre-defined constants $MinGap$, $MaxGap$ ($MinGap \leq MaxGap$) and $MinLen$, a gapped palindrome vuv^T is called *length-constrained* if it verifies $MinGap \leq |u| \leq MaxGap$ and $MinLen \leq |v|$.

Consider a word $w = w[1] \ldots w[n]$ that contains some gapped palindrome vuv^T. Assume $v = w[l'..l'']$, and $v^T = w[r'..r'']$. We use notation $w[l' : l'', r' : r'']$ for this palindrome. This palindrome is called *maximal* if its arms cannot be extended inward or outward. This means that *(i)* $w[l'' + 1] \neq w[r' - 1]$, and *(ii)* $w[l' - 1] \neq w[r'' + 1]$ provided that $l' > 1$ and $r'' < n$.

3   Long-Armed Palindromes

Let $w = w[1] \ldots w[n]$ be an input word. For technical reasons, we require that the last letter $w[n]$ does not occur elsewhere in the word. In this section, we describe a linear-time algorithm for computing all gapped palindromes occurring in w which are both maximal and long-armed.

The algorithm is based on techniques used for computing different types of periodicities in words [KK05, KK00a], namely on (an extension of) the Lempel-Ziv factorization of the input word and on longest extension functions. The variant of longest extension functions used here is defined as follows. Assume we are given two words $u[1..n]$ and $v[1..m]$ and we want to compute, for each position $j \in [1..n]$ in u, the length $LP(j)$ of the longest common prefix of $u[j..n]$ and v. Assume $m \leq n$ (otherwise we truncate v to $v[1..n]$). Then this computation can be done in time $O(n)$ (see [KK05]). If we have to compute $LP(j)$ for a subset of positions $j \in [1..N]$ for some $N \leq n$, then the time bound becomes $O(N + m)$. Similar bounds apply if we want to compute the lengths of longest common suffixes of $u[1..j]$ and v.

We now describe the algorithm. First, we compute the *reversed Lempel-Ziv factorization* of $w = f_1 f_2 \ldots f_m$ defined recursively as follows:

- if a letter a immediately following $f_1 f_2 \ldots f_{i-1}$ does not occur in $f_1 f_2 \ldots f_{i-1}$ then $f_i = a$,
- otherwise, f_i is the longest subword of w following $f_1 f_2 \ldots f_{i-1}$ which occurs in $(f_1 f_2 \ldots f_{i-1})^T$.

This factorization can be computed in time $O(n \log |A|)$, where A is the alphabet of w, by building the suffix tree for w^T with the Weiner's algorithm that processes the suffixes from shortest to longest (i.e. processes the input word from right to left) [CR94]. For $i = 1, 2 \ldots m$, we construct the suffix tree T_i of the word $(f_1 f_2 \ldots f_i)^T$, and compute f_{i+1} as the longest word that occurs immediately after $f_1 f_2 \ldots f_i$ in w and is present in T_i. If no such word exists, f_{i+1} is defined to be the letter immediately following $f_1 f_2 \ldots f_i$ in w. For each $i = 1, 2, \ldots, m$, denote $f_i = w[s_i..t_i]$ ($s_i = t_{i-1} + 1$) and $F_i = |f_i| = t_i - s_i + 1$.

After computing the reversed Lempel-Ziv factorization, we split all maximal long-armed palindromes into two categories that we compute separately: those which cross (or touch) a border between two factors and those which occur entirely within one factor. Formally, for each $i = 1, 2 \ldots m$, we define the set $P(i)$ of all maximal long-armed palindromes $w[l' : l'', r' : r'']$ that verify one of the conditions:

1. $r'' = t_{i-1}$ and $l' > s_{i-1}$, or
2. $t_{i-1} < r'' \leq t_i$ and $l' \leq s_i$.

Complementary, define $Q(i)$ to be the set of all maximal long-armed palindromes $w[l' : l'', r' : r'']$ that verify $l' > s_i$ and $r'' < t_i$.

Observe that the set $\cup_{i=1}^m P(i) \cup \cup_{i=1}^m Q(i)$ contains all maximal long-armed palindromes in w, and all sets $P(i), Q(i)$ are pairwise disjoint.

3.1   Computing $P(i)$

Each set $P(i)$ is further split into three disjoint sets $P'(i) \cup P''(i) \cup P'''(i)$. $P'(i) \subseteq P(i)$ is the set of all palindromes $w[l' : l'', r' : r'']$ which satisfy one of the conditions:

1. $r'' = t_{i-1}$ and $l' > s_{i-1}$, or
2. $t_{i-1} < r'' \leq t_i$ and $r' \leq s_i$.

$P'(i)$ are maximal long-armed palindromes with the right arm crossing (or touching from the right) the border between f_{i-1} and f_i.

$P''(i) \subseteq P(i)$ contains all palindromes $w[l' : l'', r' : r'']$ which verify both $l' \leq s_i$ and $l'' \geq t_{i-1}$. Palindromes of $P''(i)$ have their left arm crossing (or touching) the border between f_{i-1} and f_i.

Finally, $P'''(i) \subseteq P(i)$ contains all palindromes $w[l' : l'', r' : r'']$ which satisfy the conditions $l'' < t_{i-1}$ and $r' > s_i$. Palindromes of $P'''(i)$ are those for which the border between f_{i-1} and f_i falls inside the spacer.

Computing $P'(i)$. Let $w[l' : l'', r' : r'']$ be a palindrome from $P'(i)$, and let $q = r' - l'' - 1$ be the spacer length. Then the right arm $w[r'..r'']$ is a concatenation of a possibly empty prefix $u = w[r'..t_{i-1}]$ and a possibly empty suffix $v = w[s_i..r'']$. Then the left arm $w[l'..l'']$ is a concatenation of the prefix $v^T = w[l'..t_{i-1} - j]$ and suffix $u^T = w[s_i - j..l'']$ where $j = 2|u| + q$ (see Fig. 1 in the Appendix). Moreover, since the palindrome is maximal, v has to be the longest common prefix of words $w[s_i..n]$ and $w[1..t_{i-1} - j]^T$, and u has to be the longest common suffix of words $w[1..t_{i-1}]$ and $w[s_i - j..n]^T$. Since the spacer length q is no more than the arm length $|u| + |v|$, we have $q \leq |u| + |v|$, i.e. $j \leq 3|u| + |v|$.

Lemma 1. $|u| < F_{i-1}$.

Proof. If $|v| = 0$, i.e. $r'' = t_{i-1}$, then the lemma follows from the condition $l' > s_{i-1}$. If $|v| > 0$, i.e. $r'' > t_{i-1}$, then from $|u| \geq F_{i-1}$ we obtain that the prefix $w[s_{i-1}..r'']$ of $w[s_{i-1}..n]$ occurs in $(f_1 f_2 \ldots f_{i-2})^T$ as a subword of the left arm of the palindrome, which contradicts the definition of $f_{i-1} = w[s_{i-1}..r'']$ as the longest prefix of $w[s_{i-1}..n]$ that occurs in $(f_1 f_2 \ldots f_{i-2})^T$. (If f_{i-1} is a single letter that doesn't occur to the left, then we obviously have $|u| = 0$.)

From the condition $r'' \leq t_i$ we also have $|v| \leq F_i$ and then $j \leq 3|u| + |v| < 3F_{i-1} + F_i$. For all $j < 3F_{i-1} + F_i$, we compute the longest common prefix

$LP(j)$ of words $w[s_i..s_{i+1}]$ and $w[1..t_{i-1} - j]^T$ and the longest common suffix $LS(j)$ of words $w[s_{i-1}..t_{i-1}]$ and $w[s_i - j..n]^T$ (see Fig. 1). These computations can be done in time $O(F_{i-1} + F_i)$. Then each palindrome of $P'(i)$ corresponds to a value of j which satisfies the following conditions:

1. $LP(j) + 3LS(j) \geq j$,
2. if $LP(j) = 0$ then $j < F_{i-1}$,
3. $LS(j) < j/2$.

Inversely, if j satisfies the above conditions, then there exists a palindrome $w[l' : l'', r' : r'']$ for $l' = s_i - j - LP(j)$, $l'' = t_{i-1} - j + LS(j)$, $r' = s_i - LS(j)$, and $r'' = t_{i-1} + LP(j)$. Once conditions 1-3 are verified for some j, the corresponding palindrome is output by the algorithm. The whole computation takes time $O(F_{i-1} + F_i)$.

Computing $P''(i)$. Let $w[l' : l'', r' : r'']$ be a maximal long-armed palindrome from $P''(i)$, and $q = r' - l'' - 1$ be the spacer length. Then the left copy $w[l'..l'']$ is a concatenation of a possibly empty prefix $u = w[l'..t_{i-1}]$ and a possibly empty suffix $v = w[s_i..l'']$. Then the right arm $w[r'..r'']$ is a concatenation of the prefix $v^T = w[r'..t_{i-1} + j]$ and suffix $u^T = w[s_i + j..r'']$, where $j = 2|v| + q$. Moreover, v has to be the longest common prefix of words $w[s_i..n]$ and $w[1..t_{i-1} + j]^T$, and u has to be the longest common suffix of words $w[1..t_{i-1}]$ and $w[s_i + j..n]^T$ (see Fig. 2). Since the spacer length q has to be no more than the arm length $|u| + |v|$, we have that $q \leq |u| + |v|$, i.e. $j \leq |u| + 3|v|$.

Similarly to the case of $P'(i)$, we compute, for each $j = 1, 2, \ldots, F_i$, the longest common prefix $LP(j)$ of words $w[s_i..t_i]$ and $w[s_i..t_{i-1} + j]^T$ and the longest common suffix $LS(j)$ of words $w[1..t_{i-1}]$ and $w[s_i + j..s_{i+1}]^T$. Tables LP and LS are computed in time $O(F_i)$.

Each palindrome of $P''(i)$ corresponds to a value of j verifying the following conditions:

1. $3LP(j) + LS(j) \geq j$,
2. $j + LS(j) \leq F_i$,
3. $LP(j) < j/2$.

If some j satisfies the above conditions, the algorithm outputs the palindrome $w[l' : l'', r' : r'']$ where $l' = s_i - LS(j)$, $l'' = t_{i-1} + LP(j)$, $r' = s_i + j - LP(j)$, and $r'' = t_{i-1} + j + LS(j)$. The computation of $P''(i)$ is done in time $O(F_i)$.

Computing $P'''(i)$. To compute $P'''(i)$, we partition it into disjoint subsets $P_k'''(i)$ for $k = 1, 2, \ldots, \lfloor \log_2 F_i \rfloor$, where $P_k'''(i)$ is the set of all palindromes $w[l' : l'', r' : r'']$ from $P'''(i)$ such that $s_i + \lfloor \frac{F_i}{2^k} \rfloor \leq r'' < s_i + \lfloor \frac{F_i}{2^{k-1}} \rfloor$.

Lemma 2. *For any palindrome $w[l' : l'', r' : r''] \in P_k'''(i)$, we have $r' \leq s_i + \lfloor \frac{F_i}{2^k} \rfloor$.*

Proof. If $r' > s_i + \lfloor \frac{F_i}{2^k} \rfloor$, the arm length of the palindrome is no more than $\lfloor \frac{F_i}{2^k} \rfloor$, and then the spacer length is no more than $\lfloor \frac{F_i}{2^k} \rfloor$. Then, $l'' \geq r' - 1 - \lfloor \frac{F_i}{2^k} \rfloor \geq s_i$ which contradicts the definition of $P'''(i)$.

By the lemma, the right arm of the palindrome is a concatenation of a possibly empty prefix $u = w[r'..t_{i-1} + \lfloor \frac{F_i}{2^k} \rfloor]$ and suffix $v = w[s_i + \lfloor \frac{F_i}{2^k} \rfloor..r'']$. Similar to previous cases, u has to be the longest common suffix of the words $w[1..t_{i-1} + \lfloor \frac{F_i}{2^k} \rfloor]$ and $w[s_i + \lfloor \frac{F_i}{2^k} \rfloor - j..n]^T$, and v has to be the longest common prefix of the words $w[s_i + \lfloor \frac{F_i}{2^k} \rfloor..n]$ and $w[1..t_{i-1} + \lfloor \frac{F_i}{2^k} \rfloor - j]^T$, where $j = 2|u| + q$ and q is the spacer length of the palindrome (see Fig. 3).

Moreover, u and v satisfy the relations $|u| < \lfloor \frac{F_i}{2^k} \rfloor$ and $0 < |v| \le \lfloor \frac{F_i}{2^{k-1}} \rfloor - \lfloor \frac{F_i}{2^k} \rfloor$. Thus, $q \le |u| + |v| < \lfloor \frac{F_i}{2^{k-1}} \rfloor$, and then $j = 2|u| + q < 2\lfloor \frac{F_i}{2^k} \rfloor + \lfloor \frac{F_i}{2^{k-1}} \rfloor < 2\lfloor \frac{F_i}{2^{k-1}} \rfloor$. On the other hand, from the condition $l'' < t_{i-1}$ we have also $|u| < j - \lfloor \frac{F_i}{2^k} \rfloor$ which implies $j > \lfloor \frac{F_i}{2^k} \rfloor$.

Now, to compute all palindromes from $P_k'''(i)$ we apply again the same procedure: for all j such that $\lfloor \frac{F_i}{2^k} \rfloor < j < 2\lfloor \frac{F_i}{2^{k-1}} \rfloor$, we compute the longest common prefix $LP(j)$ of words $w[s_i + \lfloor \frac{F_i}{2^k} \rfloor..s_i + \lfloor \frac{F_i}{2^{k-1}} \rfloor]$ and $w[1..t_{i-1} + \lfloor \frac{F_i}{2^k} \rfloor - j]^T$, and the longest common suffix $LS(j)$ of words $w[s_i..t_{i-1} + \lfloor \frac{F_i}{2^k} \rfloor]$ and $w[s_i + \lfloor \frac{F_i}{2^k} \rfloor - j : n]^T$ (Fig. 3). Each palindrome of $P_k'''(i)$ corresponds then to a value j verifying the following conditions:

1. $LP(j) + 3LS(j) \ge j$,
2. $0 < LP(j) \le \lfloor \frac{F_i}{2^{k-1}} \rfloor - \lfloor \frac{F_i}{2^k} \rfloor$,
3. $LS(j) < \min(\lfloor \frac{F_i}{2^k} \rfloor, j - \lfloor \frac{F_i}{2^k} \rfloor)$.

If some j satisfies the above conditions, we output the palindrome $w[l' : l'', r' : r'']$, where $l' = s_i + \lfloor \frac{F_i}{2^k} \rfloor - j - LP(j)$, $l'' = t_{i-1} + \lfloor \frac{F_i}{2^k} \rfloor - j + LS(j)$, $r' = s_i + \lfloor \frac{F_i}{2^k} \rfloor - LS(j)$, and $r'' = t_{i-1} + \lfloor \frac{F_i}{2^k} \rfloor + LP(j)$.

The required functions $LP(j)$ and $LS(j)$ can be computed in time $O(\frac{F_i}{2^k})$, and then $P_k'''(i)$ can be computed in time $O(\frac{F_i}{2^k})$. Summing up over $k = 1, 2, \ldots,$ $\lfloor \log_2 F_i \rfloor$, $P'''(i)$ can be computed in time $O(F_i)$.

Thus the total time for computing of $P(i)$ is $O(F_{i-1} + F_i)$.

3.2   Computing $Q(i)$

Recall that $Q(i)$ contains all palindromes $w[l' : l'', r' : r'']$ which verify $s_i < l'$ and $r'' < t_i$, i.e. occur as a proper subword of factor f_i. Since f_i has a reversed copy in $f_1 f_2 \ldots f_{i-1}$, a reverse of each palindrome of $Q(i)$ also occurs in that copy. Therefore, it can be "copied over" from that location. Technically, this is done exactly in the same way as in the algorithm for computing maximal repetitions presented in [KK00b] (see also [KK05]). Recovering each palindrome of $Q(i)$ is done in constant time. We refer the reader to those papers for details of this procedure.

3.3   Putting All Together

Each of the sets $P'(i)$, $P''(i)$, $P'''(i)$ is computed in time $O(F_{i-1} + F_i)$, and so is $P(i)$. Summing over all i, all involved palindromes are computed in time $O(n)$. Time computed for all $Q(i)$ is $O(n + T)$, where T is the number of output palindromes. Since all sets $P(i), Q(i)$ are pairwise disjoint, we obtain the final result:

Theorem 1. *All maximal long-armed palindromes can be computed in time $O(n + S)$, where n is the length of the input word and S the number of output palindromes.*

4   Length-Constrained Palindromes

Recall that a gapped palindrome vuv^T is called *length-constrained* if $MinGap \leq |u| \leq MaxGap$ and $MinLen \leq |v|$ for some pre-defined constants $MinGap$, $MaxGap$ and $MinLen$. In this section, we are interested to compute, in a given word, all palindromes that are both length-constrained and maximal.

Note that we do not want to output palindromes that verify length constraints but are not maximal. The inward/outward extension of such a palindrome may lead to a palindrome that no longer verifies length constraints. For example, if $MinLen = 3$, $MinGap = 3$ and $MaxGap = 5$, then the palindrome ...a gtt $aaca$ ttg g... verifies length constraints but is not maximal, while its extension ...a $gtta$ ac $attg$ g... is maximal but does not verify length constraints.

First Step. Consider an input word $w = w[1..n]$. For a position i, we consider words $W(i^+) = w[i..i + MinLen - 1]$ and $W(i^-) = (w[i - MinLen..i - 1])^T$, where i^+, i^- are interpreted as start positions in forward and backward direction respectively. Consider the set $\mathcal{P} = \{i^+, i^- | i = 1..n\}$. For two positions $k_1, k_2 \in \mathcal{P}$, define the equivalence relation $k_1 \equiv k_2$ iff $W(k_1) = W(k_2)$. At the first step, we assign to each position i^-, i^+ the identifier (number) of its equivalence class under the above equivalence relation. This assignment can be done in time $O(n)$ using, e.g., the suffix array for the word $w \# w^T \$$. A simple traversal of this suffix array allows the desired assignment: two successive alphabetically-ordered suffixes belong to the same equivalence class iff the length of their common prefix is at least $MinLen$. Deciding whether position i^+ or i^- should be assigned is naturally done depending on whether the suffix starts in w or in w^T. Further details are left out. Note that the suffix array can be constructed in time $O(n)$ independent on the alphabet size [KS03].

Second Step. After the first preparatory step, the second step does the main job. Our goal is to find pairs of positions $i < j$ such that *(i)* $W(i^-) = W(j^+)$ (arm length constraint), *(ii)* $MinGap \leq j - i \leq MaxGap$ (gap length constraint), and *(iii)* $w[i] \neq w[j - 1]$ (maximality condition). Each such pair of positions corresponds to a desired palindrome. The arm length of this palindrome can then be computed by computing the longest common subword starting at positions i^- and j^+ (i.e. the longest common prefix of $(w[1..i - 1])^T$ and $w[j..n]$). This can be done in constant time using lowest common ancestor queries on the suffix tree for $w \# w^T \$$ [Gus97], but can be also done with the suffix array using the results of [KS03]. The latter solution is independent on the alphabet size.

We are now left with describing how pairs i, j are found. This is done in an online fashion during the traversal of w from left to right. For each equivalence class, we maintain the list of all "minus-positions" $(i_1)^-, (i_2)^-, \ldots, (i_k)^-$ $(i_1 < i_2 < \ldots < i_k)$ scanned so far and belonging to this equivalence class. Moreover, this list is partitioned into *runs* of consecutive list items $(i_\ell)^-, (i_{\ell+1})^-, \ldots, (i_{\ell+k_\ell})^-$ such that $w[i_\ell] = w[i_{\ell+1}] = \ldots = w[i_{\ell+k_\ell}]$ and $w[i_{\ell-1}] \neq w[i_\ell]$ and $w[i_{\ell+k_\ell}] \neq w[i_{\ell+k_\ell+1}]$ (provided that $w[i_{\ell-1}], w[i_{\ell+k_\ell+1}]$ exist in the list).

Furthermore, we maintain a pointer from each run to the next run, so that we are able to "jump", in a constant time, from the first item of the current run to the first item of the next run, avoiding the traversal of the whole run.

The list items can then be implemented by a structure with the following fields:

position: position i such that i^- belongs to the corresponding equivalence class,
NextItem: pointer to the next item in the list,
NextRun: pointer to the first item of the next run (valid only for the first item of a run).

Assume now we are processing a position j of w. First, we insert j to the list of the equivalence class of j^- and update links $NextItem$ and $NextRun$ accordingly. Then we have to find all positions i from the interval $[j - MaxGap..j - MinGap]$ such that i^- belongs to the equivalence class of j^+.

Let C be the identifier of the equivalence class of j^+. We need to check, in the list for C, those positions which belong to the interval $[j - MaxGap..j - MinGap]$. To efficiently access the corresponding fragment in the list, we remember the smallest position of the list belonging to the interval $[\ell - MaxGap..\ell - MinGap]$ for the last processed position $\ell < j$ such that ℓ^+ belongs to equivalence class C. We then start the traversal from this position looking for the positions i falling into the interval $[j - MaxGap..j - MinGap]$. This trick allows us to bound the total time for finding the starting position of segments $[j - MaxGap..j - MinGap]$ by the total size of all the lists, i.e. by $O(n)$.

For each retrieved position i, we verify if $w[i] \neq w[j - 1]$ (maximality condition). If this inequality does not hold, we jump to the first position of the next run of the list, using the run links defined above, thus avoiding consecutive negative tests and insuring that the number of those tests is proportional to the number of output palindromes. The following theorem puts together the two steps of the algorithm.

Theorem 2. *For any predefined constants MinLen, MinGap, MaxGap, all length-constrained palindromes can be found in time $O(n + S)$.*

Proof. The first step is done in time $O(n)$ using suffix array. At the second step, finding starting positions from intervals $[j - MaxGap, j - MinGap]$ in the list for class of j^+ takes time $O(n)$ overall. Testing the maximality condition and outputting the resulting palindromes takes time $O(S)$, where S is the number of output palindromes. Finally, implementing the constant-time computation of longest common subwords starting at given positions is done in time $O(n)$ independent of the alphabet size using results of [KS03].

Algorithm 1 in the Appendix presents a pseudo-code of the algorithm. Besides variables *position*, *NextItem* and *NextRun* defined previously, the algorithm uses the following variables.

$LeftClass(j)$: equivalence class of j^-,

$RightClass(i)$: equivalence class of i^+,

$LastItem(C)$: pointer to the last item in the list for class C,

$LastRun(C)$: pointer to the first item of the current last run in the list for class C,

$PreviousStartItem(C)$: pointer to the start item in the search interval for the last processed position ℓ^+ of class C, i.e. to the smallest position in the list for C belonging to the interval $[\ell - MaxGap..\ell - MinGap]$. (To avoid irrelevant algorithmic details, we assume that such a position always exists.)

$NextFirstItem(C)$: pointer to the first item in the run following the run containing $PreviousStartItem(C)$.

5   Biological Palindromes

Both algorithms presented in Sections 3 and 4 can be extended to biological palindromes, where the word reversal is defined in conjunction with the complementarity of nucleotide letters: $c \leftrightarrow g$ and $a \leftrightarrow t$ (or $a \leftrightarrow u$, in case of RNA). For example, $\ldots c \boxed{acat} aca \boxed{atgt} c \ldots$ is a maximal biological gapped palindrome.

The main part of either algorithm is extended in a straightforward way: each time the algorithm compares two letters, this comparison is replaced by testing their complementarity.

Some parts of the algorithms deserve a special attention. For the algorithm of Section 3 for computing long-armed palindromes, the computation of the reversed Lempel-Ziv factorization extends in a straightforward way too: when computing the next factor f_{i+1}, one has to use the complementarity relation. Similarly, the computation of extension functions LP and LS are also extended straightforwardly.

The algorithm of Section 4 for length-constrained palindromes requires a straightforward modification of the first step: we now need to compute the suffix array for $w\#w^T\$$, where w^T stands for the "biological inversion" (i.e. reversal together with complement). At the second step, the algorithm uses the same suffix array (or alternatively, the suffix tree for $w\#w^T\$$) in order to implement constant-time common subword queries.

6   Concluding Remarks

The algorithm for computing long-armed palindromes from Section 3 can be generalized to palindromes vuv^T verifying condition $|u| \leq c|v|$ for some constant $c \geq 1$. The resulting complexity is $O(cn + S)$.

An interesting open question is whether one can compute the reverse Lempel-Ziv factorization in time $O(n)$ independent on the alphabet size.

Acknowledgments. Part of this work was done during the stay of R. Kolpakov at Inria Lille - Nord Europe, supported by INRIA. R. Kolpakov acknowledges the support of the Russian Foundation for Fundamental Research (Grant 08-01-00863) and of the program for supporting Russian scientific schools (Grant NSh–5400.2006.1).

References

[ABCD03]   Allouche, J.-P., Baake, M., Cassaigne, J., Damanik, D.: Palindrome complexity. Theor. Comput. Sci. 292(1), 9–31 (2003)

[ABG94]   Apostolico, A., Breslauer, D., Galil, Z.: Parallel detection of all palindromes in a string. In: Enjalbert, P., Mayr, E.W., Wagner, K.W. (eds.) STACS 1994. LNCS, vol. 775, pp. 497–506. Springer, Heidelberg (1994)

[BBD⁺03]   Biedl, T., Buss, J., Demaine, E., Demaine, M., Hajiaghayi, M., Vinar, T.: Palindrome recognition using a multidimensional tape. Theor. Comput. Sci. 302(1-3), 475–480 (2003)

[BG95]   Breslauer, D., Galil, Z.: Finding all periods and initial palindromes of a string in parallel. Algorithmica 14, 355–366 (1995)

[Col69]   Cole, S.N.: Real-time computation by n-dimensional iterative arrays of finite-state machines. IEEE Transactions on Computers 18, 349–365 (1969)

[Coo71]   Cook, S.: Linear time simulation of deterministic two-way pushdown automata. In: Proceedings of the 5th World Computer Congress, IFIP 1971, Ljubljana, Yugoslavia, August 23-28, 1971, Vol. 1, pp. 75–80 (1971)

[CR94]   Crochemore, M., Rytter, W.: Text algorithms. Oxford University Press, Oxford (1994)

[DLDL05]   De Luca, A., De Luca, A.: Palindromes in Sturmian words. In: De Felice, C., Restivo, A. (eds.) DLT 2005. LNCS, vol. 3572, pp. 199–208. Springer, Heidelberg (2005)

[DP99]   Droubay, X., Pirillo, G.: Palindromes and Sturmian words. Theoret. Comput. Sci. 223, 73–85 (1999)

[Dro95]   Droubay, X.: Palindromes in the Fibonacci word. Information Processing Letters 55(4), 217–221 (1995)

[Gal78]   Galil, Z.: Palindrome recognition in real time by a multitape turing machine. Journal of Computer and System Sciences 16(2), 140–157 (1978)

[Gus97]   Gusfield, D.: Algorithms on Strings, Trees, and Sequences: Computer Science and Computational Biology. Cambridge University Press, Cambridge (1997)

[KK00a]   Kolpakov, R., Kucherov, G.: Finding repeats with fixed gap. In: Proceedings of the 7th International Symposium on String Processing and Information Retrieval (SPIRE), A Coruña, Spain, September 27-29, 2000, pp. 162–168. IEEE, Los Alamitos (2000)

[KK00b]   Kolpakov, R., Kucherov, G.: On maximal repetitions in words. Journal of Discrete Algorithms 1(1), 159–186 (2000)

[KK05]   Kolpakov, R., Kucherov, G.: Identification of periodic structures in words. In: Berstel, J., Perrin, D. (eds.) Applied combinatorics on words, Encyclopedia of Mathematics and its Applications. Lothaire books, ch.8, vol. 104, pp. 430–477. Cambridge University Press, Cambridge (2005)

[KMP77]   Knuth, D., Morris, J., Pratt, V.: Fast pattern matching in strings. SIAM J. Comput. 6, 323–350 (1977)

[KS03]   Kärkkäinen, J., Sanders, P.: Simple linear work suffix array construc-
            tion. In: Baeten, J.C.M., Lenstra, J.K., Parrow, J., Woeginger, G.J. (eds.)
            ICALP 2003. LNCS, vol. 2719, pp. 943–955. Springer, Heidelberg (2003)
[LJDL07]   Lu, L., Jia, H., Dröge, P., Li, J.: The human genome-wide distribution
            of DNA palindromes. Functional and Integrative Genomics 7(3), 221–227
            (2007)
[Man75]   Manacher, G.: A new linear-time "on-line" algorithm for finding the small-
            est initial palindrome of a string. Journ. ACM 22(3), 346–351 (1975)
[PB02]   Porto, A.H.L., Barbosa, V.C.: Finding approximate palindromes in
            strings. Pattern Recognition 35, 2581–2591 (2002)
[Sli73]   Slisenko, A.O.: Recognition of palindromes by multihead turing machines.
            In: Orverkov, V.P., Sonin, N.A. (eds.) Problems in the Constructive Trend
            in Mathematics VI, Proceedings of the Steklov Institute of Mathematics,
            vol. 129, pp. 30–202 (1973)
[Sli81]   Slissenko, A.: A simplified proof of real-time recognizability of palindromes
            on Turing machines. J. of Soviet Mathematics 15(1), 68–77 (1981); Rus-
            sian original. In: Zapiski Nauchnykh Seminarov LOMI, vol. 68, pp. 123–
            139 (1977)
[vdSSer]   van de Snepscheut, J., Swenker, J.: On the design of some systolic algo-
            rithms. J. ACM 36(4), 826–840 (1989)
[WGC+04]   Warburton, P.E., Giordano, J., Cheung, F., Gelfand, Y., Benson, G.: In-
            verted repeat structure of the human genome: The X-chromosome con-
            tains a preponderance of large, highly homologous inverted repeats that
            contain testes genes. Genome Research 14, 1861–1869 (2004)

Appendix

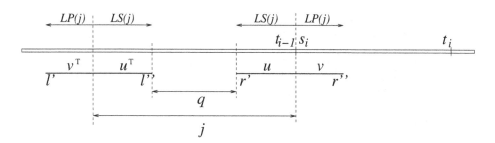

Fig. 1. Computing palindromes of $P'(i)$ (Section 3.1)

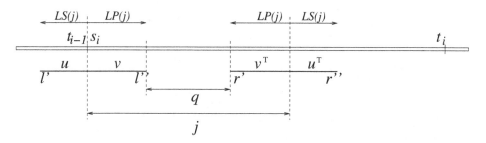

Fig. 2. Computing palindromes of $P''(i)$ (Section 3.1)

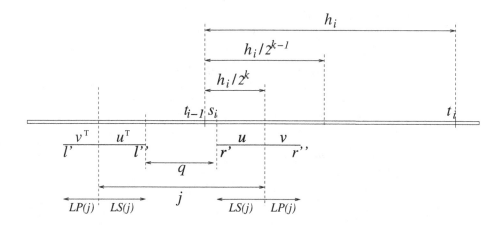

Fig. 3. Computing palindromes of $P'''(i)$ (Section 3.1)

for $j \leftarrow MinLen + 1$ **to** n **do**

    /* insert position j^- to the appropriate list   */

    begin

        $C \longleftarrow LeftClass(j)$;

        create a new item $NewItem$ to the list of class C;

        $NewItem.position \longleftarrow j$;

        $LastItem(C).NextItem \longleftarrow NewItem$;

        if $w[j] \neq w[LastItem(C).position]$ **then**

            $LastRun(C).NextRun \longleftarrow NewItem$;

            $LastRun(C) \longleftarrow NewItem$;

        end

        $LastItem(C) \longleftarrow NewItem$;

    end

    /* find all maximal length-constrained palindromes with the right
        arm starting at position j */

    begin

        $C \longleftarrow RightClass(j)$;

        /* find, in the list for class c, the first position greater
            than or equal to $(j - MaxGap)$ */

        $SearchItem \longleftarrow PreviousStartItem(C)$;

        while $SearchItem.position < j - MaxGap$ **do**

            $SearchItem \longleftarrow SearchItem.NextItem$;

            if $SearchItem = NextFirstItem(C)$ **then**

                $NextFirstItem(C) \longleftarrow SearchItem.NextRun$;

            end

        end

        $PreviousStartItem(C) \longleftarrow SearchItem$;

        /* for each position in the list for class c between
            $(j - MaxGap)$ and $(j - MinGap)$, check if there exists a
            corresponding maximal palindrome   */

        while $SearchItem.position \leq (j - MinGap)$ **do**

            if $w[SearchItem.position] \neq w[j - 1]$ **then**

                $lp \longleftarrow$ length of the longest common prefix of words

                    $w[j + MinLen..n]$ and

                    $(w[1..SearchItem.position - MinLen - 1])^T$;

                output the palindrome $w[SearchItem.position - MinLen - lp :$
                    $SearchItem.position - 1, j : j + MinLen + lp - 1]$;

                $SearchItem \longleftarrow SearchItem.NextItem$;

            end

            else

                if $SearchItem$ is the first item of a run **then**

                $SearchItem \longleftarrow SearchItem.NextRun$;

                else $SearchItem \longleftarrow NextFirstItem(C)$;

            end

        end

    end

  end

end

Algorithm 1: Step 2 of the algorithm for computing length-constrained palindromes

Parameterized Algorithms and Hardness Results for Some Graph Motif Problems

Nadja Betzler[1,*], Michael R. Fellows[2,**], Christian Komusiewicz[1,***], and Rolf Niedermeier[1]

[1] Institut für Informatik, Friedrich-Schiller-Universität Jena, Ernst-Abbe-Platz 2, D-07743 Jena, Germany
{betzler,ckomus,niedermr}@minet.uni-jena.de
[2] PC Research Unit, Office of DVC (Research), University of Newcastle, Callaghan, NSW 2308, Australia
Michael.Fellows@newcastle.edu.au

Abstract. We study the NP-complete GRAPH MOTIF problem: given a vertex-colored graph $G = (V, E)$ and a multiset M of colors, does there exist an $S \subseteq V$ such that $G[S]$ is connected and carries exactly (also with respect to multiplicity) the colors in M? We present an improved randomized algorithm for GRAPH MOTIF with running time $O(4.32^{|M|} \cdot |M|^2 \cdot |E|)$. We extend our algorithm to list-colored graph vertices and the case where the motif $G[S]$ needs not be connected. By way of contrast, we show that extending the request for motif connectedness to the somewhat "more robust" motif demands of biconnectedness or bridge-connectedness leads to W[1]-complete problems. Actually, we show that the presumably simpler problems of finding (uncolored) biconnected or bridge-connected subgraphs are W[1]-complete with respect to the subgraph size. Answering an open question from the literature, we further show that the parameter "number of connected motif components" leads to W[1]-hardness even when restricted to graphs that are paths.

1 Introduction

With the advent of network biology [1, 15] and complex network analysis in general, the study of pattern matching problems in graphs has become more and more important. In this context, the term "graph motif" plays a central role. Roughly speaking, there are two views of graph (or network) motifs. The older is the topological view where one basically ends up with certain subgraph isomorphism problems. For instance, the term "network motif" has been used to represent patterns of interconnections that occur in a network at frequencies much higher than those found in random networks [16, 18]. By way of contrast,

---

* Supported by the DFG, project DARE, GU 1023/1.
** Supported by the Australian Research Council. Work done while staying in Jena as a recipient of the Humboldt Research Award of the Alexander von Humboldt Foundation, Bonn, Germany.
*** Supported by a PhD fellowship of the Carl-Zeiss-Stiftung.

P. Ferragina and G. Landau (Eds.): CPM 2008, LNCS 5029, pp. 31–43, 2008.

the second and more recent view on graph motifs takes a more "functional approach". Here, topology is of lesser importance but the functionalities of network nodes (expressed by colors) form the governing principle. This approach has been propagated by Lacroix et al. [12] and has been followed up by Fellows et al. [9], defining the following problem.

GRAPH MOTIF: Input: A vertex-colored undirected graph $G = (V, E)$ and a multiset of colors M, with $|M| = k$. Question: Does there exist an $S \subseteq V$ such that the induced subgraph $G[S]$ is connected and there is a bijection between the colors of the vertices in S and M?

The different vertex colors are used to model different functionalities. Although originally introduced in a biological context [9, 12], it is conceivable that GRAPH MOTIF is an interesting problem not only for biological networks, but also may prove useful when studying complex social or technical networks.

Known Results. Not surprisingly, GRAPH MOTIF is a computationally hard problem. It is NP-complete even if the input multiset M actually is a set and the input graph is a tree with maximum vertex degree three [9]. Moreover, NP-completeness has also been shown for the case that M consists of only two colors and the input graph is restricted to be bipartite with maximum degree four [9]. Given the apparent hardness of GRAPH MOTIF, Fellows et al. [9] initiated a parameterized complexity analysis. Unfortunately, it turned out that GRAPH MOTIF is W[1]-hard for trees when parameterized by the number of different colors in the motif multiset M. That is, there is no hope to confine the seemingly inevitable combinatorial explosion to the number of colors. By way of contrast, there are good news for other parameterizations. First, when parameterized by the motif size $k := |M|$, GRAPH MOTIF can be solved by a color-coding algorithm running in $O(87^k \cdot k \cdot n^2)$ time on an n-vertex graph, proving its fixed-parameter tractability with respect to the motif size [9][1]. Finally, Dondi et al. [6] extended these investigations for GRAPH MOTIF by studying the case where the subgraph induced by the chosen motif vertices needs not be connected.

New Results. Our work makes two sorts of contributions. First, we present significantly faster algorithms for GRAPH MOTIF and two natural variants, now giving hope for practically useful implementations. In all these cases, the motif size is the governing parameter. Second, we further chart the range of tractability of GRAPH MOTIF by exploring natural variants that become W[1]-hard (again with respect to the parameter motif size). More specifically, we achieve the following results. On the positive side, we improve the randomized algorithm of Fellows et al. [9] running in $O(87^k \cdot k \cdot n^2)$ time and consuming $O(4^k \cdot n)$ space to a new randomized algorithm running in $O(4.32^k \cdot k^2 \cdot m)$ time and consuming $O(2.47^k \cdot n)$ space on an m-edge graph. Note that both algorithms are based on the color-coding technique due to Alon et al. [2], which has recently proven

---

[1] Fellows et al. [9] do not explicitly state the running time of the randomized version of their algorithm. Instead, they demonstrate a running time of $O(2^{5k} \cdot k \cdot n^2)$ per trial. Using k colors for color-coding, $O(e^k)$ trials are needed to achieve a sufficiently low error probability, which results in a total running time of $O(87^k \cdot k \cdot n^2)$.

practical usefulness [5, 7, 10, 14]. Both algorithms can be derandomized, but the current state of the art of derandomization techniques seems prohibitive from a practical point of view (also see [10]). We extend our fixed-parameter tractability results for GRAPH MOTIF to two variants: LIST-COLORED GRAPH MOTIF, where each chosen vertex may allow for a list of colors that it can match, and MIN-CC GRAPH MOTIF, where we specify the number of connected components the graph motif may have. On the negative side, we also provide several parameterized hardness results. First, we investigate the search for somewhat "more robust" motifs. In other words, we show that if one requires that the found motif shall not only be connected but biconnected or bridge-connected, then in both cases the corresponding GRAPH MOTIF problem becomes W[1]-complete with respect to the parameter motif size (actually, even special cases thereof do so). Since these are the two most simple demands one may pose for more robust motifs, this shows that the request for connected motifs is already a topology demand close to the border of tractability and intractability.[2] Finally, somewhat aside, we answer an open question of Dondi et al. [6] by proving that the aforementioned MIN-CC GRAPH MOTIF problem is W[1]-hard with respect to the number of components even if the input graph is restricted to be only a path. Due to the lack of space, some details are deferred to the full version.

Preliminaries. We consider only simple undirected graphs $G = (V, E)$, where $n := |V|$ and $m := |E|$ throughout the whole work. For a vertex $v \in V$, let $N(v) := \{u \mid \{u, v\} \in E\}$ denote the *open neighborhood* of v, and let $N[v] := N(v) \cup \{v\}$ denote the *closed neighborhood* of v. A *coloring* of an undirected graph $G = (V, E)$ is a function $c : V \to C$, where C is a set of colors. Unless stated otherwise, a *motif* is a multi-set of colors. In case that the motif is a set, we call the motif *colorful*. An *occurrence* of a motif M in G is a set of vertices $S \subseteq V$ such that $|S| = |M|$, $G[S]$ is connected, and there are x vertices of color c in S iff M contains c exactly x times. Let $col(v)$ denote the color of a vertex v and $col(S)$ the multiset of colors of the vertices of S. A vertex u in an undirected graph is called a *cut vertex* if there are two vertices v, w with $v \neq u$ and $w \neq u$ such that every path from v to w contains u. If an undirected graph G is connected and has no cut-vertex, then G is *biconnected*. In general, if a graph $G = (V, E)$ cannot be disconnected by deletion of any set of $p - 1$ vertices, it is called *p-connected*. A graph is called *p-edge-connected* if it cannot be disconnected by deletion of any set of $p - 1$ edges. A 2-edge-connected graph is called *bridge-connected*.

The *color-coding* technique yields randomized fixed-parameter algorithms [2]. The main idea is to randomly color the vertices of the graph, and then to solve the corresponding problem under the assumption that the subgraph that is searched for obtains a *colorful* coloring, that is, all of the vertices of the subgraph have pairwise different colors. This assumption often leads to a problem solvable more

---

[2] Our results also generalize to higher connectivity demands. Even further, they hold for uncolored graphs, where one searches for a subgraph with the specific connectivity demand, and the parameter is the number of subgraph vertices.

efficiently. The procedure of coloring and then solving the subsequent problem on the colored graph is repeated as often as necessary to obtain a sufficiently low error probability. We say that a randomized algorithm solves a problem with *error probability* ϵ if the probability that it fails to return the correct answer is at most ϵ.

Parameterized algorithmics aims at a multivariate complexity analysis of problems [8, 13]. The hope lies in accepting the seemingly inevitable combinatorial explosion for NP-hard problems, but to confine it to a parameter k. A given parameterized problem (I, k) is *fixed-parameter tractable (FPT)* with respect to the parameter k if it can be solved within running time $f(k) \cdot \text{poly}(|I|)$ for some computable function f. Downey and Fellows [8] developed a theory of parameterized intractability by means of devising a completeness program with complexity classes. The first level of (presumable) parameterized intractability is captured by the complexity class W[1]. A *parameterized reduction* reduces a problem instance (I, k) in $f(k) \cdot \text{poly}(|I|)$ time to an instance (I', k') such that (I, k) is a yes-instance if and only if (I', k') is a yes-instance and k' only depends on k but not on $|I|$. If for a given parameterized problem L there is a parameterized problem L' such that L' is W[1]-hard and there is a parameterized reduction from L' to L, then L is also W[1]-hard.

2   Fixed-Parameter Algorithms

Our accelerated algorithm for GRAPH MOTIF, as the previous one [9], is based on the color-coding technique [2]. However, we make use of the following new observation on colorful motifs.

Lemma 1. *Let* (G, M) *be a* GRAPH MOTIF *instance such that* M *is colorful. Then,* GRAPH MOTIF *can be solved in* $O(3^k \cdot m)$ *time.*

Proof. We describe a dynamic programming algorithm that finds an occurrence of M. In the dynamic programming table, entry $D_{v,C}$ stores the "minimum score" of a color set C for a vertex v, where a score of 0 means that an occurrence of C that includes v exists. We initialize the entries of the dynamic programming table with

$$D_{v,C} = \begin{cases} 0, & C = \{\text{col}(v)\}, \\ 1, & \text{otherwise.} \end{cases}$$

In the recurrence, we look for the combination of subsets of a color set such that the sum of the entries is minimum:

$$D_{v,C} = \min_{u \in N(v), \, C' \subset C} \left\{ \begin{array}{l} D_{u, C \setminus \{\text{col}(v)\}}, \\ D_{v, C' \cup \{\text{col}(v)\}} + D_{v, (C \setminus C') \cup \{\text{col}(v)\}} \end{array} \right\}.$$

Since the motif M is colorful, we can restrict attention to joining sets of vertices that have disjoint color sets. Therefore, we never join vertex sets that have vertices in common. If a colorful motif M occurs in G, then for some $v \in V$,

$D_{v,M} = 0$. Furthermore, during the dynamic programming procedure we only need to consider color sets C that are subsets of M. Therefore, we have $O(2^k)$ entries per vertex, which results in a table size of $O(2^k \cdot n)$. Overall, the first part of the recursion can be executed in $O(2^k \cdot m)$ time, since for each color set $C \subseteq M$ and for every vertex v we have to scan once through the adjacency list of v and for each neighbor the corresponding table entry can be found in constant time. The second part of the recursion can be executed in $O(3^k \cdot n)$ time overall: for each vertex v the number of combinations that have to be considered is bounded by $O(3^k)$, since we have to consider all possible subsets of M and for each subset we have to consider all possibilities to split this subset. Overall this amounts to $O(3^k)$ combinations, since there are 3^k possibilities to split a subset of size k into three disjoint subsets (in our case these subsets are $M \setminus C$, C' , and C''). For each combination the computation of the recursion can be performed in constant time. Overall, the running time amounts to $O(3^k \cdot m)$. An occurrence of the motif can be computed by traceback within the same asymptotic running time bound.   □

The above dynamic programming procedure is basically a simplified version of the procedure for the related problem of finding a minimum-weight tree of size k [14]. The main difference is that for GRAPH MOTIF, we do not have additional weights that are associated with the graph vertices.

We now show how to use Lemma 1 in order to obtain an algorithm in case that the motif is a multiset of colors. The main idea is to use the technique of color-coding [2] in order to transform any instance that has a multiset of colors as motif into an instance that has a colorful motif. To this latter instance then Lemma 1 applies. In the following, we describe this transformation in detail. Let M be the motif and let $\mathrm{occ}(c)$ denote the number of occurrences of a color c in M. For each color c with $\mathrm{occ}(c) \geq 2$ we introduce $\mathrm{occ}(c)$ new colors $c_1, c_2, \ldots, c_{\mathrm{occ}(c)}$. Then, we randomly recolor each vertex that has color c with one of the new colors, where the probability for each color is exactly $1/ \mathrm{occ}(c)$ (uniform distribution). Let M' be the set of colors that contains the colors that occurred only once in M together with the colors $\{c_1, c_2, \ldots, c_{\mathrm{occ}(c)}\}$ for every color c with $\mathrm{occ}(c) \geq 2$. Furthermore, let S be an occurrence of M. We say that S achieves a *colorful recoloring* if $\mathrm{col}(S)$ is colorful after the recoloring procedure. Clearly, if S achieves a colorful recoloring, then $\mathrm{col}(S) = M'$. An occurrence of M' can be found via dynamic programming by Lemma 1. This procedure of recoloring with subsequent dynamic programming is repeated until either an occurrence of M is found, or the probability that there is an S that has not achieved a colorful recoloring is acceptably low.

Proposition 1. GRAPH MOTIF *can be solved with error probability ϵ within* $O(|\ln(\epsilon)| \cdot 8.16^k \cdot m)$ *time.*

Proof. By Lemma 1, we can find an occurrence of a colorful motif in $O(3^k \cdot m)$ time. Therefore, the total running time of the algorithm is $O(t(\epsilon) \cdot 3^k \cdot m)$, where $t(\epsilon)$ denotes the number of trials that is needed in order to achieve a colorful recoloring of the vertices of the motif in at least one of the trials with a probability of at least $1 - \epsilon$. For each color $c \in M$, the probability P_c that the $occ(c)$ vertices in S that have color c receive a colorful recoloring is $(occ(c))!/occ(c)^{occ(c)}$, because each coloring has the same probability and $(occ(c))!$ colorings of the $occ(c)^{occ(c)}$ possible colorings are colorful. Using Stirling's approximation for factorials we can show that $occ(c)!/occ(c)^{occ(c)} > \sqrt{2 \cdot \pi \cdot occ(c)} \cdot e^{-occ(c)}$. For two colors c_1 and c_2 the probabilities P_{c_1} and P_{c_2} are independent. Therefore, the probability $P_{c_1 \wedge c_2}$ that the vertices of both color classes achieve a colorful recoloring is

$$P_{c_1 \wedge c_2} = P_{c_1} \cdot P_{c_2} > \sqrt{2 \cdot \pi \cdot (occ(c_1) + occ(c_2))} \cdot e^{-(occ(c_1) + occ(c_2))}.$$

The probability P_M that an occurrence of M receives a colorful recoloring thus is

$$P_M = \prod_{c \in col(M)} P_c > \sqrt{2 \cdot \pi \cdot k} \cdot e^{-\sum_{c \in col(M)} occ(c)} > e^{-k}.$$

After t trials the error probability, that is, the probability that a colorful recoloring was not achieved, is $(1 - P_M)^t$. Therefore, the number of trials $t(\epsilon)$ to achieve an error probability of at most ϵ is $t(\epsilon) = \lceil |\ln(\epsilon)|/\ln(1 - P_M) \rceil = |\ln(\epsilon)| \cdot O(e^k)$. Hence, the total running time of the algorithm when an error probability of at most ϵ is allowed is $O(|\ln(\epsilon)| \cdot e^k \cdot 3^k \cdot m) = O(|\ln(\epsilon)| \cdot 8.16^k \cdot m)$.   □

Applying two speed-up techniques, we can further improve the running time of the algorithm. First, as proposed by Hüffner et al. [10], we can increase the number of colors that are used for color-coding in order to increase the probability of an occurrence of M to receive a colorful recoloring[3]. Second, we can speed up the dynamic programming procedure of Lemma 1 by using the technique of *fast subset convolution*. This novel technique was developed by Björklund et al. [3], who used it to speed up several dynamic programming algorithms including the algorithm by Scott et al. [14] for computing minimum weight size k trees in signalling networks.

Let f and g be functions defined on the power set of a finite set N with $|N| = n$, that is, $f, g : \mathcal{P}(N) \to I$. For any ring over I that defines addition and multiplication on elements of I, the *subset convolution* of f and g, denoted by $f * g$, is defined for each $S \subseteq N$ as

$$f * g : \mathcal{P}(N) \to I, \quad (f * g)(S) = \sum_{T \subseteq S} f(T)g(S \setminus T).$$

To calculate the subset convolution means to determine the value of $f * g$ for all 2^n possible inputs, assuming that f and g can be evaluated in constant time (typically by being stored in a table). A naive algorithm that calculates each value independently needs $O(\sum_{i=0}^{n} \binom{n}{i} 2^i) = O(3^n)$ ring operations. The following result shows a substantial improvement.

---

[3] Increasing the number of colors has been independently examined by Deshpande et al. [5]. Hüffner et al. [10] derive a better bound on the worst-case running time.

Theorem 1 (Björklund et al. [3]). *The subset convolution over an arbitrary ring can be computed with $O(2^n \cdot n^2)$ ring operations.*

Björklund et al. [3] showed how to apply Theorem 1 to also calculate the subset convolution for the integer min-sum semiring

$$f * g : \mathcal{P}(N) \to \mathbb{Z}, \quad (f * g)(S) = \min_{T \subseteq S} f(T) + g(S \setminus T)$$

by embedding it into the standard integer sum-product ring.[4] Recall the recurrence of the dynamic programming procedure for colorful motifs:

$$D_{v,C} = \min_{u \in N(v), \, C' \subset C} \left\{ \begin{array}{l} D_{u,C \setminus \{\mathrm{col}(v)\}}, \\ D_{v,C' \cup \{\mathrm{col}(v)\}} + D_{v,(C \setminus C') \cup \{\mathrm{col}(v)\}} \end{array} \right\}.$$

The first part of the recurrence can be evaluated in $O(2^k \cdot m)$ time. For the second part we can use fast subset convolution and can thus compute the recurrence in $O(2^k \cdot k^2 \cdot n)$ time, because each ring operation can be performed in constant time, since the maximum weight that is used for the basic table entries is 1. Clearly, the graph G has an occurrence of M if there is a table entry $D_{v,M} = 0$ in the final table. The actual occurrence of the motif can be computed in $O(2^k \cdot k \cdot m)$ time by traceback. In the following theorem, we upper-bound the running time of the algorithm that is obtained from combining the two described speed-up techniques.

Theorem 2. GRAPH MOTIF *can be solved with error probability ϵ in $O(|\ln(\epsilon)| \cdot 4.32^k \cdot k^2 \cdot m)$ time.*

Proof. Hüffner et al. [10] showed that when using $1.3 \cdot k$ colors, the number of trials that is needed to obtain error probability ϵ is $O(|\ln(\epsilon)| \cdot 1.752^k)$. However, this increases the running time of the dynamic programming procedure, since now the color set has size $1.3 \cdot k$. The modified dynamic programming procedure then has a running time of $O(2^{1.3 \cdot k} \cdot k^2 \cdot m)$. Overall, the running time amounts to $O(|\ln(\epsilon)| \cdot 1.752^k \cdot 2^{1.3 \cdot k} \cdot k^2 \cdot m) = O(4.32^k \cdot k^2 \cdot m)$. □

A drawback of using $1.3 \cdot k$ colors is that the memory requirement increases from $O(2^k \cdot m)$ to $O(2^{1.3k} \cdot m) = O(2.47^k \cdot m)$. However, it was shown that the running time improvement is enormous in practice [10]. In some special cases, we need even less trials to achieve an exponentially low error probability. For example, if every color in the motif occurs at most twice, then we have to use at most two colors per vertex. Furthermore, there can be at most $k/2$ colors that appear twice in the motif. Using two colors for each color c that appears twice in M, the two vertices in an occurrence of M that have a color c receive different colors with probability $1/2$. Hence, the probability that a recoloring is a colorful recoloring is $2^{-k/2} = (\sqrt{2})^{-k}$. The number of trials needed to achieve exponentially low error probability then is $O((\sqrt{2})^k)$ and the total running time $O((\sqrt{2})^k \cdot 2^k \cdot k^2 \cdot m) = O(2.83^k \cdot k^2 \cdot m)$.

---

[4] Björklund et al. [3] also considered the variant where we do not have disjoint sets T and $S \setminus T$ but allow one element occurring in both sets (as we make use of in the following).

Two Natural Graph Motif Variants. We extend our randomized algorithm for the basic GRAPH MOTIF problem to two practically interesting problem variants. The original formulation of GRAPH MOTIF allows multiple colors per vertex [12]. This makes sense in a biological context in order to model multiple functionalities of one element. The input graph can then be formalized as a *list-colored graph*, in which a *list* of colors is attached to every vertex of the graph. In other words, for a vertex $v \in V$ of a list-colored graph, $col(v)$ denotes a set of colors instead of a single color.

LIST-COLORED GRAPH MOTIF: Input: A list-colored undirected graph $G = (V, E)$ and a multiset of colors M. Question: Does there exist a vertex subset $S \subseteq V$ such that the induced subgraph $G[S]$ is connected and there is a bijection $f : S \to M$ such that $\forall v \in S : f(v) \in col(v)$?

Unfortunately, we cannot use our above algorithm for LIST-COLORED GRAPH MOTIF. The difficulty is that in list-colored graphs we do not have a one-to-one correspondence between vertices and colors; hence, two disjoint color sets do not imply two disjoint vertex sets. However, we can apply a different color-coding procedure, partially resembling the algorithm by Fellows et al. [9].

Theorem 3. LIST-COLORED GRAPH MOTIF *can be solved with error probability ϵ in* $O(|\ln(\epsilon)| \cdot 10.88^k \cdot m)$ *time.*

Proof. We use color-coding. To avoid ambiguities, we call the random colors assigned by the color-coding procedure *labels*, and the term color only refers to the colors of the list-colored graph. Let $L = \{l_1, l_2, \ldots, l_k\}$ denote a set of k distinct labels. We randomly assign (uniformly distributed) the labels of L to the vertices of the graph and solve the problem of finding an occurrence of the motif M under the assumption that all vertices of the occurrence have received a different label. Without loss of generality, assume that M is colorful. Otherwise, we transform M and G as follows: For each color c that occurs $occ(c)$ times, we add $occ(c)$ new colors to M and completely remove c from G. Furthermore, for every vertex v in G with $c \in col(v)$, we remove c from $col(v)$ and add the $occ(c)$ new colors to $col(v)$. Let M' and G' be the thus modified motif and graph, respectively. We now solve the problem of finding an occurrence of M' in G'. Each such occurrence clearly corresponds to an occurrence of M in G.

The problem of finding a colorful occurrence of M that has the labels of L is solved by dynamic programming. First, we extend our notion of occurrence. Let $F \subseteq (L \cup M)$ be a set that contains labels as well as colors. An occurrence of F is defined as a set of vertices S such that the vertices of S have exactly the labels of $F \cap L$, and there is a bijection $f : S \to F \cap M$, such that for each vertex v $f(v) \in col(v)$. An entry $D_{v,F}$ of the dynamic programming table denotes the "score" of an occurrence of F that contains v. We initialize the table as follows:

$$D_{v,\{c,l\}} = \begin{cases} 0, & c \in col(v) \wedge l = label(v), \\ 1, & \text{otherwise.} \end{cases}$$

Furthermore, we assign weight 1 to all entries $D_{v,\{c\}}$ and $D_{\{l\}}$. The recurrence reads

$$
D_{v,F} = \min_{u \in N(v),\ c \in \mathrm{col}(v),\ F' \subset F} \left\{ \begin{array}{l} D_{u,F \setminus \{c,\mathrm{label}(v)\}}, \\ D_{v,F' \cup \{c,\mathrm{label}(v)\}} + D_{v,(F \setminus F') \cup \{c,\mathrm{label}(v)\}} \end{array} \right\}.
$$

We calculate the score for sets $F \subseteq M$ of increasing cardinality. Note that by initializing the entries $D_{v,\{c\}}$ and $D_{v,\{l\}}$ with 1, we make sure that a score of 0 of an "occurrence" of F can only be achieved when there is a one-to-one correspondence between labels and colors of the occurrence. Therefore, if there is a $v \in V$ such that $D_{v,L \cup M} = 0$, then there is an occurrence of $L \cup M$ in G. An actual occurrence then can be computed by traceback.

For the running time consider the following. Clearly $|L \cup M| = 2 \cdot k$. The recursion is similar to the recursion in the proof of Theorem 2. Hence, we can also apply subset convolution and obtain a running time of $O(2^{2 \cdot k} \cdot (2 \cdot k)^2 \cdot m) = O(4^k \cdot k^2 \cdot m)$ for the dynamic programming procedure. The number of trials that is needed to obtain a *good labelling* with probability at least $1 - \epsilon$ is $O(|\ln(\epsilon)| \cdot e^k)$. The total running time thus amounts to $O(|\ln(\epsilon)| \cdot e^k \cdot 4^k \cdot k^2 \cdot m) = O(10.88^k \cdot k^2 \cdot m)$. $\square$

Our second variant of GRAPH MOTIF has been introduced by Dondi et al. [6], who proposed a generalization of GRAPH MOTIF in which it is no longer demanded that the motif is connected.

MIN-CC GRAPH MOTIF: Input: A vertex-colored undirected graph $G = (V, E)$, a multiset of colors M with $|M| = k$, and a nonnegative integer d. Question: Does there exist an $S \subseteq V$ such that $G[S]$ has at most d components, and there is a bijection between the colors of the vertices in S and M?

Clearly, GRAPH MOTIF is MIN-CC GRAPH MOTIF with $d = 1$. Among other results, Dondi et al. [6] showed that the algorithms for GRAPH MOTIF by Fellows et al. [9] can be adapted to solve MIN-CC GRAPH MOTIF. We can also modify our GRAPH MOTIF algorithm to solve MIN-CC GRAPH MOTIF.

Theorem 4. MIN-CC GRAPH MOTIF *can be solved with error probability ϵ in* $O(|\ln(\epsilon)| \cdot 4.32^k \cdot k^2 \cdot m)$ *time.*

3   Parameterized Hardness Results

Lacroix et al. [12] motivated the study of (variants) of the GRAPH MOTIF problem by considerations comparing "topological motifs" with "functional motifs". The GRAPH MOTIF problem only poses a minimal demand on the motif topology by requiring connectedness. The natural question arises what happens if we ask for somewhat "more robust" motifs, replacing the connectedness demand by demands for biconnectivity, bridge-connectivity and the like. As we will show in this section, these seemingly small steps towards topologically more constrained motifs already lead to W[1]-completeness. Finally, the only time considering a parameter other than motif size, we answer an open question of Dondi et al. [6]

Fig. 1. An example of the transformation of a CLIQUE instance with $k = 3$ into a BICONNECTED SUBGRAPH instance with $k' = 15$. White vertices in G' belong to V_1, black vertices to V_2.

by showing that the parameter "number of connected components" in a graph motif leads to a W[1]-hard problem.

BICONNECTED GRAPH MOTIF: Input: A vertex-colored undirected graph $G = (V, E)$ and a multiset of colors M. Question: Does there exist an $S \subseteq V$ such that the induced subgraph $G[S]$ is biconnected and there is a bijection between the colors of the vertices in S and M?

We will show that BICONNECTED GRAPH MOTIF is W[1]-complete when parameterized by the size of the motif M. In fact, we prove an even stronger result. Consider the special case that M contains only one color c, $|M| = k$, and that all vertices in G have color c. Then, the remaining problem to find a biconnected subgraph of size *exactly* k is denoted as:

BICONNECTED SUBGRAPH: Input: An undirected graph $G = (V, E)$ and a non-negative integer k. Question: Does there exist an $S \subseteq V$ of size k such that the induced subgraph $G[S]$ is biconnected?

Note that looking for a biconnected subgraph of size *at least* k is solvable in linear time [17]. However, restricting the size of the biconnected subgraph to exactly k makes the problem surprisingly hard. We prove the parameterized hardness by reduction from the CLIQUE problem, which is known to be W[1]-complete [8] with respect to the size of the clique searched for.

CLIQUE: Input: An undirected graph G and a nonnegative integer k. Question: Is there a complete subgraph of size k in G?

Theorem 5. BICONNECTED SUBGRAPH *is W[1]-complete with respect to* k.

Proof. To show the W[1]-hardness, we give a parameterized reduction from CLIQUE to BICONNECTED SUBGRAPH. Let (G, k) be a CLIQUE instance. We construct a graph G' from G by replacing every edge e of G with a simple path p_e that has $\binom{k}{2} + 1$ internal new vertices. The vertex set of G' can be partitioned into two vertex sets V_1 and V_2, where V_1 contains the vertices that correspond to vertices of the original graph G and V_2 contains the new internal path vertices. An example of this reduction is shown in Figure 1.

We prove in the following that G has a clique of size k iff G' has a biconnected subgraph of size $k' = k + \binom{k}{2} \cdot (\binom{k}{2} + 1)$. If G has a clique C of size k, then the subgraph that is induced by the k vertices of C and by the vertices on the $\binom{k}{2}$ paths that were created from the $\binom{k}{2}$ clique edges of C in G has size exactly $k + \binom{k}{2} \cdot (\binom{k}{2} + 1)$. Clearly, this subgraph is also biconnected.

It remains to show that if G' has a biconnected subgraph of size $k' = k + \binom{k}{2} \cdot (\binom{k}{2} + 1)$, then G has a clique of size k. Let G' have a biconnected subgraph $G'[S]$ of size k. If S contains one vertex of a path p_e, then it must contain all vertices from p_e, because otherwise $G'[S]$ would not be biconnected. Hence, the number of vertices k' in S can be expressed as $k' = a + b \cdot (\binom{k}{2} + 1)$, where $a = |S \cap V_1|$ and b denotes the number of paths in G' that correspond to edges of G.

We distinguish two main cases. In the first case, let $a = k$. Then, $G'[S]$ must contain exactly $\binom{k}{2}$ paths that correspond to edges in G. Let $e = \{u, v\}$ be an edge of G, and let $A := S \cap V_1$. Since $G'[S]$ is biconnected, if a path p_e is contained in S, then $\{u, v\} \subseteq A$. Since $G'[S]$ contains exactly $\binom{k}{2}$ paths consisting of vertices from V_2 and each path must connect two vertices of A, all vertices of A are pairwise connected via a path of length $\binom{k}{2}$. Hence, the subgraph $G[A]$ must be a size-k clique since it contains exactly k vertices and exactly $\binom{k}{2}$ edges.

We now consider the case $a \neq k$ and show that in this case, either $a + b \cdot (\binom{k}{2} + 1) \neq k + \binom{k}{2} \cdot (\binom{k}{2} + 1) = k'$ or $G'[S]$ cannot be biconnected. Clearly, if $b = \binom{k}{2}$, then

$$a + b \cdot \left(\binom{k}{2} + 1 \right) = a + \binom{k}{2} \cdot \left(\binom{k}{2} + 1 \right) \neq k + \binom{k}{2} \cdot \left(\binom{k}{2} + 1 \right) = k'.$$

Therefore, we can assume that $b \neq \binom{k}{2}$. In the following, we list all remaining cases and show that either $a + b \cdot \binom{k}{2} \neq k'$ or $G'[S]$ is not biconnected.

Case 1: $b > \binom{k}{2}$.

$$a + b \cdot \left(\binom{k}{2} + 1 \right) \geq a + \left(\binom{k}{2} + 1 \right) \cdot \left(\binom{k}{2} + 1 \right) > k + \binom{k}{2} \cdot \left(\binom{k}{2} + 1 \right)$$

Case 2.1 : $b < \binom{k}{2}$ and $a < \binom{k}{2}$.

$$a + b \cdot \left(\binom{k}{2} + 1 \right) < \binom{k}{2} + \left(\binom{k}{2} - 1 \right) \cdot \left(\binom{k}{2} + 1 \right) < k + \binom{k}{2} \cdot \left(\binom{k}{2} + 1 \right)$$

Case 2.2 : $b < \binom{k}{2}$ and $a \geq \binom{k}{2}$.

In this case, $G'[S]$ cannot be biconnected: S consists of at least $a \geq \binom{k}{2}$ vertices from V_1 and less than $\binom{k}{2}$ paths that correspond to edges of G. Therefore, at least one of the $\binom{k}{2}$ vertices from V_1 is connected to at most one path. By construction, vertices in V_1 may only be adjacent to vertices in V_2. Hence, $G'[S]$ is not biconnected.

Summarizing, G has a clique of size k iff G' has a biconnected subgraph of size $k \cdot \binom{k}{2} \cdot (\binom{k}{2} + 1)$. The reduction can be clearly performed in polynomial time. We omit the proof for containment in W[1]. □

A second natural way to heighten the robustness demands for GRAPH MOTIF is to search for bridge-connected motifs. We define BRIDGE-CONNECTED SUB-GRAPH in complete analogy to BICONNECTED-CONNECTED SUBGRAPH, simply replacing the demand for biconnectivity by the demand for bridge-connectivity. The reduction from CLIQUE as used in the proof of Theorem 5 works also for bridge-connected subgraphs.

Theorem 6. BRIDGE-CONNECTED SUBGRAPH *is W[1]-complete with respect to k (number of subgraph vertices).*

Further, we can generalize the hardness results to graph motifs of higher connectivity. To this end, consider the following problem.

p-(EDGE) CONNECTED SUBGRAPH: Input: An undirected graph G and a nonnegative integer k. Question: Does there exist an $S \subseteq V$ of size k such that the induced subgraph $G[S]$ is p-(edge) connected?

Theorem 7. p-(EDGE) CONNECTED SUBGRAPH *is W[1]-complete with respect to k (number of subgraph vertices).*

The following theorem answers an open question of Dondi et al. [6].

Theorem 8. MIN-CC GRAPH MOTIF *restricted to paths is W[1]-hard with respect to the parameter "number of components".*

Proof. (Construction) We reduce from the W[1]-complete PERFECT CODE [4] problem: Given an undirected graph $G = (V, E)$ and a positive integer k, is there is a size-k-subset $V' \subseteq V$ such that for every vertex $v \in V$ there is exactly one vertex in $N[v] \cap V'$. Given a PERFECT CODE instance $(G = (V, E), k)$, we construct a MIN-CC GRAPH MOTIF instance consisting of a path P and a motif M. It asks for the existence of a solution consisting of k connected components. The vertex set of P consists of $|N[v]|$ vertices with color c_v for every $v \in V$, $n-1$ "separator" vertices with color s each, and $2 \cdot n$ "end" vertices with color e each. Now, we describe the order of the vertices in the path P. For this, let a "subpath" of P denote a connected path that is part of P. Then, for every vertex $v \in V$ there is a subpath containing $|N[v]|$ vertices colored by $\{c_w \mid w \in N[v]\}$ in an arbitrary order. At both ends of every subpath we add an end vertex with color e. Finally, we connect all subpaths in an arbitrary order such that two neighboring subpaths are connected through a separator vertex with color s. The motif set M consists of $2 \cdot k$ times the color e and $\{c_v \mid v \in V\}$.

We omit to show that G has a perfect code of size k iff the there are k subpaths $P_1, \ldots, P_k$ such that there is a bijection between the colors of their vertices and the colors of M. □

4   Conclusion

GRAPH MOTIF and its variants are natural graph-theoretic pattern matching problems with prospective applications. Our positive algorithmic results should support implementation and experimental work, similarly to previous positive experiences with color-coding based graph algorithms [5, 7, 10, 11, 14]. It is particularly interesting whether the recently introduced subset convolution technique [3], which so far has been studied purely from a theoretical point of view, also yields a significant speed-up in practice.

Acknowledgments. We are grateful to Jiong Guo (hinting to Theorem 8) and Frances Rosamond for helpful comments.

References

[1] Alm, E., Arkin, A.P.: Biological networks. Curr. Opin. Struc. Biol. 13(2), 193–202 (2003)

[2] Alon, N., Yuster, R., Zwick, U.: Color-coding. J. ACM 42(4), 844–856 (1995)

[3] Björklund, A., Husfeldt, T., Kaski, P., Koivisto, M.: Fourier meets Möbius: fast subset convolution. In: Proc. 39th STOC, pp. 67–74. ACM, New York (2007)

[4] Cesati, M.: Perfect code is W[1]-complete. Inform. Process. Lett. 81, 163–168 (2002)

[5] Deshpande, P., Barzilay, R., Karger, D.R.: Randomized decoding for selection-and-ordering problems. In: Proc. NAACL HLT 2007. Association for Computational Linguistics, pp. 444–451 (2007)

[6] Dondi, R., Fertin, G., Vialette, S.: Weak pattern matching in colored graphs: Minimizing the number of connected components. In: Proc. 10th ICTCS. WSPC, vol. 4596, pp. 27–38. World Scientific, Singapore (2007)

[7] Dost, B., Shlomi, T., Gupta, N., Ruppin, E., Bafna, V., Sharan, R.: QNet: A tool for querying protein interaction networks. In: Speed, T., Huang, H. (eds.) RECOMB 2007. LNCS (LNBI), vol. 4453, pp. 1–15. Springer, Heidelberg (2007)

[8] Downey, R.G., Fellows, M.R.: Parameterized Complexity. Springer, Heidelberg (1999)

[9] Fellows, M.R., Fertin, G., Hermelin, D., Vialette, S.: Sharp tractability borderlines for finding connected motifs in vertex-colored graphs. In: Arge, L., Cachin, C., Jurdziński, T., Tarlecki, A. (eds.) ICALP 2007. LNCS, vol. 4596, pp. 340–351. Springer, Heidelberg (2007)

[10] Hüffner, F., Wernicke, S., Zichner, T.: Algorithm engineering for color-coding to facilitate signaling pathway detection. In: Proc. 5th APBC. Advances in Bioinf. and Comput. Biol., vol. 5, pp. 277–286. Imperial College Press (2007); Extended version to appear in Algorithmica

[11] Hüffner, F., Wernicke, S., Zichner, T.: FASPAD: fast signaling pathway detection. Bioinformatics 23(13), 1708–1709 (2007)

[12] Lacroix, V., Fernandes, C.G., Sagot, M.-F.: Reaction motifs in metabolic networks. In: Casadio, R., Myers, G. (eds.) WABI 2005. LNCS (LNBI), vol. 3692, pp. 178–191. Springer, Heidelberg (2005)

[13] Niedermeier, R.: Invitation to Fixed-Parameter Algorithms. Oxford University Press, Oxford (2006)

[14] Scott, J., Ideker, T., Karp, R.M., Sharan, R.: Efficient algorithms for detecting signaling pathways in protein interaction networks. J. Comput. Biol. 13(2), 133–144 (2006)

[15] Sharan, R., Ideker, T.: Modeling cellular machinery through biological network comparison. Nat. Biotechnol. 24, 427–433 (2006)

[16] Shen-Orr, S., Milo, R., Mangan, S., Alon, U.: Network motifs in the transcriptional regulation network of escherichia coli. Nat. Genet. 31(1), 64–68 (2002)

[17] Tarjan, R.E.: Depth first search and linear graph algorithms. SIAM J. Comp. (1), 146–160 (1972)

[18] Wernicke, S.: Efficient detection of network motifs. IEEE ACM T. Comput. Bi. 3(4), 347–359 (2006)

Finding Largest Well-Predicted Subset of Protein Structure Models

Shuai Cheng Li[1], Dongbo Bu[1,3], Jinbo Xu[2], and Ming Li[1]

[1] David R. Cheriton School of Computer Science, University of Waterloo, Canada
{scli,dbu,mli}@cs.uwaterloo.ca
[2] Toyota Technological Institute at Chicago, USA
j3xu@tti-c.org
[3] Institute of Computing Technology, Chinese Academy of Sciences, China

Abstract. [1] How to evaluate the quality of models is a basic problem for the field of protein structure prediction. Numerous evaluation criteria have been proposed, and one of the most intuitive criteria requires us to find a *largest well-predicted subset* — a maximum subset of the model which matches the native structure [12]. The problem is solvable in $O(n^7)$ time, albeit too slow for practical usage. We present a $(1 + \epsilon)d$ distance approximation algorithm that runs in time $O(n^3 \log n/\epsilon^5)$ for general protein structures. In the case of globular proteins, this result can be enhanced to a randomized $O(n \log^2 n)$ time algorithm with probability at least $1 - O(1/n)$. In addition, we propose a $(1 + \epsilon)$-approximation algorithm to compute the minimum distance to fit all the points of a model to its native structure in time $O(n(\log \log n + \log 1/\epsilon)/\epsilon^5)$. We have implemented our algorithms and results indicate our program finds much more matched pairs with less running time than TMScore, which is one of the most popular tools to assess the quality of predicted models.

1 Introduction

Quite a number of protein structure prediction methods are available, and each method produces a large number of models for a given sequence. Evaluation of the quality of these models is a difficult and fundamental subject which has been intensively studied in structural bioinformatics, and is still under active research [12]. Among the proposed techniques, Root Means Square Deviation (RMSD) is the most popular one [4]. However, using RMSD has a few short-comings. For example, RMSD fails to identify the quality of a model when only a substructure is predicted correctly. RMSD was first proposed to handle noisy data in which the error is small, and does not perform well when the overall structures have a large distance. RMSD is also not equivalent between targets of different lengths. For example, the quality of a model of 10 residues with an RMSD of 3Å is considered bad, while the quality of a model of 100 residues with an RMSD of 3Å is considered an accurate model. To overcome these difficulties,

---

[1] We have implemented a package named ApproxSub. The source code is available upon requests.

P. Ferragina and G. Landau (Eds.): CPM 2008, LNCS 5029, pp. 44–55, 2008.

measurements such as MaxSub [12], Global Distance Test (GDT), Local/Global Alignment (LGA) [14] and TMScore [15] have been proposed. For a more comprehensive review, we refer readers to [10]. Most of these methods use RMSD as a subroutine, and are heuristic.

Though intensive studies have been conducted to attack the similarity evaluation problem, most of the proposed methods are heuristic and do not have theoretical performance bound. In [12], the problem is formulated as to compute the largest 'well-predicted' subset (LWPS) to overcome the shortcomings of RMSD. The evaluation criteria by LWPS is elegant and intuitive, but it was believed this problem was NP-hard and heuristic approaches are proposed [12,10]. We provide a tool in this paper to address this problem more directly. The problem is actually polynomial solvable by the techniques from the field of computational geometry called *largest common point sets under approximate congruence*, with a distance threshold d (d-LCP) (under bottleneck matching measure). We also propose a $O(n^3 \log n/\epsilon^5)$ time algorithm to solve the $(1 + \epsilon)d$ distance approximation problem for general protein structures. In the case of globular proteins, this result can be enhanced to a randomized $O(n \log^2 n)$ time algorithm with probability at least $1 - O(1/n)$. In addition, we propose a $(1 + \epsilon)$-approximation algorithm to compute the minimum distance to fit all the corresponding points of a model and its native structure in time $O(n(\log \log n + \log 1/\epsilon)/\epsilon^5)$. Furthermore, our algorithms are simple to implement.

2   Preliminaries

2.1   Problem Definition and Notations

A protein structure A consists of an ordered set of n points in three-dimensional (3D) space, i.e., $A = (a_1, a_2, \ldots, a_n)$, $a_i \in \mathbb{R}^3$. Biologically, each point represents a $C\alpha$ atom. These $3D$ coordinates have some special characteristics. For example, A is bounded within a sphere of radius R_A. It is known that for general proteins, $R_A = O(n)$ and for globular proteins, $R_A = cn^{1/3}$, for some constant c [9]. For notation simplicity, we omit the leading constant of R_A for globular protein structure. The distance between any two points ($C\alpha$ atoms) in a protein structure cannot be too small due to steric clashes. Furthermore, the distance between any two non-consecutive points is no less than 4Å. The distance between any two consecutive points is about 3.8Å. Due to such distance constraints, the maximum number of points that can be encapsulated in a given sphere with radius r is proportional to the volume of the sphere. For notation simplicity, we use r^3 and the number of points that can be encapsulated in the sphere exchangeably, when the context is clear.

Given a set of indices M' and a point set $P = \{p_1, \ldots, p_n\}$, denote $P[M'] = \{p_i | i \in M\}$.

The predicted model B of the protein consists of an ordered set of n points, i.e., $B = \{b_1, b_2, \ldots, b_n\}$. We assume B has the geometry properties of protein structure. Given a threshold d and a rigid transformation $\mathcal{I}$, if $|a_i - \mathcal{I}(b_i)| \leq d$, we say a_i *matches* b_i, or b_i *fits into* a_i under $\mathcal{I}$. Denote $M_{\mathcal{I}} = \{i | |a_i - \mathcal{I}(b_i)| \leq d\}$

and $M_{\mathcal{I}}$ is referred to as a *match set*. We also refer to $A[M_{\mathcal{I}}]$ and $B[M_{\mathcal{I}}]$ as the match set when the context is clear.

We now define the problems that will be studied in this paper:

LARGEST WELL-PREDICTED SUBSET PROBLEM. Given a protein structure A, a model B and a threshold d, the *largest well-predicted subset problem*, or $LWPS$ (A, B, d), is to identify a maximum match set $M_{opt} \subseteq \{1, 2, \ldots, n\}$ and a corresponding rigid transformation $\mathcal{I}_{opt}$ (a rotation and translation) [12,10]. d is called the *bottleneck distance*. Denote $A_{opt} = A[M_{opt}]$ and $B_{opt} = B[M_{opt}]$

Also, we are interested in solving the minimum bottleneck distance problem.

MINIMUM BOTTLENECK DISTANCE PROBLEM. Given a protein structure A and a model B, find the smallest distance d_{opt} and a corresponding rigid transformation $\mathcal{I}_{opt}$ such that $\forall i, 1 \leq i \leq n, |a_i - \mathcal{I}_{opt}(b_i)| \leq d_{opt}$.

With a careful examination of the algorithm for d-LCP in [3,5], one can see that the LWPS problem has a polynomial time solution in $O(n^7)$, which contradicts to the claim in [12] that the problem is NP-hard.

Theorem 1 ([5]). *The largest well-predicted subset can be solved in $O(n^7)$ time under general transformations.*

In this paper, we are interested in the following types of approximation algorithms:

1. Distance Approximation for $LWPS(A, B, d)$. Find a transformation $\mathcal{T}$ to bring a subset $B' \subseteq B$ of size at least B_{opt} such that $\forall b_i \in B', ||a_i - \mathcal{T}(b_i)|| \leq (1 + \epsilon)d$, ϵ is some constant.
2. Bottleneck Distance Approximation. Find a transformation $\mathcal{T}$, such that $\forall b_i \in A, ||a_i - \mathcal{T}(b_i)|| \leq (1 + \epsilon)d_{opt}$, ϵ is some constant.

3 A Discretization of the Rigid Transformation

We now start with basic notations. The *d-sphere* of a point p is the sphere of radius d centered at p. Given a point p and a point set P, p' form a *radial point in P with respect to p* iff p' is the furthest point in P from p. Points $p, p' \in P$ is a *radial pair* $\langle p, p' \rangle$ of P iff p' is a radial point w.r.t. p. (Note that $\langle p, p' \rangle$ is a radial pair of P does not imply that $\langle p', p \rangle$ is a radial pair of P.)

A rigid transformation consists of a rotation and a translation. In this paper, we consider how to approximate a rigid transformation $\mathcal{T}$ on a point set P by considering $\mathcal{T}$ as being composed of the following two components: (1) an initial transformation T that transforms an arbitrary chosen radial pair $\langle p_1, p_2 \rangle$ in P into their positions under $\mathcal{T}$, i.e. $T(p_1) = \mathcal{T}(p_1)$ and $T(p_2) = \mathcal{T}(p_2)$; (2) a rotation R around the axis along $\overrightarrow{T(p_1)T(p_2)}$ such that $\forall p \in P, R(T(p)) = \mathcal{T}(p)$.

Let T be an initial transformation, as in (1), of $\mathcal{T}$ on a given point set P, and let $\langle p_1, p_2 \rangle$ be the radial pair in P on which T is derived. Suppose we are able to find an approximation T' for T such that $T'(p_1) \approx \mathcal{T}(p_1)$ and $T'(p_2) \approx \mathcal{T}(p_2)$, then we claim that there exists a rotation R around the axis along $\overrightarrow{T'(p_1)T'(p_2)}$, that transforms every point $p \in T'(P)$ to some point near $\mathcal{T}(p)$. Formally,

Lemma 1. *Given a point set P, rigid transformations $\mathcal{T}$ and T', let $\langle p_1, p_2 \rangle$ be a radial pair of P, if $|\mathcal{T}(p_1) - T'(p_1)| \leq \epsilon$ and $|\mathcal{T}(p_2) - T'(p_2)| \leq \epsilon$, then there exists a rotation R around the axis along $\overrightarrow{T'(p_1)T'(p_2)}$, such that $\forall p \in P$, $\|R(T'(p)) - \mathcal{T}(p)\| \leq 3\epsilon$.*

Proof. Denote $p_1' = T'(p_1)$ and $p_2' = T'(p_2)$. With two points fixed, the only degree of freedom for a rigid transformation T' on P are rotations around the axis along $\overrightarrow{p_1' p_2'}$. Therefore, we just need to show that there exists a transformation T'' which transforms $\mathcal{T}(P)$ such that $T''(\mathcal{T}(p_1))$ coincides with p_1', and $T''(\mathcal{T}(p_2))$ coincides with p_2', and $\forall p \in P$, $\|T''(\mathcal{T}(p)) - \mathcal{T}(p)\| \leq 3\epsilon$.

We consider T'' as follows, in two steps. First, we translate $\mathcal{T}(P)$ with translation t such that $\mathcal{T}(p_1)$ coincides with $T'(p_1)$ and let $\mathcal{T}(P) - t$ be the 3D point set of $\mathcal{T}(P)$ with translation t. Second, we rotate $\mathcal{T}(P) - t$ with rotation axis as the line which passes though p_1' and is orthogonal to the plane defined by points p_1', $\mathcal{T}(p_2) - t$ and p_2', with rotation angle as the angle formed by $\mathcal{T}(p_2) - t$, p_1' and p_2', where p_1' is the vertex. Denote this rotation as R'' and the rotation angle as α. It can be verified that $p_1' = T''(\mathcal{T}(p_1))$ and $p_2' = T''(\mathcal{T}(p_2))$. With rotation R'', we move $\mathcal{T}(p_2) - t$ to coincide with p_2'.

By translation t, we know that $\forall p \in P$, $\|\mathcal{T}(p) - (\mathcal{T}(p) - t)\| = \|t\| = \|\mathcal{T}(p_1) - p_1'\| \leq \epsilon$. As $\|\mathcal{T}(p_2) - T'(p_2)\| \leq \epsilon$, we have $\|(\mathcal{T}(p_2) - t) - p_2'\| \leq 2\epsilon$. Consider the angle formed by $\mathcal{T}(p) - t$, p_1' and $R(\mathcal{T}(p) - t)$, $p \in P$, where p_1' is the vertex, we know that (1) the angle formed by points p_1', $\mathcal{T}(p) - t$ and $R(\mathcal{T}(p) - t)$ with p_1' as the vertex is at most α; and (2) $\|p_1 - p\| \leq \|p_1 - p_2\|$. By these two properties, we have $\|(\mathcal{T}(p) - t) - R(\mathcal{T}(p) - t)\| \leq \|(\mathcal{T}(p_2) - t) - p_2'\| \leq 2\epsilon$. Therefore, by triangle inequality we have $\|T''(p) - \mathcal{T}(p)\| \leq 3\epsilon$. The statement holds. $\qquad\square$

3.1 Match a Radial Pair Approximately

If we know a radial pair $\langle b_i, b_j \rangle$ of B_{opt} matches to a pair $\langle a_i, a_j \rangle$ of A, then we have $\|a_i - \mathcal{T}_{opt}(b_i)\| \leq d$ and $\|a_j - \mathcal{T}_{opt}(b_j)\| \leq d$. Thus we can discretize the $(1 + \epsilon)d$-spheres of a_i and a_j with a grid of side length $1/3\epsilon d$, and exhaustively try every grid pair for possible positions of b_i and b_j. This will result in an error at most ϵd according to Lemma 1.

As shown in Fig 1, we partition the $(1 + \epsilon)d$ sphere of a_i with 3D grids of side length $1/3\epsilon d$. The number of grid points to partition the $(1 + \epsilon)d$ sphere of a_i is bounded by $O((d + \epsilon)^3/(1/3\epsilon d)) = O(1/\epsilon^3)$. Here we can try all the grid positions for b_i.

Once we have fixed b_i at a grid point, all the possible positions for b_j fitting into the $(1 + \epsilon)$-sphere a_j form a sphere cap centered at b_i with radius $\|b_j - b_i\|$ and contained in the $(1 + \epsilon)d$ sphere of a_j. The spherical cap has an area of $O(d^2)$. We partition the sphere cap with grids of resolution size $1/\epsilon$. This can be approximated by creating the smallest cube which encapsulates the sphere (the one the sphere cap belongs to) and create grids of side length $O(1/\epsilon)$ of the six faces of the cube. Then we can use the grid on the cube to partition the sphere cap — a common trick used in computation geometry to round the

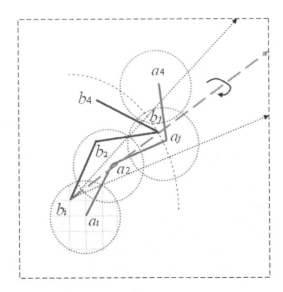

Fig. 1. Approximating $\mathcal{T}_{opt}(b_i)$ and $\mathcal{T}_{opt}(b_i)$. *First, we match a radial axis to approximated positions (the grid points), then we rotate B around this radial axis to find a maximum match. A and B both consists of four points in this example. We match b_i to a_i approximately. Then we discretize the possible directions for $\langle b_i, b_j \rangle$. Last, we rotate B around $\langle b_i, b_j \rangle$ to find a maximum match.*

directions [1]. Note that we do not need to create the grid explicitly. It is easy to show that only $O(1/\epsilon^2)$ grid points are necessary to partition the sphere cap.

Combining with Lemma 3, to be shown later, we have the following result.

Lemma 2. *If we know a radial pair $\langle b_i, b_j \rangle$ of B_{opt} matching to a pair $\langle a_i, a_j \rangle$ of A, there are $O(1/\epsilon^5)$ possible choices to transform $\langle b_i, b_j \rangle$ such that at least one of the transformations results in error at most ϵd for each $b \in B$ from their optimal positions.*

Note that we do not know which pair of points is a radial pair of B_{opt}. However, this can be overcome by enumerating all the possible $O(n^2)$ combinations, as we next discuss.

3.2   Exact Algorithm for Restricted Rotation Axis

Suppose all the points of B must be rotated around a given axis, we want to identify an angle $\theta \in [0, 2\pi)$ such that the number of matched pairs is maximized. If we represent the interval $[0, 2\pi)$ as a unit circle, then it is not difficult to see that the angle that moves b_i into the $(1 + \epsilon)d$-sphere of a_i form an arc of the circle. Totally, there are $O(n)$ arcs. Each arc consists of two endpoints, and the circle is subdivided into $O(n)$ circular intervals. Each of these intervals consists of a set of equivalent rotation angles and we can simply pick up an angle contained in the interval to represent the the interval. The problem is equivalent to that of finding a point on the circle covered by the maximum number of arcs and it can be solved by the algorithm used in [2,6] with a plane-sweep approach.

Lemma 3. *The LWPS(A, B, d) problem can be solved in time $O(n \log n)$ when rotations are allowed only on a given rotation axis.*

3.3   Distance Approximating of the LWPS Problem

As we do not know which pair is a radial pair of B_{opt}, we enumerate all the possible cases. Totally there are $O(n^2)$ possible cases. For two pairs $\langle b_i, b_j \rangle$ and $\langle a_i, a_j \rangle$, we have $O(1/\epsilon^5)$ ways to match them. For each discretizaion we need time $O(n \log n)$ to find the best match by Lemma 3. Therefore we have the following result.

Theorem 2. *The $LWPS(A, B, d)$ can be solved in time $O(n^3 \log n/\epsilon^5)$ with a $(1 + \epsilon)d$ distance approximation algorithm.*

4   An Efficient Randomized Algorithm for Globular Protein Structure

The distance approximation algorithm proposed in Section 3.3 has a time complexity of $O(n^3 \log n)$, which is still inefficient. If we know a radial pair $\langle b_i, b_j \rangle$ of B_{opt}, then we can solve the problem in time $O(n \log n/\epsilon^5)$. This observation inspires us to improve the algorithm by identifying a radial pair $\langle b_i, b_j \rangle$ or some pair good enough to approximate a radial pair. This section presents an efficient method to identify such a pair with high probability for *meaningful* models of globular proteins.

A model is *meaningful* if the TMScore is greater than 0.4 [15]. Here TMScore is defined as:

$$TM(A, B) = 1/n \sum_{i \in M'} \frac{1}{1 + (d_i/d)^2}$$

where d_i is the Euclidean distance between a_i and b_i under the transformation, d has a similar meaning as in present paper, which is a predefined threshold, and M' is a subset of $[1, n]$.

Immediately, we can prove that M' has a size of at least $0.4n$. A careful analysis of the TMScore paper will show that M' has a subset of at least $0.1n$ matched pairs with distance less than d for a meaningful match. Therefore, the following assumption is reasonable:

Assupmption 1. *A meaningful prediction B of structure A has $|LWPS(A, B, d)| \geq \alpha n$, for some constant α.*

We call a pair of points b_i and b_j a *pseudo radial* pair if $|b_i - b_j| \geq (1/2\alpha n)^{1/3}$. We create grids of side length $1/3(1/2\alpha)^{1/3}\epsilon d$, recall that for globular proteins

$R_B = n^{1/3}$. If we use pseudo radial pairs as a radial pair, the error introduced at each point in the match set is less than:

$$3\frac{n^{1/3}}{(1/2\alpha n)^{1/3}} \times 1/3(1/2\alpha)^{1/3}\epsilon d = \epsilon d$$

Therefore, we can make the following statement:

Lemma 4. *Given a globular protein* P *, rigid transformations* $\mathcal{T}$ *and* $\mathcal{T}'$, *let* $\langle p_1, p_2 \rangle$ *be a pseudo radial pair of* P, *if* $|\mathcal{T}(p_1) - \mathcal{T}'(p_1)| \leq \epsilon$ *and* $|\mathcal{T}(p_2) - \mathcal{T}'(p_2)| \leq \epsilon$, *then there exists a rotation* R *around the axis along* $\overrightarrow{\mathcal{T}'(p_1)\mathcal{T}'(p_2)}$, *such that* $\forall p \in P$, $\|R(\mathcal{T}'(p)) - \mathcal{T}(p)\| \leq 3c\epsilon$, *where* c *is some constant.*

The proof is omitted. Thus any pseudo radial pair gives us a $(1 + \epsilon)$ distance approximation algorithm.

Theorem 3. *There exists a probabilistic* $(1 + \epsilon)d$ *distance approximation algorithm for LWPS for globular proteins of meaningful models with probability at least* $1 - O(1/n)$ *in time* $O(n \log^2 n/\epsilon^5)$.

We first prove that there exist *enough* radial pairs.

Lemma 5. B_{opt} *contains at least* $1/2|B_{opt}|^2$ *pairs* b_i *and* b_j *such that* $|b_i - b_j| \geq (1/2\alpha n)^{1/3}$.

Proof. The number points confined in the ball centered at p with radius $(1/2\alpha n)^{1/3}$, $p \in P$ is bounded by $1/2\alpha n$. This implies that the number of points in B_{opt} has a distance at least $(1/2\alpha n)^{1/3}$ is at least $|B_{opt}| - 1/2\alpha n$. Thus the statement holds.   □

Since there are at least $1/2(\alpha n)^2$ pseudo radial pairs, randomly sampling $\lceil 1/\alpha^2 \log n \rceil$ pairs from B yields a randomized distance approximate algorithm. Note that each pair has a probability $\frac{1/2(\alpha/n)^2}{1/2n(n-1)} = O(1)$ of being a pseudo radial pair. Given that there are $\lceil 1/\alpha^2 \log n \rceil$ pairs, the probability that none of them is a pseudo radial pair is $1 - O(1/n)$. In addition, calculation for each pair needs time $O(n \log n/\epsilon^5)$, thus the total time complexity is $O(n \log^2 n/\epsilon^5)$.

5   Approximating the Bottleneck Distance

In some cases, we need to compute the minimum distance d^* such that each point b_i in B can fit into the corresponding d^*-sphere of a_i, $a_i \in A$. Techniques in [2] can be used to attack this problem; however, these techniques suffer from high time complexity. We present in this section an efficient one.

First, we investigate the problem if we have some d' such that $d' \leq d_{opt} \leq 2d'$. We have the following fact:

Lemma 6. *If* $d' \leq d_{opt} < 2d'$, *we can approximate* d_{opt} *with ratio* $(1 + \epsilon)$ *in time* $O(\frac{n \log 1/\epsilon}{\epsilon^5})$.

Proof. We subdivide interval $[d', 2d']$ into intervals of length $0.5\epsilon d'$ (assume 1 is divisible by 0.5ϵ). Totally there are $2/\epsilon$ such intervals. For each interval $\lambda_i = [d'(1 + 0.5i\epsilon), d'(1 + 0.5(i + 1)\epsilon)]$ $(0 \leq i \leq 2/\epsilon - 1)$, we build grids of side length $1/3\epsilon d'$ for $(1 + 0.5(i + 1)\epsilon)d'$-sphere of a_i, $1 \le i \le n$. Then we check if there is a transformation specified by such grids to fit all the points. For two consecutive intervals λ_i and λ_{i+1}, if there is a feasible solution for interval $i + 1$, and it is infeasible for interval $i + 1$, then we know that $d'(1 + 0.5i\epsilon) \le d_{opt} \le d'(1 + 0.5(i+2)\epsilon) + \epsilon d'$. This yields us an $(1 + \epsilon)d_{opt}$ algorithm immediately. We can use a binary search to find such i, and we need $O(\log 1/\epsilon)$ search operations.

In addition, for each search operation, it will be expensive if we employ the enumerating techniques proposed in the previous sections. Instead, we notice that any radial pair of B can be used as we want to fit all the points. Totally, there are $O(1/\epsilon^5)$ possible choices for a given radial pair. Given a rotation axis, the angle to fit b_i into a_i can be modelled as an arc on a circle as previously, and we just need to check if there is a point on the circle covered by n arcs, this can be done in $O(n)$.

Thus each search operation can be performed in time $O(n/\epsilon^5)$.   □

Now the remaining difficulty is to find a d' meeting the requirement $d' \le d_{opt} < 2d'$. We make use of RMSD to achieve this goal. RMSD can be computed in linear time [4]. RMSD is defined as the minimal root mean square distance over all the possible transformation I, i.e.,

$$RMSD(A, B) = \min_I \sqrt{\frac{\sum_{i=1}^n ||a_i - I(b_i)||^2}{n}}$$

Let $\mathcal{D} = RMSD(A, B)$, according to the definition of RMSD, we can prove:

Lemma 7. $\mathcal{D} \le d_{opt} \le \sqrt{n}\mathcal{D}$

Proof. First, we prove that $\mathcal{D} \le d_{opt}$. Suppose $\mathcal{D} > d_{opt}$, and let I' be the transformation to obtain d_{opt}, then we have: $\sqrt{\frac{\sum_{i=1}^n ||a_i - I'(b_i)||^2}{n}} < \sqrt{\frac{\sum_{i=1}^n d_{opt}^2}{n}} = d_{opt}$. This contradicts the definition of $RMSD$.

Second, we prove $d_{opt} \le \sqrt{n}\mathcal{D}$, let I^* be the transformation to obtain the RMSD distance:

$$d_{opt}^2 \le \max_{i=1}^n \{||a_i - I^*(b_i)||^2\} \le \sum_{i=1}^n ||a_i - I^*(b_i)||^2 = n\mathcal{D}^2$$

□

We subdivide interval $[\mathcal{D}, n^{1/2}\mathcal{D}]$ into intervals $[2^i \times \mathcal{D}, 2^{i+1} \times \mathcal{D}]$, $0 \le i \le 1/2\log n - 1$ (assume $1/2\log n$ is an integer, WLOG). For each interval, we build grids of side length $\frac{1}{3}2^i \times \mathcal{D}\epsilon$ and sphere of radius $2^{i+1} \times \mathcal{D} + 2^i \times \mathcal{D}\epsilon$. If there is a feasible solution under such grids, then we know that $d_{opt} \le 2^{i+1} \times \mathcal{D} + 2^i \times \mathcal{D}\epsilon$. We can perform a binary similar as previous to find such i.

Thus we have the following results:

Theorem 4. *The bottleneck distance can be approximated with ratio $(1 + \epsilon)d_{opt}$ in time $O(n(\log \log n + \log 1/\epsilon)/\epsilon^5)$.*

6   Results

We have implemented the algorithm in Section 3.3 and named as ApproxSub. The program has been implemented carefully to avoid redundant computations. First, given pairs $\langle b_i, b_j \rangle$ and $\langle a_i, a_j \rangle$, let $d_{i,j} = |b_j - b_i| - |a_j, a_i|$. If $|d_{i,j}| < 2d$ or $|d_{i,j}| > 2d$, we simply preclude $\langle b_i, b_j \rangle$ as a radial pair candidate, as it is impossible for b_i match a_i and b_j match a_j simultaneously. Second, given a radial pair candidate $\langle b_i, b_j \rangle$, we compute an upper bound for all the $O(\epsilon^5)$ axis under pair $\langle b_i, b_j \rangle$ by employing the approximation algorithm in [6]. If the bound is smaller than the best solution that we have found so far, we do not need to explore any further for pair $\langle b_i, b_j \rangle$. Third, we try to explore the pairs which probably yield us the best solution first. We have employ some other rules to accelerate our program, such that it is practical and efficient.

MaxSub and TMScore are two popular programs used for protein quality assessment. Our program are not directly comparable to MaxSub and TMScore. However, the number of matched pairs is used as one of indicators to show the quality of the models in both methods. We compare against this indicator.

We compare our method to the MaxSub and TMScore of finding number of matched pairs. To compare with MaxSub, we use the default setting for MaxSub with a distance threshold 3.5Å. We set $d = 3.2$Å and $\epsilon d = 0.3$Å for ApproxSub. To compare with TMScore, we use the default settings for TMScore, with distance threshold 5.0Å (which is used to compute the RAW TMScore). We set $d = 4.5$Å and $\epsilon d = 0.5$Å in ApproxSub, such that any matched pairs has a distance no more than 5.0Å. We normalize the matched pairs by dividing it by n.

We use the six proteins 1fc2, 1enh, 2gb1, 2cro, 1ctf and 4icb to evaluate the AppproxSub. These six proteins are commonly used to test the quality of structural prediction methods based on fragment assembly [13,8,7]. We randomly picked up 1000 models generated by ROSETTA [13]. Fig. 2 shows the results. Fig.2(a) is the comparison between ApproxSub and MaxSub. Each point stands for a model, x is the ratio of matched pairs by MaxSub, and y is the ratio of matched pairs by ApproxSub. Fig.2(b) is the comparison between ApproxSub and TMScore. Both figures display that ApproxSub found much more matched pairs. Also we notice that MaxSub and TMScore are poor to find matched pairs when the ratio of matched pairs are higher than 60%. This is mainly due to the heuristic nature of the two methods, and in which, they extend the match by superimposing a local fragment of A and B first. This may trap them at a local minimum.

Time efficiency is the main concern for PTAS as the leading constant is generally large. The MaxSub is implemented in TMScore. So we just compare the running time against TMScore, since TMScore and MaxSub have the same running time if we use the package from [15]. The setting is the same as previous, except we use more models of various lengths that generated by prediction methods for CASP7 targets[11]. Totally, there are 92 targets. We use 100 models for each target, and take the average running time for comparison. The reason that

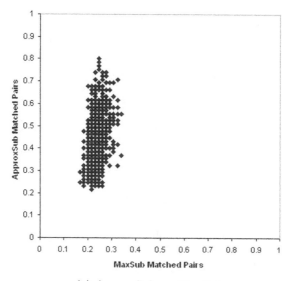

(a) ApproxSub vs. MaxSub

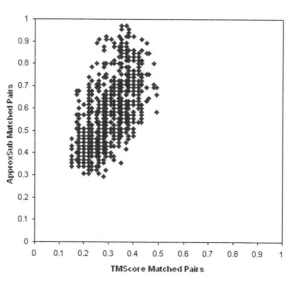

(b) ApproxSub vs. TMScore

Fig. 2. Ratio of Matched Pairs

we choose CASP7 targets rather than the above six proteins is due to that CASP7 target provides enough length variation of the proteins.

Fig. 3(a) shows the running time of ApproxSub against TMScore. The x-coordinate is the length for protein structures. The y-coordinate is the CPU time in seconds. Our program is significantly faster.

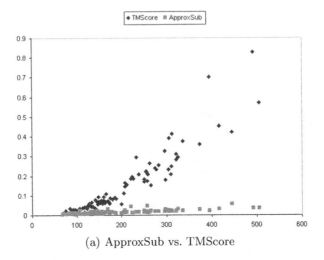

(a) ApproxSub vs. TMScore

Fig. 3. Running time of ApproxSub vs. MaxSub

7   Discussion

We have implemented our algorithms, it is efficient and practical. Our package can be used to qualify the heuristic approaches, and to identify all the well-predicted sub-structures. Furthermore, our package has a theoretical bound.

Our techniques can also be applied to *protein structure alignment* problems. However, it will yield theoretical results rather than practical tools due to high time complexity.

References

1. Agarwal, P.K., Matoušek, J., Suri, S.: Farthest neighbors, maximum spanning trees and related problems in higher dimensions. Comput. Geom. Theory Appl. 1(4), 189–201 (1992)
2. Alt, H., Mehlhorn, K., Wagener, H., Welzl, E.: Congruence, similarity, and symmetries of geometric objects. In: SCG 1987: Proceedings of the third annual symposium on Computational geometry, pp. 308–315. ACM Press, New York (1987)
3. Ambühl, C., Chakraborty, S., Gärtner, B.: Computing largest common point sets under approximate congruence. In: Paterson, M. (ed.) ESA 2000. LNCS, vol. 1879, pp. 52–64. Springer, Heidelberg (2000)
4. Arun, K.S., Huang, T.S., Blostein, S.D.: Least-squares fitting of two 3-d point sets. IEEE Trans. Pattern Anal. Mach. Intell. 9(5), 698–700 (1987)
5. Choi, V., Goyal, N.: A combinatorial shape matching algorithm for rigid protein docking. In: Sahinalp, S.C., Muthukrishnan, S.M., Dogrusoz, U. (eds.) CPM 2004. LNCS, vol. 3109, pp. 285–296. Springer, Heidelberg (2004)
6. Choi, V., Goyal, N.: An efficient approximation algorithm for point pattern matching under noise. In: Correa, J.R., Hevia, A., Kiwi, M. (eds.) LATIN 2006. LNCS, vol. 3887, pp. 298–310. Springer, Heidelberg (2006)

7. Hamelryck, T., Kent, J.T., Krogh, A.: Sampling Realistic Protein Conformations Using Local Structural Bias. PLoS Computational Biology 2(9), e131 (2006)
8. Kolodny, R., Koehl, P., Guibas, L., Levitt, M.: Small libraries of protein fragments model native protein structures accurately. J. Mol. Biol. 323, 297–307 (2002)
9. Kolodny, R., Linial, N.: Approximate protein structural alignment in polynomial time. Proc. Natl. Acad. Sci. 101, 12201–12206 (2004)
10. Lancia, G., Istrail, S.: Protein structure comparison: Algorithms and applications. In: Mathematical Methods for Protein Structure Analysis and Design, pp. 1–33 (2003)
11. Moult, J., Fidelis, K., Rost, B., Hubbard, T., Tramontano, A.: Critical assessment of methods of protein structure prediction (casp):round 6. Proteins: Struct. Funct. Genet. 61, 3–7 (2005)
12. Siew, N., Elofsson, A., Rychlewski, L., Fischer, D.: Maxsub: an automated measure for the assessment of protein structure prediction quality. Bioinformatics 16(9), 776–785 (2000)
13. Simons, K.T., Kooperberg, C., Huang, E., Baker, D.: Assembly of Protein Tertiary Structures from Fragments with Similar Local Sequences using Simulated Annealing and Bayesian Scoring Functions. J. Mol. Biol. 268 (1997)
14. Zemla, A.: LGA: a method for finding 3D similarities in protein structures. Nucl. Acids Res. 31(13), 3370–3374 (2003)
15. Zhang, Y., Skolnick, J.: Scoring function for automated assessment of protein structure template quality. Proteins: Structure, Function, and Bioinformatics 57(4), 702–710 (2004)

HP Distance Via Double Cut and Join Distance

Anne Bergeron[1], Julia Mixtacki[2], and Jens Stoye[3]

[1] Dépt. d'informatique, Université du Québec à Montréal, Canada
bergeron.anne@uqam.ca
[2] International NRW Graduate School in Bioinformatics and Genome Research,
Universität Bielefeld, Germany
julia.mixtacki@uni-bielefeld.de
[3] Technische Fakultät, Universität Bielefeld, Germany
stoye@techfak.uni-bielefeld.de

Abstract. The genomic distance problem in the Hannenhalli-Pevzner theory is the following: Given two genomes whose chromosomes are linear, calculate the minimum number of inversions and translocations that transform one genome into the other. This paper presents a new distance formula based on a simple tree structure that captures all the delicate features of this problem in a unifying way.

1 Introduction

The first solution to the genomic distance problem was given by Hannenhalli and Pevzner [6] in 1995. Their distance formula, called the general *HP distance*, requires preprocessing steps such as *capping* and *concatenation* and involves seven parameters. In the last decade, different authors pointed to problems in the original formula and in the algorithm given by Hannenhalli and Pevzner. Their algorithm was first corrected by Tesler [9]. In 2003, Ozery-Flato and Shamir [8] found a counter-example and modified one of the parameters of the distance formula. Very recently, another correction was presented by Jean and Nikolski [7]. Unfortunately, the last two recent results have not resulted in simpler presentations of the material, nor are they implemented in software tools yet. The only available tool is *GRIMM* implemented by Glenn Tesler [10].

In contrast to this rather complicated distance measure, Yancopoulos *et al.* [11] presented a general genome model that includes linear and circular chromosomes and introduced a new operation called *double cut and join* (or shortly DCJ) operation. In addition to inversions and translocations, the DCJ operation also models transpositions and block-interchanges. Beside the simple distance computation, the sorting algorithm is also basic and efficient [4].

In this paper we will show how the rearrangement model considered in the HP theory can be integrated in the more general DCJ model. Specifically, the HP distance can be expressed as

$$d_{HP} = d_{DCJ} + t$$

P. Ferragina and G. Landau (Eds.): CPM 2008, LNCS 5029, pp. 56–68, 2008.

where t represents the extra cost of not resorting to unoriented DCJ operations. The extra cost can easily be computed by a tree data structure associated to a genome.

The next section recalls the results on the DCJ distance. In Section 3, we establish the conditions under which the two distances are equal. The general case is treated in Section 4, where we introduce the basic concepts and the tree needed for the computation of the HP distance, and we give a new proof and formula for the Hannenhalli-Pevzner theorem. Section 5 presents the conclusion.

2   The Double Cut and Join (DCJ) Model

Let A and B be two linear multi-chromosomal genomes on the same set of N genes. A linear chromosome will be represented by an ordered sequence of signed genes, flanked by two unsigned telomere markers:

$$(\circ, g_1, \ldots, g_n, \circ).$$

An *interval* $(l, \ldots, r)$ in a genome is a set of consecutive genes or telomere markers within a chromosome; the set $\{l, -r\}$ is the set of *extremities* of the interval – note that $\circ = -\circ$. An *adjacency* is an interval of length 2, an adjacency that contains a telomere marker is called a *telomere*. Each gene g is the extremity of two adjacencies, one as $+g$, and one as $-g$, in both genomes A and B. This remark yields the following basic construct:

Definition 1. *The* adjacency graph $AG(A, B)$ *is a graph whose vertices are the adjacencies of genomes A and B. Each gene g defines two edges, one connecting the two adjacencies of genome A and B in which g appears as extremity $+g$, and one connecting the two adjacencies in which g appears as extremity $-g$.*

Since adjacencies that are telomeres have only one gene, the vertices of the adjacency graph will have degree one or two, thus the graph is a union of *paths* and *cycles*. Paths of odd length, called *odd paths*, connect telomeres of different genomes, and paths of even length, the *even paths*, connect telomeres of the same genome. For example, the adjacency graph of the genomes $A = \{(\circ, 3, 2, 1, 4, \circ), (\circ, 6, 5, \circ)\}$ and $B = \{(\circ, 1, 2, 3, 4, \circ), (\circ, 5, 6, \circ)\}$ has two odd paths, one cycle and two even paths:

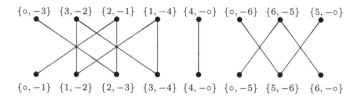

A *DCJ operation* applied to two adjacencies of the same genome disconnects the incident edges of the adjacency graph, and reconnects them in one of the possible other ways. The *DCJ distance* between genomes A and B, $d_{DCJ}(A, B)$, is the minimum number of DCJ operations necessary to transform genome A into genome B. We have:

Theorem 1 ([4]). *Let A and B be two genomes defined on the same set of N genes, then we have*

$$d_{DCJ}(A, B) = N - (C + I/2)$$

where C is the number of cycles and I the number of odd paths in $AG(A, B)$. An optimal sorting sequence can be found in $O(N)$ time.

A DCJ operation that reduces the DCJ distance by 1 is called *DCJ-sorting*. Using Theorem 1, we have the following property of DCJ-sorting operations, using the fact that a DCJ operation acts on at most two paths or cycles, and produces at most one new path or cycle:

Corollary 1. *A DCJ-sorting operation acts on a single path or cycle, or on two even paths of the adjacency graph.*

Some DCJ operations can create intermediate circular chromosomes, even if both genomes A and B are linear, and we will want to avoid them in the HP model. The following definition is a generalization of a classical concept in rearrangement theory, *oriented operations*:

Definition 2. *A DCJ-sorting operation is* oriented *if it does not create circular chromosomes.*

For two linear genomes, oriented operations are necessarily inversions, translocations, fusions or fissions. These operations are also called *HP operations*, and the HP distance between two genomes $d_{HP}(A, B)$ is the minimum number of HP operations needed to transform genome A into genome B. Since DCJ operations are more general than HP operations, we always have the following lower bound:

Proposition 1. *For two linear genomes A and B, we have that $d_{DCJ}(A, B) \leq d_{HP}(A, B)$.*

3 Components and Oriented Sorting

In this section, we introduce the notion of *components*. They roughly correspond to the classical concept of components, but in the context of adjacency graphs, we prove that they are unions of paths and cycles.

3.1 Basic Definitions

Definition 3. *Given two genomes A and B, an interval $(l, \ldots, r)$ of genome A is a* component *relative to genome B if there exists an interval in genome B:*
   a) with the same extremities,
   b) with the same set of genes, and
   c) that is not the union of two such intervals.

Example 1. Let

$$A = \{(\circ, 2, 1, 3, 5, 4, \circ), (\circ, 6, 7, -11, -9, -10, -8, 12, 16, \circ), (\circ, 15, 14, -13, 17, \circ)\},$$
$$B = \{(\circ, 1, 2, 3, 4, 5, \circ), (\circ, 6, 7, 8, 9, 10, 11, 12\circ), (\circ, 13, 14, 15, \circ), (\circ, 16, 17, \circ)\}.$$

The components of genome A relative to genome B are: $(\circ, 2, 1, 3)$, $(3, 5, 4, \circ)$, $(\circ, 6)$, $(6, 7)$, $(-11, -9, -10, -8)$, $(7, -11, -9, -10, -8, 12)$, $(\circ, 15, 14, -13)$ and $(17, \circ)$.

Note that components of length 2 are the same adjacencies in both genomes, possibly up to flipping of a chromosome. These are called *trivial components*.

Two components are *nested* if one is included in the other and their extremities are different. As the following lemma shows, two components cannot share a telomere:

Lemma 1. *If* $(\circ, \ldots, r_1)$ *and* $(\circ, \ldots, r_1, \ldots, r_2)$ *are two components, then* $r_1 = r_2$, *and if* $(l_1, \ldots, l_2, \ldots, \circ)$ *and* $(l_2, \ldots, \circ)$ *are two components then* $l_1 = l_2$.

Proof. Suppose that $(\circ, \ldots, r_1)$ and $(\circ, \ldots, r_1, \ldots, r_2)$ are two components. Since the corresponding intervals in genome B, $(\circ, \ldots, r_1)$ and $(\circ, \ldots, r_2)$, share the same gene content, the interval $(r_1, \ldots, r_2)$ shares the same gene content in both genomes, thus $(r_1, \ldots, r_2)$ is a component, and $(\circ, \ldots, r_1, \ldots, r_2)$ is the union of two components, a contradiction. The other statement has a similar proof. □

It is further known that two components can not properly overlap on two or more elements. We thus have the following generalization of a statement from [5]:

Proposition 2. *Two components are either disjoint, nested, or overlap on exactly one gene.*

Proposition 2 implies that components can be partially ordered by inclusion, and that overlapping components will have the same parent. An adjacency *properly belongs* to the smallest component that contains it.

Definition 4. *The adjacency graph of a component C is the subgraph of the adjacency graph of genomes A and B induced by the adjacencies that properly belong to C.*

An important property of the adjacency graph is the following:

Proposition 3. *The adjacency graph of a component is a union of paths and cycles of the adjacency graph of genomes A and B.*

Proof. Let $C = (l, \ldots, r)$ be a component. Since it has the same gene content and the same extremities as the corresponding interval in genome B, all edges of the adjacency graph that are within the interval $(l, \ldots, r)$ in genome A will also be within the interval $(l, \ldots, r)$ in genome A. Thus all these edges form a union of paths and cycles of the adjacency graph of genomes A and B.

Each component that is nested in C is also a union of paths and cycles of the adjacency graph of genomes A and B, and none of them contains an adjacency that properly belongs to C. We can thus remove them without compromising the connectivity of the adjacency graph of C. □

3.2 Oriented Sorting

Since orientation of genes is relative, we can always assume that all genes in a chromosome of genome B are positive and in increasing order. The proper adjacencies of a component $\mathcal{C} = (l, \dots, r)$ induce a block partition in the corresponding chromosomes of genomes A and B. If we label the blocks in the chromosome of genome B with numbers from 1 to k, the corresponding blocks of the chromosome in genome A will be a signed permutation $(p_1, \dots, p_k)$ of these integers $\{1, \dots, k\}$. We will call this permutation – or it reverse – the *permutation associated to the component* $\mathcal{C}$.

Consider for example the following two genomes

$$A = \{(\circ\ 5, 1, 3, -2, 4, 6, -10, 9, 8, -7, 11\ \circ)\}$$
$$B = \{(\circ\ 1,\ 2,\ 3,\ 4,\ 5,\ 6,\ 7,\ 8,\ 9, 10, 11\ \circ)\}$$

The associated components can easily be seen in the following diagram:

The component $(\circ, \dots, 6)$ consists of three blocks: the gene 5, the block $(1, \dots, 4)$ and the gene 6. Thus, the permutation associated to the component $(\circ, \dots, 6)$ is $(2, 1, 3)$. For the other three non-trivial components, the associated permutations are $(1, 3, -2, 4)$, $(-4, 3, 2, -1)$ and $(1, -2, 3)$.

When the permutation associated to a component has both positive and negative signs, then it is well known from the sorting by inversion theory that the component can be optimally sorted by DCJ-sorting inversions. Components whose associated permutations have only positive elements can sometimes be optimally sorted by DCJ-sorting inversions. For example, consider the pair of genomes:

$$A = (\circ, 4, 3, 2, 1, \circ) \text{ and } B = (\circ, 1, 2, 3, 4, \circ),$$

whose associated permutation is $(4, 3, 2, 1)$. Its DCJ distance is 4, and it can be optimally sorted by inverting each of the four genes. However, we have:

Lemma 2. *If all elements of the permutation associated to a component have the same sign, then no inversion acting on one of its paths or cycles can create a new cycle.*

Proof. By eventually flipping the chromosome, we can assume that all the elements of the permutation are positive. Suppose that an inversion is applied to two adjacencies $(+i, +j)$ and $(+k, +l)$ in a single path or cycle of the component, and that this creates a new cycle. The new adjacencies will be $(+i, -k)$ and $(-j, +l)$, where at most one of $+i$ and $+l$ can be a telomere. If both of these new adjacencies belong to the same path or cycle, there was no creation of a new cycle. Suppose that the adjacency $(+i, -k)$ belongs to the new cycle, then all other adjacencies of this cycle existed in the original component, and are composed of positive elements. This, however, is impossible by the construction of the adjacency graph. $\qquad\square$

Definition 5. *A component is* oriented *if there exists an oriented DCJ-sorting operation that acts on vertices of its adjacency graph, otherwise it is* unoriented.

Oriented components are characterized by the following:

Proposition 4. *A component is oriented if and only if either its associated permutation has positive and negative elements, or its adjacency graph has two even paths.*

Proof. If the associated permutation has positive and negative elements, then there is at least one change of signs between blocks labeled by consecutive integers. There thus exists an inversion that creates an adjacency in genome B, thus a new cycle, and the inversion is DCJ-sorting. If there are two even paths, then one must be a path from genome A to genome A, and the other must be a path from genome B to genome B. An inversion in genome A that acts on one adjacency in each path creates two odd paths, thus is DCJ-sorting.

In order to show the converse, suppose that all elements of the associated permutation are positive, and all paths are odd. By Corollary 1, a DCJ-sorting operation must act on a single path or cycle. This operation cannot be a translocation or a fusion since all paths and cycles of a component are within a chromosome. This operation cannot be an inversion, since inversions that create new cycles are ruled out by Lemma 2, inversions acting on a single odd path cannot augment the number of odd paths, and inversions acting on cycles never create paths. Finally, this operation cannot be a fission: a fission acting on a cycle creates an even path; and a fission acting on an odd path must circularize one of the chromosome parts in order to be DCJ-sorting, otherwise it would be split into an even path and an odd path. □

Proposition 4 implies that, in the presence of unoriented components, we have $d_{DCJ}(A, B) < d_{HP}(A, B)$, since all DCJ-sorting operations will create circular chromosomes. On the other hand, well known results from the Hannehalli-Pevzner theory show that, when all components admit a sorting inversion, then it is possible to create a new cycle at each step of the sorting process with HP operations, without creating unoriented components. The same type of result can be obtained in this context, and we give it in the Appendix. Thus we have:

Theorem 2. *Given two linear genomes A and B, $d_{HP}(A, B) = d_{DCJ}(A, B)$ if and only if there are no unoriented components.*

4   Computing the General HP Distance

In this section we will show that, given the DCJ distance d_{DCJ}, one can express the Hannenhalli-Pevzner distance d_{HP} in the form

$$d_{HP} = d_{DCJ} + t,$$

where t represents the additional cost of not resorting to unoriented DCJ operations. First, we describe how to destroy unoriented components in Section 4.1 and after that, in Section 4.2, we compute the additional cost from the inclusion and linking tree of the unoriented components.

4.1   Destroying Unoriented Components

Destroying unoriented components is done by applying a DCJ operation either on one component in order to orient it, or on two components in order to merge them, and possibly others, into a single oriented component. By using the nesting and linking relationship between components, one can minimize the number of operations necessary to destroy unoriented components.

When two components overlap on one element, we say that they are *linked*. Successive linked components form a *chain*. A chain that cannot be extended to the left or right is called *maximal*. We represent the nesting and linking relations between components of a chromosome in the following way:

Definition 6. *Given a chromosome X of genome A and its components relative to genome B, define the forest F_X by the following construction:*

1. *Each non-trivial component is represented by a round node.*
2. *Each maximal chain that contains non-trivial components is represented by a square node whose (ordered) children are the round nodes that represent the non-trivial components of this chain.*
3. *A square node is the child of the smallest component that contains this chain.*

Now, we define a tree associated to the components of a genome by combining the forests of all chromosomes into one rooted tree:

Definition 7. *Suppose genome A consists of chromosomes $\{X_1, X_2, \ldots, X_K\}$. The tree T associated to the components of genome A relative to genome B is given by the following construction:*

1. *The root is a round node.*
2. *All trees of the set of forests $\{F_{X_1}, F_{X_2}, \ldots, F_{X_K}\}$ are children of the root.*

The round nodes of T are *painted* according to the following classification:

1. The root and all nodes corresponding to oriented components are painted *black*.
2. Nodes corresponding to unoriented components that do not contain telomeres are painted *white*.
3. Nodes corresponding to unoriented components that contain one or two telomeres are painted *grey*.

The tree associated to the components of the genomes A and B of Example 1 is shown in Fig. 1. Note that grey nodes are never included into other components.

The following two propositions are general remarks on components and are useful to show how to destroy unoriented components.

Proposition 5. *A translocation acting on two (unoriented) components cannot create new (unoriented) components.*

Proposition 6. *An inversion acting on two (unoriented) components $\mathcal{A}$ and $\mathcal{B}$ creates a new component $\mathcal{D}$ if and only if $\mathcal{A}$ and $\mathcal{B}$ are included in linked components.*

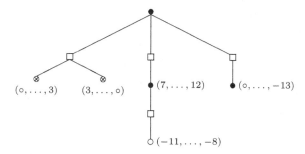

Fig. 1. The tree T associated to the genomes A and B of Example 1 has two grey leaves, one white leaf and one black leaf

Now, we have all necessary results to get rid of unoriented components. The following two propositions are straightforward generalizations of well-known results from the inversion theory [2]. We will start by looking at one single unoriented component.

Proposition 7. *If a component C is unoriented, any inversion between adjacencies of the same cycle or the same path of C orients C, and leaves the number of cycles and paths of the adjacency graph of C unchanged.*

Orienting a component as in Proposition 7 is called *cutting* the component. Note that this operation leaves the DCJ distance unchanged, and does not create new components.

It is possible to destroy more than one unoriented component with a DCJ operation acting on two unoriented components. The following proposition describes how to *merge* several components, and the relations of this operation to paths in the tree T.

Proposition 8. *A DCJ operation acting on adjacencies of two different unoriented components $\mathcal{A}$ and $\mathcal{B}$ destroys, or orients, all components on the path from $\mathcal{A}$ to $\mathcal{B}$ in the tree T, without creating new unoriented components.*

If the DCJ operation acts on two odd paths, thus on grey components, then merging the two components can be done without changing the number of odd paths, and the DCJ distance is unchanged. If the DCJ operation involves at least one cycle, then merging two components decreases the number of cycles by one, and the DCJ distance will increase by 1 in the resulting pair of genomes.

4.2   Unoriented Sorting

Let T be the tree associated to the components of genome A relative to genome B, and let T' be the smallest subtree of T that contains all the unoriented components, that is, the white and grey nodes.

Definition 8. *A* cover *of T' is a collection of paths joining all the unoriented components, such that each terminal node of a path belongs to a unique path.*

A path that contains two or more white or grey components, or one white and one grey component, is called a *long* path. A path that contains only one white or one grey component, is a *short* path.

The *cost* of a cover is defined to be the sum of the costs of its paths, where the cost of path is the increase in DCJ distance caused by destroying the unoriented components along the path. Using the remarks following Propositions 7 and 8, we have:

1. The cost of a short path is 1.
2. The cost of a long path with just two grey components is 1.
3. The cost of all other long paths is 2.

An *optimal* cover is a cover of minimal cost. Define t as the cost of any optimal cover of T'. We first establish that t is the difference between the two distances, using the following terminology:

Definition 9. *Given genomes A and B, we call a DCJ operation applied to genome A*

- *proper, if it decreases $d_{DCJ}(A, B)$ by one, i.e. $\Delta(C + I/2) = 1$,*
- *improper, if $d_{DCJ}(A, B)$ remains unchanged, i.e. $\Delta(C + I/2) = 0$, and*
- *bad, if it increases $d_{DCJ}(A, B)$ by one, i.e. $\Delta(C + I/2) = -1$.*

Theorem 3. *If t is the cost of an optimal cover of T', the smallest subtree of T that contains all the unoriented components of genome A relative to genome B, then:*

$$d_{HP}(A, B) = d_{DCJ}(A, B) + t.$$

Proof. First, we will show that $d_{HP}(A, B) \leq d_{DCJ}(A, B) + t$. Consider any cover of the tree T'. Let

- ww be the number of long paths with only white components,
- wg be the number of long paths with white and grey components,
- gg be the number of long paths with only grey components,
- w be the number of short paths with one white component,
- g be the number of short paths with one grey component.

Clearly, we have that the cost t' of this cover is $t' = 2ww + 2wg + gg + w + g$.

Suppose that the adjacency graph $AG(A, B)$ has C cycles and I odd paths. Applying $ww + wg$ bad DCJ operations and $gg + w + g$ improper DCJ operations yields a genome A'. Since each bad DCJ operation merges two cycles or one cycle and a path, the number of cycles in $AG(A', B)$ is $C - ww - wg$. Note that the number of odd paths remains unchanged. Therefore, by Theorem 2, we have that

$$d_{HP}(A, B) \leq d_{HP}(A', B) + ww + wg + gg + w + g$$
$$= N - (C + \frac{I}{2}) + 2ww + 2wg + gg + w + g$$
$$= d_{DCJ}(A, B) + t'.$$

Thus, since the above equation is true for any cover, we have: $d_{HP}(A, B) \leq d_{DCJ}(A, B) + t$.

The fact that $d_{HP}(A, B) \geq d_{DCJ}(A, B) + t$ is a consequence of the fact that an optimal sorting with HP operation necessarily induces a cover of T' since all unoriented components are eventually destroyed. □

It remains to establish a closed formula for t. A first easy but significant result on the size of t is the following lower bound. Let w be the number of white leaves and g be the number of grey leaves in T'. Since destroying a white leaf costs at least 1 and destroying a grey leaf costs at least $1/2$, and t is an integer, we have:

$$w + \left\lceil \frac{g}{2} \right\rceil \leq t.$$

It is quite remarkable, as was observed in the original paper on HP distance [6], that this bound is at most within one rearrangement operation from the optimal solution.

A branch in a tree is called a *long branch* if it has two or more unoriented components. A tree is called a *fortress* if it has an odd number of leaves, all of them on long branches. A standard theorem of the sorting by inversion theory states that the minimal cost to cover a tree that is not a fortress is ℓ, the number of leaves of the tree, and $\ell + 1$ in the case of a fortress [2].

We have first:

Theorem 4. *Let w be the number of white leaves and g be the number of grey leaves in T', the smallest subtree of T that contains all the unoriented components of genome A relative to genome B. If the root of T' has more than one child with white leaves, then the minimal cost of a cover of T' is:*

$t = w + \lceil \frac{g}{2} \rceil$ *if the smallest subtree T'' that contains all the white leaves*
           of T' is not a fortress, or g is odd,
$t = w + \lceil \frac{g}{2} \rceil + 1$ *otherwise.*

Proof. If the subtree T'' is not a fortress then it admits a cover of cost w, and pairing the maximum number of grey nodes yields a cover of T' costing $w + \lceil \frac{g}{2} \rceil$. If the subtree T'' is a fortress, then one of its white leaves is not paired with another leaf since the number of leaves is odd. A cover of T' can be obtained by pairing this white leaf with a grey leaf, which exists if g is odd. The resulting cost will be again $w + \lceil \frac{g}{2} \rceil$ which equals the lower bound and thus the cover is optimal.

If the subtree is a fortress and g is even, we can construct a cover costing $(w+1) + g/2$, using the cover of the fortress and pairing the grey nodes. To show that this cost is minimal, suppose that k grey nodes are paired with k white nodes, the remaining white and grey paired separately. If k is even, then the cost of such a cover would be $(w-k+1) + (g-k)/2 + 2k$, which is greater than or equal to $(w+1) + g/2$. If k is odd, then the cost of this cover is $(w-k) + (g-k+1)/2 + 2k$, which is again greater than or equal to $(w+1) + g/2$. □

When all the white leaves belong to a single child of the root, the situation is more delicate. Define a *junior fortress* as a tree with an odd number of white leaves, all of them on long branches, except one that is alone on its branch, called the *top* of the fortress. We have the following:

Theorem 5. *Let $w > 0$ be the number of white leaves and $g > 0$ be the number of grey leaves in T', the smallest subtree of T that contains all the unoriented components of genome A relative to genome B. If the root of T' has only one child c with white leaves then the minimal cost of a cover of T' is:*

$$t = w + \lceil \tfrac{g}{2} \rceil \quad \text{if } g \text{ is odd and the subtree } T_c \text{ that is rooted at } c$$
$$\text{is neither a fortress nor a junior fortress,}$$
$$t = w + \lceil \tfrac{g}{2} \rceil + 1 \; \text{otherwise.}$$

Proof. Suppose first that $g = 1$, then the only grey leaf either belongs to T_c or not. In the first case, this grey leaf must be the child c implying that T_c is not a junior fortress. If T_c is not a fortress, then there exists a cover with minimal cost equal to the number of leaves of T_c, which is given by $w + \lceil \tfrac{g}{2} \rceil$, since $g = 1$. If T_c is a fortress, then the minimal cost of a cover is $w + \lceil \tfrac{g}{2} \rceil + 1$.

In the other case, i.e. the grey leaf does not belong to T_c, then if T_c is a fortress or a junior fortress, the whole tree T' is a fortress with $w + \lceil \tfrac{g}{2} \rceil$ leaves, yielding a cost of $w + \lceil \tfrac{g}{2} \rceil + 1$. Otherwise, if T_c is neither a fortress nor a junior fortress, then T' can not be a fortress, and hence can be destroyed with cost $w + 1 = w + \lceil \tfrac{g}{2} \rceil$.

The same argumentation holds for any $g > 1$ if g is odd.

Now, we consider the case $g = 2$. If T_c is a fortress, two of the white leaves in T_c can be paired with the two grey leaves outside T_c at cost 4. This eliminates the two grey leaves, two of the long white branches, and the branch containing c. The remaining $w - 2$ long branches are paired at cost $w - 2$. Together, this gives a cover of cost $4 + w - 2 = w + \lceil \tfrac{g}{2} \rceil + 1$. This is optimal since the cost of T' is the same as for T_c. If T_c is not a fortress, we do not need to pair white and grey leaves. T_c can be covered with cost $w + 1$ and the g grey leaves are paired with cost $\lceil \tfrac{g}{2} \rceil$, giving again a total cost of $w + \lceil \tfrac{g}{2} \rceil + 1$.

If $g > 2$ and g is even, it is always possible to pair the grey leaves, as long as there are more than two left, and then apply the case $g = 2$. This gives the same cost $w + \lceil \tfrac{g}{2} \rceil + 1$. $\qquad\square$

For example, the genomes A and B of Example 1 have $N = 17$ genes. The adjacency graph $AG(A, B)$ has $C = 3$ cycles and $I = 6$ odd paths. After removing the dangling black leaf, the tree T' has $g = 2$ grey leaves and $w = 1$ white leaf (see Fig. 1). Therefore, by Theorem 5, we have $t = 2$ and thus

$$d_{HP}(A, B) = N - (C + \frac{I}{2}) + t = 17 - (3 + 3) + 2 = 13.$$

5   Conclusion

In this paper, we have given a simpler formula for the Hannenhalli-Pevzner genomic distance equation. It requires only a few parameters that can easily be computed directly from the genomes and from simple graph structures derived from the genomes. Traditionally used concepts that were sometimes hard to access, like weak-fortresses-of-semi-real-knots, are bypassed.

References

1. Bergeron, A.: A very elementary presentation of the hannenhalli-pevzner theory. In: Amir, A., Landau, G.M. (eds.) CPM 2001. LNCS, vol. 2089, pp. 106–117. Springer, Heidelberg (2001)
2. Bergeron, A., Mixtacki, J., Stoye, J.: Reversal distance without hurdles and fortresses. In: Sahinalp, S.C., Muthukrishnan, S.M., Dogrusoz, U. (eds.) CPM 2004. LNCS, vol. 3109, pp. 388–399. Springer, Heidelberg (2004)
3. Bergeron, A., Mixtacki, J., Stoye, J.: On sorting by translocations. J. Comput. Biol. 13(2), 567–578 (2006)
4. Bergeron, A., Mixtacki, J., Stoye, J.: A unifying view of genome rearrangements. In: Bücher, P., Moret, B.M.E. (eds.) WABI 2006. LNCS (LNBI), vol. 4175, pp. 163–173. Springer, Heidelberg (2006)
5. Bergeron, A., Stoye, J.: On the similarity of sets of permutations and its applications to genome comparison. In: Warnow, T., Zhu, B. (eds.) COCOON 2003. LNCS, vol. 2697, pp. 68–79. Springer, Heidelberg (2003)
6. Hannenhalli, S., Pevzner, P.A.: Transforming men into mice (polynomial algorithm for genomic distance problem). In: Proceedings of FOCS 1995, pp. 581–592. IEEE Press, Los Alamitos (1995)
7. Jean, G., Nikolski, M.: Genome rearrangements: a correct algorithm for optimal capping. Inf. Process. Lett. 104, 14–20 (2007)
8. Ozery-Flato, M., Shamir, R.: Two notes on genome rearrangements. J. Bioinf. Comput. Biol. 1(1), 71–94 (2003)
9. Tesler, G.: Efficient algorithms for multichromosomal genome rearrangements. J. Comput. Syst. Sci. 65(3), 587–609 (2002)
10. Tesler, G.: GRIMM: Genome rearrangements web server. Bioinformatics 18(3), 492–493 (2002)
11. Yancopoulos, S., Attie, O., Friedberg, R.: Efficient sorting of genomic permutations by translocation, inversion and block interchange. Bioinformatics 21(16), 3340–3346 (2005)

A   Proof of Theorem 2

Components whose both extremities are genes, often called *real* components, are well studied in the context of sorting a single chromosome with the same flanking genes [1]. We have the following:

Proposition 9 ([1]). *An oriented real component has an oriented DCJ-sorting operation that does not create new unoriented components.*

Components that contain one or two telomere are called *semi-real*.

Proposition 10. *An oriented semi-real component whose associated permutation is oriented can be sorted with oriented DCJ-sorting operations.*

Proof. We will show that such components can be embedded in oriented real components with the same DCJ distance. Then, we can sort the component with oriented DCJ-sorting operations. The basic idea is the following: if the component has one telomere, add an extra gene 0 or k to the associated permutation. This transforms the – only – odd path into a cycle and preserves the DCJ distance. If the component has two telomeres – it spans a whole chromosome – flip the chromosome as necessary in order to "close" each odd path into a cycle. It is then easy to show that the DCJ distance is preserved. $\square$

Proposition 11. *A semi-real component whose adjacency graph has even paths can be sorted with oriented DCJ-sorting operations.*

Proof. First, note that the semi-real component $C = (l, \dots, r)$ has two even paths. Consider the permutation $(p_1, \dots, p_k)$ associated to component C. If the permutation is oriented, then it is possible to sort the component with oriented DCJ-sorting operations by Proposition 10.

Now, if the permutation is unoriented, then all genes p_1 to p_k have the same sign. There exist two possible fissions: fission F_1 creating telomere $(k, \circ)$ and fission F_2 creating $(\circ, 1)$. It can be shown that one of these two fissions does not create new unoriented components. $\square$

Definition 10. *A DCJ operation creating the adjacency (a, b) of B, where a and b are genes, is called* interchromosomal, *if (a, x) and (y, b) belong to different chromosomes in A.*

1. *If $x \neq \circ$ and $y \neq \circ$, the DCJ operation is a translocation.*
2. *If $x = \circ$ or $y = \circ$, the DCJ operation is a semi-translocation.*
3. *If $x = \circ$ and $y = \circ$, the DCJ operation is a fusion.*

The next proposition is the key, it says that for any interchromosomal DCJ operation that creates an unoriented component there always exists an alternative interchromosomal DCJ-sorting operation that does not. This statement, already proven in the context of sorting by translocations in [3], can be shown similarly for the general case.

Proposition 12. *Given two linear genomes A and B, if an interchromosomal DCJ operation creates an unoriented component, then there exists another interchromosomal DCJ-sorting operation that does not.*

Theorem 2. *Given two linear genomes A and B, $d_{HP}(A, B) = d_{DCJ}(A, B)$ if and only if there are no unoriented components.*

Proof. The "if" part comes from the fact that we can sort a genome without unoriented components with DCJ-sorting operations (Propositions 10, 11, 12), adding the fact that semi-real components whose graphs have even paths can be "destroyed" by fissions. The "only if" part comes from the fact that if there are unoriented components, then $d_{DCJ}(A, B) < d_{HP}(A, B)$, since we showed in Proposition 4 that all DCJ-sorting operations create circular chromosomes in these cases. $\square$

Fixed Parameter Tractable Alignment of RNA Structures Including Arbitrary Pseudoknots

Mathias Möhl[1,*], Sebastian Will[2,*], and Rolf Backofen[2]

[1] Programming Systems Lab, Saarland University, Saarbrücken, Germany
mmohl@ps.uni-sb.de
[2] Bioinformatics, Institute of Computer Science, Albert-Ludwigs-Universität,
Freiburg, Germany
{will,backofen}@informatik.uni-freiburg.de

Abstract. We present an algorithm for computing the edit distance of two RNA structures with arbitrary kinds of pseudoknots. A main benefit of the algorithm is that, despite the problem is NP-hard, the algorithmic complexity adapts to the complexity of the RNA structures. Due to fixed parameter tractability, we can guarantee polynomial run-time for a parameter which is small in practice. Our algorithm can be considered as a generalization of the algorithm of Jiang *et al.* [1] to arbitrary pseudoknots. In their absence, it gracefully degrades to the same polynomial algorithm. A prototypical implementation demonstrates the applicability of the method.

Keywords: RNA alignment, pseudoknots, fixed parameter tractability.

1 Introduction

Over the last years, numerous discoveries attribute to RNA a central role that goes far beyond being a messenger and comprises regulatory as well as catalytic functions [2]. The turn of focus from purely sequence based analysis, as largely applied for DNA and proteins, to structure based analysis, as required for RNA, imposes a challenge to bioinformatics.

For this reason, RNA sequence/structure alignment is a rich and active field of research [1,3,4,5,6]. Almost all current approaches rely on the assumption that the pseudoknot-free representation of RNA structures suffices to obtain reasonable alignments. This is justified, algorithmically, since this restriction allows for an efficient treatment, as well as biologically, since the function of an RNA-molecule is mainly determined by its pseudoknot-free, secondary structure, which is usually more conserved than its sequence. Recent findings at least question the assumption that pseudoknots can be neglected. Today, it is known that many natural RNA molecules not only contain pseudoknots, but that these pseudoknots have diverse and important functions in the cell [7] and are therefore highly conserved [8]. Moreover, the concrete alignment of the pseudoknot region

---

* These authors contributed equally.

P. Ferragina and G. Landau (Eds.): CPM 2008, LNCS 5029, pp. 69–81, 2008.

is of major interest, since pseudoknots often occur at the functional centers of RNAs.

Many problems associated with the prediction or alignment of structures with arbitrary pseudoknots are NP-hard [1,9]. To overcome the limitation to pseudoknot-free structures, but still maintain a complexity that is affordable in practice, one has several alternatives. A first approach is to consider only a restricted class of pseudoknots, which allows a polynomial algorithm [10,11,12]. Second, there are heuristic approaches which are usually fast, but which are not guaranteed to find the optimal structure or do not give a performance guarantee, or both [6]. Here we will follow a third direction, namely to design an algorithm that can align arbitrary pseudoknots, always computes optimal structures and nevertheless has a performance guarantee in terms of fixed parameter tractability. Whereas polynomial runtime cannot be guaranteed in general for NP-hard problems, unless P=NP, fixed parameter tractability allows to guarantee polynomial runtime if some parameter, which is usually small on practical instances, is considered as constant.

We present an algorithm that computes the optimal alignment of two RNA structures with respect to their edit distance. The parameter determining the exponential runtime depends on how complex the crossing stems are arranged and is small in practice. As a nice property, the algorithm gracefully degrades to the algorithm of Jiang *et al.* [1] for the simpler class of structures handled by their algorithm.

Related Work. Most of the algorithms for RNA sequence structure alignment are not able to align pseudoknots [13,5,3,1].

Among the algorithms supporting pseudoknots, there are several grammar-based approaches for motif finding, which try to align a sequence with given structure to a sequence with unknown structure (usually a genomic sequence) [14,15]. In these approaches, the class of supported pseudoknots depends on the expressivity of the underlying grammar formalism.

Concerning the alignment of two pseudoknotted structures, Evans [9] developed a fixed parameter tractable algorithm that computes the longest arc preserving common subsequence of two sequences with arbitrary kinds of pseudoknots. This problem is related to edit distance. However, on input classes where our algorithm guarantees polynomial run-time due to the fixed parameter, the run-time of Evan's algorithm is not polynomially bounded.

Another algorithm by Evans [12] finds the maximum common ordered substructure of two RNA structures in polynomial time (more precisely in $O(n^{10})$ time and $O(n^8)$ space, where n denotes the length of the sequences), but only for a restricted class of pseudoknots.

Bauer *et al.* [6] give an algorithm based on integer linear programming with Lagrangian relaxation that aligns two sequences with arbitrary pseudoknots. As a heuristic approach, it works usually well in practice but gives no guarantees on performance and may even fail to yield optimal results.

Furthermore, there is a fixed parameter tractable algorithm by Blin *et al.* [16] for protein design involving RNA, which shares an important idea with our approach, namely the bipartitioning of the complete set of base pairs into an efficiently tractable subset and the remaining "hard" base pairs (in our case, pairs of base pairs).

2 Preliminaries

An *arc-annotated sequence* is a pair (S, P), where S is a string over the set of bases $\{A, U, C, G\}$ and P is a set of arcs (l, r) with $1 \leq l < r \leq |S|$ representing bonds between bases, such that each base is adjacent to at most one arc, i.e. $\forall (l, r) \neq (l', r') \in P : l \neq l' \wedge l \neq r' \wedge r \neq l' \wedge r \neq r'$. We denote the i-th symbol of S by $S[i]$. For an arc $p = (l, r)$, we denote its left end l and right end r by p^{L} and p^{R}, respectively.

For an arc-annotated sequence (S, P), an arc $p \in P$ is called *crossing* if there is an arc $p' \in P$ such that $p^{\mathrm{L}} < p'^{\mathrm{L}} < p^{\mathrm{R}} < p'^{\mathrm{R}}$ or $p'^{\mathrm{L}} < p^{\mathrm{L}} < p'^{\mathrm{R}} < p^{\mathrm{R}}$. In the first case, p is called *right crossing*, in the second case *left crossing*; p and p' form a *pseudoknot*. An arc-annotated sequence (S, P) containing crossing arcs is called *crossing*, otherwise *non-crossing* or *nested*.

For two arc annotated sequences (S_1, P_1) and (S_2, P_2), we define χ, ψ_1, ψ_2:

$$\chi(i, j) := \text{ if } S_1[i] \neq S_2[j] \text{ then } 1 \text{ else } 0,$$
$$\psi_k(i) := \text{ if } \exists j : (i, j) \in P_k \text{ or } (j, i) \in P_k \text{ then } 1 \text{ else } 0 \quad (\text{for } k = 1, 2).$$

An *alignment* A of two arc-annotated sequences (S_1, P_1) and (S_2, P_2) is a set $A \subseteq [1..|S_1|] \times [1..|S_2|]$ of *alignment edges* such that for all $(i, j), (i', j') \in A$ holds 1.) $i > i'$ implies $j > j'$ and 2.) $i = i'$ if and only if $j = j'$. For an alignment A and i, i', j, j' such that neither (i, j) nor (i', j') cuts any alignment edge (formally $A \cap [i..i'] \times [1..|S_2|] = A \cap [1..|S_1|] \times [j..j']$), we define the *subalignment* $A(i, i'; j, j')$ of A by $A \cap [i..i'] \times [j..j']$. An *arc pair* is a pair of arcs $a = (p_1, p_2) \in P_1 \times P_2$. We call $a = (p_1, p_2)$ *realized by* A if and only if $(p_1^{\mathrm{L}}, p_2^{\mathrm{L}}), (p_1^{\mathrm{R}}, p_2^{\mathrm{R}}) \in A$, i.e. when the arcs p_1 and p_2 are *matched by* A. The *set* $\mathrm{OA}(A; i, i'; j, j')$ of *open arc pairs* of a subalignment $A(i, i'; j, j')$ in A is the set of arc pairs (p_1, p_2) that are realized by A and where either $p_1^{\mathrm{L}} < i \leq p_1^{\mathrm{R}} \leq i'$ and $p_2^{\mathrm{L}} < j \leq p_2^{\mathrm{R}} \leq j'$ or $i \leq p_1^{\mathrm{L}} \leq i' < p_1^{\mathrm{R}}$ and $j \leq p_2^{\mathrm{L}} \leq j' < p_2^{\mathrm{R}}$. In Fig. 1a), we show realized arc pairs and a subalignment; its set of open arc pairs is $\{(A, D), (C, F)\}$.

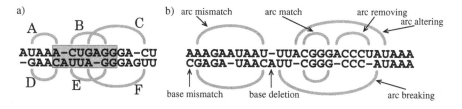

Fig. 1. a) Realized arc pairs (A,D), (B,E), and (C,F). (A,D) and (C,F) are open for the highlighted subalignment. b) Edit operations (cf. [1]).

Each alignment has an associated cost based on an edit distance with two classes of operations. The operations are illustrated in Fig. 1b). Base operations (mismatch and insertion/deletion) work solely on positions that are not incident to an arc. *Base mismatch* replaces a base with another base and has associated cost w_m. A *base insertion/deletion* removes or adds one base and costs w_d. The second class consists of operations that involve at least one position that is incident to an arc. An *arc mismatch* replaces one or both of the bases incident to an arc. It costs $\frac{w_{am}}{2}$ if one base is replaced or w_{am} if both are replaced. An *arc breaking* removes one arc and leaves the incident bases unchanged. The associated cost is w_b. *Arc removing* removes one arc and both incident bases and costs w_r. Finally, *arc altering* removes one of the two bases that are incident to an arc and costs w_a.

An alignment A has a corresponding minimal sequence of edit operations. The cost of A is defined as the sum of the cost of these edit operations.

3   A Fixed Parameter Tractable Algorithm

The algorithm we present computes the minimum cost alignment of two arc annotated sequences (S_1, P_1) and (S_2, P_2) containing arbitrary pseudoknots. In terms of Jiang *et al.* [1], we solve EDIT(CROSSING,CROSSING) for their class of *reasonable* scoring schemes. These schemes are restricted by $w_a = \frac{w_b}{2} + \frac{w_r}{2}$.

The central idea of the algorithm is to partition the set of arc pairs $P_1 \times P_2$ into a set NC of "non-crossing" arc pairs and a set of "crossing" arc pairs CR $= P_1 \times P_2 - $ NC such that the algorithm can interleave a polynomial alignment method for the arc pairs in NC with an exponential method for the arc pairs in CR. The exact requirement for such a partition is made precise in the definition of "valid partition".

The immediate result is a fixed parameter tractable algorithm whose parameter is loosely understood as the number of arc pairs in CR that cover a common base match. The presented algorithm further reduces this factor substantially by precomputing the alignment of stems of arcs in CR.[1]

3.1   Partition into Crossing and Non-crossing Arc Pairs

Two arcs p and p' of a sequence cross, iff $p^L < p'^L < p^R < p'^R$ or $p'^L < p^L < p'^R < p^R$. To generalize this from arcs to arc pairs, we define the left and right end point of an arc pair as

$$\nwarrow(p_1, p_2) = (p_1^L, p_2^L) \qquad \text{and} \qquad \searrow(p_1, p_2) = (p_1^R, p_2^R),$$

respectively. On those points we consider the partial order $\prec$ defined as $(x_1, y_1) \prec (x_2, y_2)$ if and only if $x_1 < x_2$ and $y_1 < y_2$.

---

[1] In principle, the idea can be extended from stems to arbitrary non-crossing sub-structures that are, like stems, closed by an inner and an outer arc. At the cost of precomputation this lowers the exponential factor of the algorithm further.

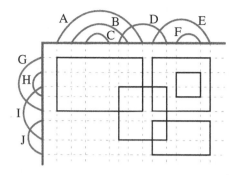

Fig. 2. Visualization of the arc pairs of two sequences. The first sequence has arcs A to E, the second sequence arcs G to J. To maintain readability, only some of the arc pairs are visualized.

Two arc pairs a and a' *cross*, iff $\seararrow a \prec \searrow a' \prec \searrow a \prec \searrow a'$ or $\searrow a' \prec \searrow a \prec \searrow a' \prec \searrow a$. Figure 2 represents arc pairs as rectangles in the plane whose dimensions correspond to the two sequences. If two arc pairs cross, the rectangles partially overlap, but note that the converse implication does not hold. In Fig. 2 for example, (D, I) and (E, J) cross, whereas (D, I) and (E, G) do not cross.

Definition 1 (valid partition). *A (bi-)partition of $P_1 \times P_2$ into NC and CR is valid if and only if for all $a, a' \in NC$ it holds that a and a' do not cross.*

A valid partition of $P_1 \times P_2$ can be lifted from a partition of the arcs of P_1 and P_2 by choosing appropriate sets $CR_1 \subseteq P_1$ and $CR_2 \subseteq P_2$ such that $P_1 - CR_1$ and $P_2 - CR_2$ are non-crossing and set $CR = CR_1 \times CR_2$. However, this does not work for arbitrary non-crossing sets $P_1 - CR_1$ and $P_2 - CR_2$. For example, in Fig. 2 choosing $CR = \{A, B, E\} \times \{I\}$ is not valid, since it contains none of the two crossing arc pairs (A, G) and (D, I). Valid partitions are obtained, if CR_1 and CR_2 contain all left crossing edges.

Lemma 1 (sufficient criterion for a partition). *The partition of $P_1 \times P_2$ into $CR = \{ p_1 \in P_1 \mid p_1$ is left crossing$\} \times \{ p_2 \in P_2 \mid p_2$ is left crossing$\}$ and $NC = P_1 \times P_2 - CR$ is valid.*

The claim holds since for two arbitrary crossing arc pairs one of them is in CR: for arc pairs a, a' with $\searrow a \prec \searrow a' \prec \searrow a \prec \searrow a'$ the two arcs of a' are left crossing. Analogously, a valid partition is obtained, if CR_1 and CR_2 contain all right crossing arcs.

Since our algorithm handles arc pairs in NC more efficient than arc pairs in CR, the partition into NC and CR is crucial for the runtime. A good partition should be minimal in the sense that it becomes invalid, if any element is removed from CR. Finding the best partition among these local minima involves balancing several parameters, since not only the cardinality of CR influences the complexity. Thinking of the arc pairs in CR as rectangles (as indicated in Fig. 2),

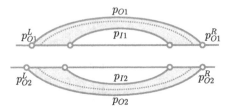

Fig. 3. Stem pair $(a_O, a_I) = ((p_{O1}, p_{O2}), (p_{I1}, p_{I2}))$, which covers the dotted arc pair

both the area of the rectangles and the number of rectangles that overlap in a common point influence the runtime.

The partition according to Lem. 1 is not yet aware of these aspects and sometimes does not lead to a local minimum. As an example, in Fig. 2 we would have $\mathrm{CR} = \{D, E\} \times \{I, J\}$, but (E, J) can safely be added to NC, since it only crosses with $(D, I) \in \mathrm{CR}$. The fact that no other arc pair containing arc E or J can be removed from CR indicates that this is a general limitation of partitions of arc pairs that are lifted from partitions of arcs. Instead of using these partitions directly, they could serve as a starting point for further optimization at the level of arc pairs with techniques like stochastic local search or genetic algorithms.

In the following we assume a valid partition of the arc pairs into CR and NC.

3.2   Precomputation of Stem Pairs

In order to align whole stems in one step, we group arc pairs in CR into pairs of stems. We define a stem Q in P (for $P \in \{P_1, P_2\}$) as a set of arcs $\{p_1, \dots, p_k\} \subseteq P$ with $p_1^{\mathrm{L}} < \cdots < p_k^{\mathrm{L}} < p_k^{\mathrm{R}} < \cdots < p_1^{\mathrm{R}}$ such that no end of arcs in $P - Q$ is in one of the intervals $[p_1^{\mathrm{L}}..p_k^{\mathrm{L}}]$ or $[p_k^{\mathrm{R}}..p_1^{\mathrm{R}}]$. In Fig. 2, for example, $\{A, B\}$ is a stem, but $\{A, B, C\}$ is not, since the left end of D is between the right endpoints of B and C. Note that, according to this notion, stems are allowed to include bulges and internal loops and do not need to be maximal.

The *stem pair* of two stems $Q_1 \subseteq P_1$ and $Q_2 \subseteq P_2$ is characterized by the pair (a_O, a_I) of arc pairs, where $a_O = (p_{O1}, p_{O2})$ is the pair of the outermost arcs and $a_I = (p_{I1}, p_{I2})$ is the pair of the innermost arcs of Q_1 and Q_2, i.e. Q_k consists of the arcs $P_k \cap [p_{Ok}^{\mathrm{L}}..p_{Ik}^{\mathrm{L}}] \times [p_{Ik}^{\mathrm{R}}..p_{Ok}^{\mathrm{R}}]$ $(k = 1, 2)$ (cf. Fig. 3). The stem pair *covers* an arc pair a iff $a \in Q_1 \times Q_2$. A stem pair is *realized in an alignment* A if and only if a_O and a_I are realized in A.

We write the set of all stem pairs (a_O, a_I) where $\{a_O, a_I\} \subseteq \mathrm{CR}$ as $\mathrm{ST}_{\mathrm{CR}}$.[2] A stem pair (a_O, a_I) is *open for a subalignment* $A(i, i'; j, j')$ in A if and only if a_O and a_I are open for $A(i, i'; j, j')$ in A. The *set of maximal open stem pairs of* $A(i, i'; j, j')$ in A is the smallest set M of open stem pairs of $A(i, i'; j, j')$ in A such that each $a \in \mathrm{OA}(A; i, i'; j, j')$ is covered by a stem pair in M.

---

[2] We assume that the arc pairs of a stem pair are either completely contained in CR or completely contained in NC, since minimal partitions (as well as partitions according to Lem. 1) satisfy this property.

$$S'(i, i'; j, j'; a_I) =$$

$$\min \begin{cases} S'(i+1, i'; j, j'; a_I) + w_d + \psi_1(i)(\frac{w_r}{2} - w_d) & \text{(gap)} \\ S'(i, i'; j+1, j'; a_I) + w_d + \psi_2(j)(\frac{w_r}{2} - w_d) & \text{(gap)} \\ S'(i, i'-1; j, j'; a_I) + w_d + \psi_1(i')(\frac{w_r}{2} - w_d) & \text{(gap)} \\ S'(i, i'; j, j'-1; a_I) + w_d + \psi_2(j')(\frac{w_r}{2} - w_d) & \text{(gap)} \\ S'(i+1, i'; j+1, j'; a_I) + \chi(i,j)w_m + (\psi_1(i) + \psi_2(j))\frac{w_b}{2} & \text{(align bases)} \\ S'(i, i'-1; j, j'-1; a_I) + \chi(i', j'l)w_m + (\psi_1(i') + \psi_2(j'))\frac{w_b}{2} & \text{(align bases)} \\ \text{if } ((i,i'), (j,j')) \in \text{CR} \\ \quad S'(i+1, i'-1; j+1, j'-1; a_I) + (\chi(i,j) + \chi(i',j'))\frac{w_{am}}{2} & \text{(align arcs)} \end{cases}$$

Fig. 4. Recursion equation to compute S' items

For each $(a_O, a_I) \in \text{ST}_{\text{CR}}$, we precompute the cost to align the respective stem pair as the value of an item $S(a_O, a_I)$. More precisely, for $a_O = (p_{O1}, p_{O2})$ and $a_I = (p_{I1}, p_{I2})$, the value of $S(a_O, a_I)$ is the cost to align $S_1[p_{O1}^L] \ldots S_1[p_{I1}^L]$ to $S_2[p_{O2}^L] \ldots S_2[p_{I2}^L]$ and simultaneously $S_1[p_{I1}^R] \ldots S_1[p_{O1}^R]$ to $S_2[p_{I2}^R] \ldots S_2[p_{O2}^R]$. In this sense, an S item describes the cost of two subalignments that are not independent of each other due to arcs in CR.

The computation of S items is based on temporary items $S'(i, i'; j, j'; a_I)$ that correspond to $S(((i,i'), (j,j')); a_I)$ if $((i,i'), (j,j'))$ is an arc pair, but are not limited to this case. $S'(i, i'; j, j'; ((i_a, i_a'), (j_a, j_a')))$ is invalid if $i > i_a$, $i' < i_a'$, $j > j_a$ or $j' < j_a'$. The alignment of the innermost arc is computed as $S'(i, i'; j, j'; ((i, i'), (j, j'))) = (\chi(i,j) + \chi(i',j'))\frac{w_{am}}{2}$ and step by step enlarged with the recursions given in Fig. 4, where implicitly recursive cases relying on invalid items are skipped.

By the recursion for S', only the arc pair a_I is guaranteed to be realized in the precomputed optimal stem alignments. However, we want to consider in the core dynamic programming algorithm only items $S(a_O, a_I)$ where a_I *and* a_O are realized, in order to avoid ambiguity in the recursion. Therefore, we define items where a_O is not realized as invalid. In consequence, cases referring to these items are skipped in the core algorithm.

3.3   Core of the Algorithm

The main part of the algorithm recursively computes costs of subalignments. The recursions are given in Fig. 6 and an illustration is provided in Fig. 5.

The subalignment costs are represented by items $D(i, i'; j, j'|M)$ where i, i', j, and j' specify the range of the subalignment and $M \subseteq \text{ST}_{\text{CR}}$ is its set of maximal open stem pairs. The precise semantics is that the value of $D(i, i'; j, j'|M)$ is the minimal cost among all subalignments $A(i, i'; j, j')$ of all alignments A that satisfy (a) M is the set of maximal open stem pairs of $A(i, i'; j, j')$ in A and (b) for all $(a_O, a_I) \in M$ the precomputed subalignment corresponding to $S(a_O, a_I)$ is a subalignment of A. A helpful intuition of M in the D items is that one end

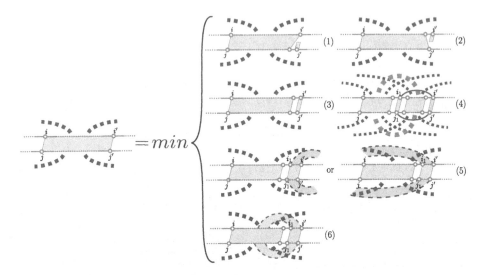

Fig. 5. Illustration of the recursion for computing items $D(i, i'; j, j'|M)$. The red dotted arcs represent the set of open stem pairs M. Case (4) recurses to $D(i, i_1 - 1; j, j_1 - 1|M_1)$ and $D(i_1 + 1, i' - 1; j_1 + 1, j' - 1|M_2)$. There, the green dotted arcs represent the set of stem pairs shared between the two alignment fragments (i.e. $M_1 \cap M_2$) and the red dotted arcs represent the remaining elements of $M_1 \cup M_2$, which make up M. Cases (5) and (6) show concrete stem pairs in light red (in M) and light green (not in M), respectively.

of the stem pairs in M is aligned and the other half is required to be aligned outside of the range $(i, i'; j, j')$.

The semantics is reasonable only for a restricted class of items, which we call *valid items*. $D(i, i'; j, j'|M)$ is valid if $i' \geq i-1, j' \geq j-1$, and there is an alignment A such that M is the set of maximal open stem pairs of $A(i, i'; j, j')$ in A.

Given the semantics of D items, the cost of the entire alignment is the value of $D(1, |S_1|; 1, |S_2| \ |\emptyset)$. It is computed following the recursion in Fig. 6 with base cases $D(i, i-1; j, j-1|\emptyset) = 0$ (for all i, j). Implicitly, in each recursive step the cases involving invalid items are skipped.

We will take a closer look at the cases of the recursion. First note that for $CR = \emptyset$, only items of the form $D(i, i'; j, j'|\emptyset)$ are valid. Then, the cases (5) and (6) are always skipped and the recursion degenerates to the recursion of Jiang *et al.* [1], shown in Fig. 7, where the items $D(i, i'; j, j'|\emptyset)$ directly correspond to matrix entries $DP(i, i'; j, j')$.[3]

In the absence of crossing arcs, the recursion in Fig. 7 is correct since the case distinction is exhaustive and each case assigns the correct cost. To see this assume an optimal alignment A with a subalignment $A(i, i'; j, j')$ without open arc pairs. Considering the positions i' and j', there are exactly the following cases directly corresponding to the recursion. (1) i' is not aligned in A. If i' is

---

[3] Note that the restrictions $i > i_1, j > j_1$ in case 4 of the recursion in Fig. 7 is implicit in our recursion by skipping cases with invalid items; here $D(i, i_1 - 1; j, j_1 - 1|M_1)$ is invalid.

$$D(i, i'; j, j'|M) = \min$$

$$\begin{cases}
D(i, i'-1; j, j'|M) + w_d + \psi_1(i')(\frac{w_r}{2} - w_d) & (1) \\[2mm]
D(i, i'; j, j'-1|M) + w_d + \psi_2(j')(\frac{w_r}{2} - w_d) & (2) \\[2mm]
D(i, i'-1; j, j'-1|M) + \chi(i', j') \cdot w_m + (\psi_1(i') + \psi_2(j'))\frac{w_b}{2} & (3) \\[2mm]
\text{if there exist some } i_1, j_1 \text{ with } ((i_1, i'), (j_1, j')) \in \text{NC} \\
\min \left\{ \begin{array}{l|l} D(i, i_1 - 1; j, j_1 - 1|M_1) \\ +D(i_1+1, i'-1; j_1+1, j'-1|M_2) & M_1, M_2 \subseteq \text{ST}_{\text{CR}}, \text{ where} \\ +(\chi(i_1,j_1) + \chi(i', j'))\frac{w_{am}}{2} & M = (M_1 \cup M_2) - (M_1 \cap M_2) \end{array} \right\} & (4) \\[2mm]
\text{if there exists some } (a_O, a_I) \in M \text{ with} \\
\quad \searrow a_O = (i_1, j_1) \wedge \searrow a_I = (i', j') \text{ or } \searrow a_I = (i_1, j_1) \wedge \searrow a_O = (i', j') \\
\quad D(i, i_1 - 1; j, j_1 - 1|M - \{(a_O, a_I)\}) + \frac{S(a_O, a_I)}{2} & (5) \\[2mm]
\min \left\{ \begin{array}{l|l} D(i, i_1-1; j, j_1-1|M \cup \{(a_O, a_I)\}) & (a_O, a_I) \in \text{ST}_{\text{CR}}, \text{ where} \\ +\dfrac{S(a_O, a_I)}{2} & \searrow a_O = (i', j') \text{ and} \\ & \searrow a_I = (i_1, j_1) \end{array} \right\} & (6)
\end{cases}$$

Fig. 6. Recursion equation to compute D items

$$DP(i, i'; j, j') = \min$$

$$\begin{cases}
DP(i, i'-1; j, j') + w_d + \psi_1(i')(\frac{w_r}{2} - w_d) & (1) \\[2mm]
DP(i, i'; j, j'-1) + w_d + \psi_2(j')(\frac{w_r}{2} - w_d) & (2) \\[2mm]
DP(i, i'-1; j, j'-1) + \chi(i', j') \cdot w_m + (\psi_1(i') + \psi_2(j'))\frac{w_b}{2} & (3) \\[2mm]
\text{if there exist some } i < i_1, j < j_1 \text{ with } ((i_1, i'), (j_1, j')) \in P_1 \times P_2 \\
\quad DP(i, i_1 - 1; j, j_1 - 1) + DP(i_1+1, i'-1; j_1+1, j'-1) & (4) \\
\quad +(\chi(i_1, j_1) + \chi(i', j'))\frac{w_{am}}{2}
\end{cases}$$

Fig. 7. Recursion equation for the algorithm of Jiang *et al.*

not adjacent to an arc this is due to a base deletion with cost w_d. Otherwise, the arc is either removed or altered, which causes cost $w_r/2$. (2) j' is not aligned in A, analogously. (3) $(i', j') \in A$, but A realizes no arc pair involving (i', j'). All adjacent arcs are broken, each causing cost $w_b/2$. If $S_1[i']$ and $S_2[j']$ mismatch this causes additional cost w_m. (4) $(i', j') \in A$ and A realizes an arc pair $((i_1, i'), (j_1, j'))$ with right end (i', j'). Then, the cost of the whole arc pair is charged and the subalignment is decomposed into a subalignment before the arc pair and the subalignment inside the arc pair.

Note that due to the assumption that $A(i, i'; j, j')$ has no open arc pairs the case where $(i', j') \in A$ and A realizes an arc pair with left end (i', j') can not occur. Furthermore, all cases only decompose into subalignments without open arc pairs. In particular in case (4), the two subalignments can not have open arc pairs since such arc pairs would be open arc pairs of both subalignments and then cross the arc pair $((i_1, i'), (j_1, j'))$.

The key to understand the recursion in Fig. 6 (illustrated in Fig. 5) is that it maintains the decomposition of the algorithm of Jiang *et al.* as much as possible in the presence of crossing arc pairs. Then, Jiang's case (4) is no more exhaustive and has to be extended to consider all cases where the recursive subalignments contain open arc pairs. To achieve this, we make the open arc pairs explicit via the additional component M of our items. In principle, it suffices to directly represent the set of open arc pairs in M. For efficiency reasons, we combine open arc pairs into open stem pairs in order to keep the sets M small. As a direct consequence of making the open stem pairs explicit, we can exhaustively minimize over all alternatives in case (4). In particular, these include the cases where M_1 and M_2 contain open stem pairs that are not contained in M, namely those where $M_1 \cap M_2 \neq \emptyset$.

For cases (1) to (3), M is invariant since no arc pairs are realized in these cases. In consequence, the generalization of these cases is straightforward.

In order to make our case distinction exhaustive for CR $\neq \emptyset$, we need additional cases (5) and (6) that cover the situations where an arc pair a in CR has left or right end in (i', j') (recall that only the arc pairs in NC are handled in case (4)). There are two such cases: either a is open in $A(i, i'; j, j')$ or not. In the first case, the maximal open stem pair that covers a is contained in M and hence uniquely determined. We can therefore decompose into (the respective subalignment of) this maximal open stem pair and the remaining subalignment, where this stem pair is no more open (case (5)). In the second case, we minimize over all possible maximal open stem pairs that cover a. Each time, we decompose again into (the respective subalignment of) this maximal open stem pair and the remaining subalignment, where now the stem pair is open in this remaining subalignment (case (6)). Note that we distribute the cost of the precomputed stem pairs equally among the two subalignments. This is correct, since it is guaranteed that each alignment contains either both subalignments or none of them. Further note that, when descending in the recursion, open stem pairs are introduced via cases (4) or (6) and are removed again via case (5).

When the cost of the alignment is determined, the actual alignment can be constructed by the usual backtracing techniques.

3.4   Complexity

Let n be $\max(|S_1|, |S_2|)$, let s and s' be the maximal number of arcs and bases in a crossing stem, respectively. For an item $S(a_O, a_I)$ we have $O(n^2 s^2)$ possible instances: for a_O, we can freely choose among the $O(n^2)$ arc pairs in CR and for a_I we have $O(s^2)$ possible choices, since the arcs of a_O and a_I must belong to the same stems. Analogously, for the S' items we need $O(n^2 s'^4)$ space. Since each of these items can be computed in constant time, the time complexity coincides with the required space.

An item $D(i, i'; j, j'|M)$ has $O(n^4)$ possible instances of i, i', j, j', but analogously to the algorithm of Jiang *et al.* [1] only $O(n^2)$ of them need to be maintained permanently. To measure the number of instances of M, we need the notion of the *crossing number of a point* $(x, y) \in [1..|S_1|] \times [1..|S_2|]$, defined as $C(x, y) = |\{ (a_O, a_I) \in \text{ST}_{\text{CR}}^{\text{MAX}} | \diagdown a_I \prec (x, y) \prec \diagdown a_I \}|$, where $\text{ST}_{\text{CR}}^{\text{MAX}}$ denotes the subset of ST_{CR} that only contains pairs of maximal stems (with respect to set inclusion). We denote the maximal crossing number with k. Since each maximal stem pair has $O(s^4)$ fragments, there are at most $O((s^4)^{C(i,j)+C(i'j')}) = O(s^{8k})$ possible instances of M for fixed i, j, i', j'.[4] Hence we need to compute $O(n^4 s^{8k})$ D items and maintain $O(n^2 s^{8k})$ of them in memory at the same time.

The computation of a D item needs only for the recursive alternatives (4) and (6) of Fig. 6 more than constant time. In alternative (6), iteration over all $O(s^2)$ possible instances of a_I is necessary and in alternative (4) we need to iterate over all possible instances of M_1 and M_2. Since M_2 is uniquely determined by M and M_1, there are $O(s^{8k})$ of these instances. The computation of all the $O(n^4 s^{8k})$ D items hence requires $O(n^4 s^{8k} \cdot s^{8k}) = O(n^4 s^{16k})$ time.

If at least one of two sequences does not contain pseudoknots, the only minimal partition is CR $= \emptyset$ and NC $= P_1 \times P_2$. In this case the algorithm gracefully degrades to the one of Jiang *et al.* [1] requiring $O(n^4)$ time and $O(n^2)$ space.

4   Practical Evaluation

We implemented a prototype of the algorithm in C++ to evaluate its applicability in practice. With the prototype, we computed pairwise alignments of some RNA structures of the tmRNA database [17]. For our evaluation we have chosen the longest tmRNA sequence (*Mycobacteriophage Bxz1*, MB), the shortest sequence (*Cyanidium caldarium*, CC), the sequence that contains the largest crossing stems (*Ureaplasma parvum*, UP), and a nested version (UPnest) of the latter, where we removed all left crossing arcs.

We were able to compute the pairwise alignments of these sequences with 1 GB of memory with one exception using 2 GB. Table 1 shows that the runtime scales well with the complexity of the involved pseudoknots. As we suggested, the exponential factor k is small on all instances. Whereas alignments of sequences with large pseudoknots take several hours, sequences with small pseudoknots can be aligned in a few minutes. In contrast, sequence length has a much smaller impact on runtime, as in particular the alignments with UPnest show.

For the results in Table 1 we partitioned into NC and CR according to the left crossing stem criterion (see Lem. 1). However, the runtime can depend heavily on the partition into NC and CR. For example the alignment of *Ureaplasma parvum* and *Mycobacteriophage Bxz1* took less than three hours if we chose CR to contain the pairs of left crossing arcs, but more than 6 hours if we chose the right crossing arcs instead. Notably, in this case the better partitioning can be identified in advance by comparing the parameters k and s; k is equal for

---

[4] This is a coarse estimate, that counts many invalid requirement sets, in particular those, where some stem pairs cannot be realized in the same alignment.

Table 1. Runtime of the alignments (on a single Xeon 5160 processor with 3.0 GHz) and the properties of the aligned structures (n=sequence length, s=max. number of arcs in crossing stem, pk=number of pseudoknots, k=fixed parameter of the algorithm) for left crossing partitioning

aligned sequences	n	s	k	pk	memory	runtime
UP / UP	413/413	10/10	1	4/4	$\leq$ 2 GB	726m 52s
UP / MB	413/437	10/7	1	4/2	$\leq$ 1 GB	172m 53s
UP / CC	413/254	10/2	1	4/1	$\leq$ 1 GB	11m 51s
UP / UPnest	413/413	10/0	0	4/0	$\leq$ 1 GB	4m 43s
MB / MB	437/437	7/7	1	2/2	$\leq$ 1 GB	43m 20s
MB / CC	437/254	7/2	1	2/1	$\leq$ 1 GB	3m 56s
MB / UPnest	437/413	7/0	0	2/0	$\leq$ 1 GB	3m 27s
CC / CC	254/254	2/2	1	1/1	$\leq$ 1 GB	1m 11s
CC / UPnest	254/413	2/0	0	1/0	$\leq$ 1 GB	2m 6s
UPnest/UPnest	413/413	0/0	0	0/0	$\leq$ 1 GB	4m 21s

both cases, s is 10/7 for the left crossing and 12/12 for the right crossing case. This comparison indicates that a more sophisticated partitioning into crossing and nested arc pairs, e.g. greedy or stochastic local optimization, may result in significant speed-ups in practice.

Finally note that the efficiency could be improved further by heuristic optimizations as utilized in many existing alignment tools. For example, skipping the computation of items that are unlikely to contribute to the optimal alignment can significantly reduce computation time.

5   Conclusion

We have presented an algorithm that is able to align RNA structures with arbitrary pseudoknots using a general edit distance for reasonable scoring schemes. The algorithm is fixed parameter tractable and our prototypical implementation shows its applicability in practice.

A central insight due to our method is that pseudoknots can be effectively handled by partitioning the RNA structure into a set of "easy" and "difficult" interactions. Then, expensive, exponential computation can be restricted to the "difficult" part, whereas state-of-the art polynomial methods can be applied to the "easy" part. Furthermore, since for alignment the dynamic programming recursions operate on pairs of sequences even more effective partitionings can be obtained on the level of arc pairs instead of lifting partitions on single arcs.

The idea of partitioning and making this level of abstraction explicit in the algorithm offers possibilities for further optimization. First, since the concrete partition strongly impacts the run-time, optimizing the partition is worth investigating. Second, one obtains heuristic versions of our algorithm by filtering out unlikely arc pairs.

Acknowledgments. We thank Marco Kuhlmann and the anonymous reviewers for useful comments. M. Möhl is funded by the German Research Foundation.

References

1. Jiang, T., Lin, G., Ma, B., Zhang, K.: A general edit distance between RNA structures. J. Comput. Biol. 9(2), 371–388 (2002)
2. Couzin, J.: Breakthrough of the year. Small RNAs make big splash. Science 298(5602), 2296–2297 (2002)
3. Siebert, S., Backofen, R.: MARNA: multiple alignment and consensus structure prediction of RNAs based on sequence structure comparisons. Bioinformatics 21(16), 3352–3359 (2005)
4. Havgaard, J.H., Torarinsson, E., Gorodkin, J.: Fast pairwise structural RNA alignments by pruning of the dynamical programming matrix. PLoS Comput. Biol. 3(10), 1896–1908 (2007)
5. Will, S., Reiche, K., Hofacker, I.L., Stadler, P.F., Backofen, R.: Inferring non-coding RNA families and classes by means of genome-scale structure-based clustering. PLoS Comput. Biol. 3(4), 65 (2007)
6. Bauer, M., Klau, G.W., Reinert, K.: Accurate multiple sequence-structure alignment of RNA sequences using combinatorial optimization. BMC Bioinformatics 8, 271 (2007)
7. Staple, D.W., Butcher, S.E.: Pseudoknots: RNA structures with diverse functions. PLoS Biol. 3(6), 213 (2005)
8. Theimer, C.A., Blois, C.A., Feigon, J.: Structure of the human telomerase RNA pseudoknot reveals conserved tertiary interactions essential for function. Mol. Cell 17(5), 671–682 (2005)
9. Evans, P.A.: Finding common subsequences with arcs and pseudoknots. In: CPM 1999: Proceedings of the 10th Annual Symposium on Combinatorial Pattern Matching, London, UK, pp. 270–280. Springer, Heidelberg (1999)
10. Rivas, E., Eddy, S.R.: A dynamic programming algorithm for RNA structure prediction including pseudoknots. J. Mol. Biol. 285(5), 2053–2068 (1999)
11. Reeder, J., Giegerich, R.: Design, implementation and evaluation of a practical pseudoknot folding algorithm based on thermodynamics. BMC Bioinformatics 5, 104 (2004)
12. Evans, P.A.: Finding common RNA pseudoknot structures in polynomial time. In: Lewenstein, M., Valiente, G. (eds.) CPM 2006. LNCS, vol. 4009, pp. 223–232. Springer, Heidelberg (2006)
13. Sankoff, D.: Simultaneous solution of the RNA folding, alignment and protosequence problems. SIAM J. Appl. Math. 45(5), 810–825 (1985)
14. Matsui, H., Sato, K., Sakakibara, Y.: Pair stochastic tree adjoining grammars for aligning and predicting pseudoknot RNA structures. Bioinformatics 21(11), 2611–2617 (2005)
15. Dost, B., Han, B., Zhang, S., Bafna, V.: Structural alignment of pseudoknotted RNA. In: Apostolico, A., Guerra, C., Istrail, S., Pevzner, P.A., Waterman, M. (eds.) RECOMB 2006. LNCS (LNBI), vol. 3909, pp. 143–158. Springer, Heidelberg (2006)
16. Blin, G., Fertin, G., Hermelin, D., Vialette, S.: Fixed-parameter algorithms for protein similarity search under mRNA structure constraints. In: Kratsch, D. (ed.) WG 2005. LNCS, vol. 3787, pp. 271–282. Springer, Heidelberg (2005)
17. Zwieb, C., Gorodkin, J., Knudsen, B., Burks, J., Wower, J.: tmRDB (tmRNA database). Nucleic Acids Res. 31(1), 446–447 (2003)

Faster Algorithm for the Set Variant of the String Barcoding Problem

Leszek Gąsieniec[1], Cindy Y. Li[2,*], and Meng Zhang[3,*]

[1] Department of Computer Science, University of Liverpool, Liverpool, UK
L.A.Gasieniec@liverpool.ac.uk
[2] Histocompatibility and Immunogenetics Laboratory, National Blood Service, Bristol, UK
Ying.Li@nbs.nhs.uk
[3] College of Computer Science and Technology, Jilin University, Changchun, China
zhangmeng@jlu.edu.cn

Abstract. A *string barcoding problem* is defined as to find a minimum set of substrings that distinguish between all strings in a given set of strings S. In a biological sense the given strings represent a set of genomic sequences and the substrings serve as probes in a hybridisation experiment. In this paper, we study a variant of the string barcoding problem in which the substrings have to be chosen from a particular set of substrings of cardinality n. This variant can be also obtained from more general *test set problem*, see, e.g., [1] by fixing appropriate parameters. We present almost optimal $O(n|S| \log^3 n)$-time approximation algorithm for the considered problem. Our approximation procedure is a modification of the algorithm due to Berman *et al.* [1] which obtains the best possible approximation ratio $(1 + \ln n)$, providing $NP \not\subseteq DTIME(n^{\log \log n})$. The improved time complexity is a direct consequence of more careful management of processed sets, use of several specialised graph and string data structures as well as tighter time complexity analysis based on an amortised argument.

1 Introduction

The string barcoding problem, discussed by Rash and Gusfield [9], is used for identification of *genomic sequences (targets)*, such as viruses or bacteria, from among a set of known targets. Applications of this technique range from efficient pathogen identification in medical diagnosis to monitoring of microbial communities in environmental studies [2]. The wide range of applications lead to the same methodological problem which is to determine the presence or absence of one target in a biological sample. Targets identification is performed by synthesising the Watson-Crick complements of the *probes* on a microarray [5], then hybridising to the array the fluorescently labelled DNA extracted from the unknown target. Under the assumption of perfect hybridisation stringency, the hybridisation pattern can be viewed as a string of 0's and 1's where 1 represents a probe hybridises to a target. This $0/1$ pattern is referred to as a *barcode* of the target. For unambiguous identification, probes must be selected such that each genomic

---

* This work was done while Cindy Y. Li was a PhD student and Meng Zhang was a research visitor at University of Liverpool.

P. Ferragina and G. Landau (Eds.): CPM 2008, LNCS 5029, pp. 82–94, 2008.

sequence has a distinct barcode. The problem is to compute a minimal set of probes that distinguishes the target in the sample during performed hybridisation experiment.

Borneman et al. [2] were among the first to study the string barcoding problem. They proposed two efficient algorithms based on simulated annealing and Lagrangian relaxation for selecting a minimal probe set to be used in the oligonucleotide fingerprinting of rDNA clones by hybridisation experiments on DNA microarrays. However, the running time of these algorithms does not scale well with the number and the length of the genomic sequences mainly because of its requirement of large memory space. The string barcoding problem has been popularised by Rash and Gusfield [9]. In their paper, they proposed an *integer programming* approach to express the minimisation problem and represented strings by using suffix trees. They also stated that the constrained version of string barcoding, where the maximum length of each probe is bounded by a constant with the alphabet size at least 3, is NP-hard. They also stated the approximability of string barcoding as an open problem. In [8], Lancia and Rizzi showed that the string barcoding problem is as hard to approximate as the *set cover*. Furthermore, they showed that the constrained version of string barcoding with probes of bounded length is also hard to approximate. Finally they proved that both constrained and unconstrained string barcoding are NP-complete even for binary alphabets. Klau et al. [7] presented an approach to select a minimal probe set for the case of non-unique probes in the presence of a small number of multiple targets in the sample. Their approach is based on Integer Programming mixed with a branch-and-cut algorithm. Their preliminary implementation is capable of separating all pairs of targets optimally in a reasonable time and achieves a considerable reduction on the numbers of probes needed compared to previous greedy algorithms. DasGupta et al. [4,3] proposed a greedy algorithm for robust string barcoding. Their method enabled probe selection based on whole genomic sequences of hundreds of microorganisms of up to bacterial size on well-equipped work station. Berman et al. [1] proposed an $O(n^2|\mathcal{S}|)$ time approximation algorithm for test set problem (TS) with approximation ratio $(1+\ln n)$. The approximability results in [1] holds for general test set problems which includes the set variant of the string barcoding as a special case. The algorithm proposed in [1] is a greedy procedure where the test set to be added at each step is determined by a certain *entropy* function.

Our Results. In this paper, we improve the time complexity $O(n^2|\mathcal{S}|)$ of the approximation algorithm for the test set problem proposed by Berman et al. [1], which solves also the variant of the string barcoding problem adopted here. Our algorithm works in almost optimal time $O(n|\mathcal{S}|\log^3 n)$ in view of the input size $\Omega(|\mathcal{S}|n)$ (the input is very often expressed as the binary matrix with $|\mathcal{S}|$ rows and n columns, where the entry (i,j) set to 1 (0) means that substring j (does not) belongs to string i, see Table 1). The improved time complexity is a direct consequence of more careful management of processed sets, use of several specialised graph and string data structures and tighter time complexity analysis based on an amortised argument.

The paper is organised as follows. We present first short description of the studied variant of the string barcoding problem, see Section 2.1. Further in Section 2.2, we provide basic notation and definitions used later in the paper. In Section 2.3, we highlight implementation differences between our algorithm and its counterpart from [1]. This is followed by detail description of specialised data structures used by our algorithm, see Section 2.4. The analysis of the time complexity based on amortised argument is presented in Section 2.5. The conclusion and discussion of further work is available in Section 3.

2   The Problem and the Method

We start this section with a short introduction to the set variant of the string barcoding problem.

2.1   String Barcoding Problem

Given a set of $|\mathcal{S}|$ genomic sequences (targets), $\mathcal{S} = \{s_1, s_2, ..., s_{|\mathcal{S}|}\}$. The objective is to find as small as possible set of elements (probes) $\mathcal{T} = \{p_1, p_2, ..., p_m\}$ from a given set of substrings of strings in $\mathcal{S}$, such that, for any pair of strings $s_i, s_j \in \mathcal{S}$, there is at least one probe $p \in \mathcal{T}$ where p is a substring of s_i or s_j, but not both of them. We say $\mathcal{T}$ distinguishes $\mathcal{S}$ if this property holds. The hybridisation pattern can be viewed as a string of m zeros and ones, referred to as the barcode of each target sequence in $\mathcal{S}$.

Let $\mathcal{S}$ be the set of targets $\{AGGT, ACCTGA, TGGAT, GCA, CGCGATT, GTTAC\}$. Then, the set $\mathcal{T} = \{AC, C, GA\}$ gives a set of valid barcodes (shown in Table 1) for the input sequences in $\mathcal{S}$.

Table 1. Targets and probes: an entry (s_i, p_j) has value 1 if and only if p_j hybridises to s_i

	AC	C	GA
$AGGT$	0	0	0
$ACCTGA$	1	1	1
$TGGAT$	0	0	1
GCA	0	1	0
$CGCGATT$	0	1	1
$GTTAC$	1	1	0

2.2   Notation and Definitions

We use $[i..j]$ to denote the set of consecutive integers $\{i, i+1, ..., j-1, j\}$ and $P(T)$ to denote the power set of a set $T \subseteq [0..n-1]$. The complement $[0..n-1] \setminus T$ of T is denoted by $\overline{T}$. The cardinality of a set T is represented by $|T|$. We say that a set T *distinguishes* two elements $x, y \in [0..n-1]$ where $x \neq y$, if $|\{x, y\} \cap T| = 1$.

Definition 1 (*Test set problem TS*)
INSTANCE: $(n, \mathcal{S})$ where $\mathcal{S} \subseteq P([0..n-1])$.
VALID SOLUTION: *A collection* $\mathcal{T} = \{T_0, T_1, ..., T_{|\mathcal{T}|-1}\} \subseteq \mathcal{S}$ *such that for every pair of distinct integers* $x, y \in [0..n-1]$, *there exists* $T \in \mathcal{T}$ *that distinguishes* x *and* y.
OBJECTIVE: *minimise* $|\mathcal{T}|$.

Example 1: Let $n = 3$ and $\mathcal{S} = \{\{0\}, \{1\}, \{0,1\}\}$. Then, $\mathcal{T} = \{\{0\}, \{0,1\}\}$ is a valid solution since $\{0,1\}$ distinguishes 0 from 2 ($|\{0,2\} \cap \{0,1\}| = 1$) as well as 1 from 2 ($|\{1,2\} \cap \{0,1\}| = 1$) while $\{0\}$ distinguishes 0 from 1 ($|\{0,1\} \cap \{0\}| = 1$).

Definition 2 (*Equivalence relation* $\overset{\mathcal{T}}{\equiv}$). *Assume that $\mathcal{T}$ is a collection of subsets of the set $[0..n - 1]$. An equivalence relation $\overset{\mathcal{T}}{\equiv}$ is defined on $[0..n - 1]$ and the collection $\mathcal{T}$, where for any $i, j \in [0..n - 1]$, $i \overset{\mathcal{T}}{\equiv} j$ if and only if $\forall T \in \mathcal{T}$, either $\{i,j\} \subseteq T$ or $\{i,j\} \cap T = \emptyset$. The equivalence relation $\overset{\mathcal{T}}{\equiv}$ partitions $[0..n - 1]$ into m equivalence classes, where the lth equivalence class is denoted by $E(\mathcal{T}, l)$, for $l = 0, 1, ..., m - 1$ and let $E(\mathcal{T}) = \{E(\mathcal{T}, 0), E(\mathcal{T}, 1), ..., E(\mathcal{T}, m - 1)\}$.*

Note that each non-trivial equivalence class $E(\mathcal{T}, l)$ of $\overset{\mathcal{T}}{\equiv}$ is a product of the form $T_0^* \cap T_1^* ... \cap T_{|\mathcal{T}|-1}^*$, where $T_k^* = T_k$ or $T_k^* = \overline{T_k}$, for $k = 0, 1, ..., |\mathcal{T}| - 1$.

Definition 3 (*Entropy function $H_{\mathcal{T}}$, adopted from [1]*). *The entropy function $H_{\mathcal{T}}$ is defined as $H_{\mathcal{T}} = \log_2(\Pi_{l=0}^{m-1} |E(\mathcal{T}, l)|!)$ where m is the number of equivalence classes in $E(\mathcal{T})$.*

Example 2: Let $\mathcal{T} = \{T_0\}$, where $T_0 = \{2, 3, 4\}$ and $n = 8$. Then, $E(\mathcal{T}, 0) = T_0 = \{2, 3, 4\}$, $E(\mathcal{T}, 1) = \overline{T_0} = \{0, 1, 5, 6, 7\}$. So, $E(\mathcal{T}) = \{\{2, 3, 4\}, \{0, 1, 5, 6, 7\}\}$ and $H_{\mathcal{T}} = \log_2((3!)(5!)) \approx 9.492$.

Example 3: Let $\mathcal{T} = \{T_0, T_1\}$ where $T_0 = \{2, 3, 4\}$, $T_1 = \{1, 3, 5, 7\}$ and $n = 8$. Then, $E(\mathcal{T}, 0) = T_0 \cap T_1 = \{3\}$, $E(\mathcal{T}, 1) = T_0 \cap \overline{T_1} = \{2, 4\}$, $E(\mathcal{T}, 2) = \overline{T_0} \cap T_1 = \{1, 5, 7\}$, $E(\mathcal{T}, 3) = \overline{T_0} \cap \overline{T_1} = \{0, 6\}$. We get, $E(\mathcal{T}) = \{\{3\}, \{2, 4\}, \{1, 5, 7\}, \{0, 6\}\}$ and $H_{\mathcal{T}} = \log_2((1!)(2!)(3!)(2!)) \approx 4.585$.

Definition 4 (*Combination of equivalence relations $\otimes$*). *A combination of two equivalence relations $\overset{\mathcal{T}}{\equiv}$ and $\overset{\mathcal{T}'}{\equiv}$ on the set $[0..n-1]$ is defined as $E(\mathcal{T}) \otimes E(\mathcal{T}') = E(\mathcal{T} \cup \mathcal{T}')$.*

Definition 5 (*Basic block*). *A basic block $B(i, l)$ is a set of consecutive integers $[(i - 1) \times n/2^l .. i \times n/2^l - 1]$, where $1 \leq i \leq 2^l$ and $0 \leq l \leq \log n$.*

2.3   Two Algorithms

In this paper, we propose asymptotically more efficient implementation of the algorithm due to Berman, DasGupta and Kao [1] for the TS problem. As mentioned earlier, Berman *et al.* [1] proposed an $O(n^2|\mathcal{S}|)$ time approximation algorithm for TS with the approximation ratio $(1 + \ln n)$. In each round of their algorithm, where rounds correspond to consecutive iterations of loop while in Algorithm 1, they compute combinations $E(\mathcal{T}) \otimes E(\{T_j\})$ for all $T_j \in \mathcal{S} \setminus \mathcal{T}$. They also select T_j that minimises the entropy function $H_{\mathcal{T} \cup \{T_j\}}$ and then move T_j (from $\mathcal{S} \setminus \mathcal{T}$) to the collection $\mathcal{T}$. Since for each remaining T_j, a naive computation of $E(\mathcal{T}) \otimes E(\{T_j\})$ and $H_{\mathcal{T} \cup \{T_j\}}$ takes time $\Omega(n)$, and since for most of rounds, $|\mathcal{S} \setminus \mathcal{T}| = \Omega(|\mathcal{S}|)$, each round requires time $\Omega(n|\mathcal{S}|)$. Algorithm 1 is executed in at most $n - 1$ rounds because n integers can be separated by at most $n - 1$ sets from $\mathcal{S}$ in the worst case. Therefore, the total

---
Algorithm 1. Berman, DasGupta and Kao [1]

1: $\mathcal{T} = \emptyset$;
2: **while** $H_{\mathcal{T}} \neq 0$ **do**
3:     select a $T_j \in \mathcal{S} \setminus \mathcal{T}$ that minimises $H_{\mathcal{T} \cup \{T_j\}}$;
4:     $\mathcal{T} = \mathcal{T} \cup \{T_j\}$;
5: **end while**

---

complexity of Algorithm 1 is $O(n^2|\mathcal{S}|)$. Note also that in Algorithm 1, no structural information about $E(\mathcal{T}) \otimes E(\{T_j\})$ and the entropy $H_{\mathcal{T} \cup \{T_j\}}$ is kept for future use in later rounds (apart from T_j that minimises the entropy function).

In our paper, the main idea is to store and to utilise information about all previously computed combinations $E(\mathcal{T}) \otimes E(\{T_j\})$ together with the history of their entropy computation (see Algorithm 2). This is to reduce the overall time complexity. Let $T(i-1)$ be the selected set that minimises the entropy function at the end of round $i-1$. Let $\mathcal{T}(i) = \{T(0), T(1), ..., T(i-1)\}$ represent the collection of sets selected as part of the solution in rounds $0, 1, ..., i$. Since we keep records on all combinations $E(\mathcal{T}(i-1)) \otimes E(\{T_j\})$ obtained in round $i-1$, later during round i, we can compute $E(\mathcal{T}(i) \cup \{T_j\})$ applying $E(\mathcal{T}(i-1) \cup \{T_j\}) \otimes E(\mathcal{T}(i-1) \cup \{T(i-1)\})$ rather than via direct computation of $E(\mathcal{T}(i) \cup \{T_j\})$ as it is done in Berman *et al.* algorithm. We introduce a new concept of hierarchical data structure (see Section 2.4) that allows to represent and manipulate equivalence classes $E(\mathcal{T}(i) \cup \{T_j\})$ and $E(\mathcal{T}(i) \cup \{T(i-1)\})$ efficiently. Moreover, we make use of a directed acyclic graph to compute the history of the entropy values (see Section 2.4).

2.4   Data Structures

We introduce a hierarchical data structure $\mathcal{H}$ to represent, compare and process efficiently a dynamic collection of sets $\mathcal{C}$ of small integers, i.e., subsets of $[0..n-1]$. Initially $\mathcal{C} = \mathcal{S}$, and later it contains all (including intermediate) subsets of $[0..n-1]$

Algorithm 2. Our algorithm

1: $\mathcal{T}(0) = \emptyset$;
2: **for** $j = 1, 2, ..., |\mathcal{S}|$ **do**
3:     Compute $E(\{[0..n-1]\} \cup \{T_j\})$;
4: **end for**
5: $T(0) = T_j$ such that T_j minimises $H_{\{T_j\}}$;
6: **for** ($i = 1; H_{\mathcal{T}(i-1)} \neq 0; i + +$) **do**
7:     /* i is the number of current round */
8:     $\mathcal{T}(i) = \mathcal{T}(i-1) \cup \{T(i-1)\}$;
9:     **for** $j = 1, 2, ..., |\mathcal{S}|$ **do**
10:        Compute $E(\mathcal{T}(i) \cup \{T_j\})$ by applying $E(\mathcal{T}(i-1) \cup \{T_j\}) \otimes E(\mathcal{T}(i-1) \cup \{T(i-1)\})$;
11:    **end for**
12:    $T(i) = T_j$ such that T_j minimises $H_{\mathcal{T}(i) \cup \{T_j\}}$;
13: **end for**
14: **return** $\mathcal{T}(i-1)$;

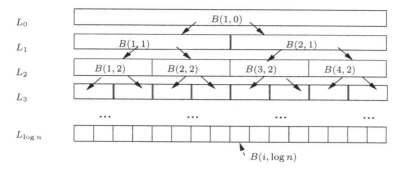

L_0

$B(1,0)$

L_1

$B(1,1)$ $B(2,1)$

L_2

$B(1,2)$ $B(2,2)$ $B(3,2)$ $B(4,2)$

L_3

$\cdots$ $\cdots$ $\cdots$ $\cdots$

$L_{\log n}$

$B(i,\log n)$

Fig. 1. Binary tree representation of a set

corresponding to all considered equivalence classes generated by Algorithm 2. The new data structure allows equality tests on two sets from the collection to be performed in constant time. Moreover, single element insertions to and deletions from any set from the collection are implemented in poly-logarithmic time.

Binary Tree Representation of a Set. In principle, each set S in $\mathcal{C}$ is represented by a binary tree structure D_s (of pointers) defined as follows. In each tree, there are exactly $\log n + 1$ levels enumerated from 0 (root level) to $\log n$ (leaf level). At the level l there are 2^l nodes. Each internal node v in the tree is the parent of two children, the left child $l(v)$ and the right child $r(v)$. Moreover, each node of the tree representing S stores information about the content of S projected on a specific basic block, chosen according to the following rule. The root of the tree stores information about the content of S projected on $B(1,0)$. And later if a parent node v on level l stores the information about the content of S projected on $B(i,l)$, then $l(v)$ and $r(v)$ store information about the contents of S projected on $B(2i-1,l+1)$ and $B(2i,l+1)$ respectively. For example, the leaves store information about the content of set S projected on consecutive basic blocks $B(1,\log n), B(2,\log n),\ldots, B(i,\log n),\ldots, B(n,\log n)$ which are either empty sets or singletons (see Figure 1).

Hierarchical Data Structure for a Collection of Sets. The nodes of binary tree structures representing sets from the collection $\mathcal{C}$ whose contents refer to the same basic block *correspond* to each other and we say that they belong to the same *group*. The binary tree structures representing sets in $\mathcal{C}$ are stored in the hierarchical data structure $\mathcal{H}$ in a compact form, where two corresponding nodes (associated with the same basic block) in different trees with the same content (the same subset of $[0..n-1]$) are represented by a single node in $\mathcal{H}$ (see Figure 2). In order to create $\mathcal{H}$ (from the trees) and further manipulate it efficiently, we propose a new application of the naming method (for definition see, e.g., [6]).

Naming Method. The naming method adopted here requires application of a system of *counters* and *balanced binary search trees*. Each group of nodes based on a specific basic block $B(i,l)$, for $0 \le i \le n-1$ and $0 \le l \le \log n$, requires a separate counter $C_{i,l}$ (that is used to generate new names within the group of nodes) and a balanced binary

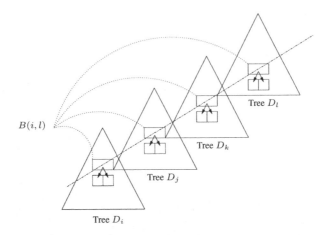

Fig. 2. Hierarchical data structure for a collection of sets

search tree $T_{i,l}$ (that keeps all used names in the group of nodes indexed by the pair of children's names). In each tree of the collection embedded in $\mathcal{H}$, the nodes get integer names, level by level, starting from the lowest level $\log n$.

The counters are initialised to value 0 and the balanced binary search trees are set to be empty. At the bottom level ($\log n$), the i-th leaf, for any $0 \leq i \leq n - 1$, in the tree representing S (embedded in $\mathcal{H}$) is given name 0 if $\{i\} \cap S = \emptyset$ and 1 otherwise. Above that, at each consecutive higher level $1 \leq l \leq \log n$ in every tree D_s, for all $S \in \mathcal{C}$, and at every internal node $v \in D_s$ associated with some $B(i,l)$, we first check whether the pair of names $(N(l(v)), N(r(v)))$ already occurs as the pair of names of children of some corresponding node w in some other tree $D_{s'}$. This can be done in time $O(\log n)$ by searching for the pair $(N(l(v)), N(r(v)))$ as the key in the balanced binary search tree $T_{i,l}$. If the pair $(N(l(v)), N(r(v)))$ does not occur as the key in $T_{i,l}$, we increase the counter $C_{i,l}$ by 1 and assign its new value as the name of v, i.e., $N(v) = ++C_{i,l}$. Moreover we insert to $T_{i,l}$ a new record with the content $N(v)$ and the key $(N(l(v)), N(r(v)))$. This is also done in time $O(\log n)$. Otherwise, if there already exists a node w such that $(N(l(v)), N(r(v))) = (N(l(w)), N(r(w)))$, v adopts the name of w, i.e., $N(v) = N(w)$, and v is represented by the node w in $\mathcal{H}$.

Lemma 1. *The initialisation of the hierarchical structure $\mathcal{H}$ takes time $O(n|\mathcal{S}| \log n)$.*

Proof. The nodes in the hierarchical structure $\mathcal{H}$ get integer names, level by level, starting from the lowest level $\log n$. As explained above, the computation of a single name including manipulation of respective data structure in $\mathcal{H}$ requires time $O(\log n)$. At each level l for $0 \leq l \leq \log n$, there are at most $2^l |\mathcal{S}|$ nodes to be named. Thus, to generate names of nodes at level l in $\mathcal{H}$ requires time $O(2^l |\mathcal{S}| \log n)$. In conclusion, the initial computation of the names of all nodes in $\mathcal{H}$ is done in time $O(\sum_{l=0}^{l=\log n} 2^l |\mathcal{S}| \log n)$, which is $O(n|\mathcal{S}| \log n)$.

Set Operations. In our amortised analysis argument provided in section 2.5, we use three operations performed on sets from the dynamic collection $\mathcal{C}$. Namely, *equality test* $Eq(S, S')$ for the contents of two sets $S, S' \in \mathcal{C}$, i.e., whether $S = S'$, *deletion operation* $Delete(S, x)$ that removes x from S, i.e., $S = S \setminus \{x\}$ and *insertion operation* $Insert(S, x)$ that adds x to S, i.e., $S = S \cup \{x\}$. When we perform equality test $Eq(S, S')$ on two sets from $\mathcal{C}$, we only need to compare the names of nodes representing sets S and S' in $\mathcal{H}$. This can be done in constant time. When we remove an element x from a set S ($Delete(S, x)$), we change the name from 1 to 0 of the appropriate node v representing x in S located at the bottom level in $\mathcal{H}$ and then we update the names of all nodes on the path from the node v to the node representing the whole set S at the top level of $\mathcal{H}$. Since there are $O(\log n)$ names to be changed at different levels in $\mathcal{H}$ and the computation of the name of a node in $\mathcal{H}$ requires time $O(\log n)$ as explained in Section 2.4, the deletion operation takes time $O(\log^2 n)$. The insertion operation ($Insert(S, x)$) is implemented analogously (where we change name from 0 to 1 at the bottom level of $\mathcal{H}$) to the deletion operation. As a result, we get the following lemma.

Lemma 2. *The structure $\mathcal{H}$ provides a mechanism for $O(1)$-time equality test for two sets in $\mathcal{C}$ and $O(\log^2 n)$-time single element removal from and insertion to a set in $\mathcal{C}$.*

Efficient Cross-examination of Equivalence Classes. Note that in any advanced round i of Algorithm 2 each equivalence relation $E(\mathcal{T}(i-1) \cup \{T_j\})$ may potentially have $\Omega(n)$ equivalence classes. Thus a naive cross-examination with all classes in $E(\mathcal{T}(i-1) \cup \{T(i-1)\})$ (see line 10 in Algorithm 2) may lead to $\Omega(|\mathcal{S}|n)$ comparisons during each round. And since the number of rounds may be as large $\min(|\mathcal{S}|, n)$ we would see no improvement in the time complexity in comparison with the algorithm presented in [1].

$E(\mathcal{T}(i-1) \cup \{T_j\})$	$\left[\; E_1^{L(j)} \;\right]$	$\left[\; E_1^{R(j)} \;\right]$	$[\qquad\qquad]$	$[\ldots]$	$\left[\; E_c^{L(j)} \;\right]$	$\left[\; E_c^{R(j)} \;\right]$
$E(\mathcal{T}(i-1))$	$\left[\; E_1 = E(\mathcal{T}(i-1),1) \;\right.$	$\left[\; E_2 = E(\mathcal{T}(i-1),2) \;\right]$	$[\ldots]$		$\left[\; E_c = E(\mathcal{T}(i-1),c) \;\right]$	
$E(\mathcal{T}(i-1) \cup \{T(i-1)\})$	$\left[\; E_1^L \;\right]$	$\left[\; E_1^R \;\right]\left[\; E_2^L \;\right]$	$\left[\; E_2^R \;\right]$	$[\ldots]$	$\left[\; E_c^L \;\right]$	$\left[\; E_c^R \;\right]$

Fig. 3. Cross-examination of equivalence classes

In order to reduce the number of cross-examined equivalence classes we provide another data structure SL (structured list of equivalence classes) based on unique names of classes available in the hierarchical structure $\mathcal{H}$ and defined as follows. Assume that during round i we have an equivalence relation $E(\mathcal{T}(i-1))$ formed of c equivalence classes $E_1 = E(\mathcal{T}(i-1), 1), \ldots, E_c = E(\mathcal{T}(i-1), c)$, s.t., each class $E_x = E(\mathcal{T}(i-1), x)$ is potentially split into two classes E_x^L and E_x^R (possibly empty) in $E(\mathcal{T}(i)) = E(\mathcal{T}(i-1) \cup \{T(i-1)\})$ (see Figure 3). Also each equivalence relation $E(\mathcal{T}(i-1) \cup \{T_j\})$ potentially bears two subclasses $E_x^{L(j)}$ and $E_x^{R(j)}$ (possibly empty) for each $E(\mathcal{T}(i-1), x) \in E(\mathcal{T}(i-1))$. We assume that at the beginning of round i

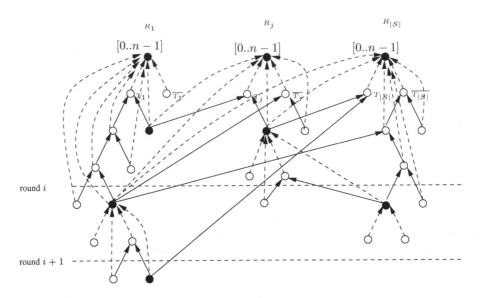

Fig. 4. Compute entropy function by using $\vec{G}$. Express link is shown by dash line.

the structure SL is formed of c lists $L_1, L_2, .., L_c$, s.t., each L_x contains all different pairs of subclasses (multiplicities are discarded to avoid dummy cross-examinations) of E_x present both in $E(\mathcal{T}(i-1) \cup \{T(i-1)\})$ and in each $E(\mathcal{T}(i-1) \cup \{T_j\})$. On the conclusion of round i each list L_x associated with E_x in SL is split (if needed) into two lists associated with two equivalence classes E_x^L and E_x^R, where each of these lists contains now all pairs of different subclasses in new $E(\mathcal{T}(i) \cup \{T_j\})$. Finally note that since every cross-examination of two different pairs of sub-classes results in creation of new equivalence classes (at least one split) the number of all cross-examined pairs can be bounded by $O(|\mathcal{S}| \cdot n)$. The total cost of all handling the structured list of equivalence classes has to be multiplied by the factor of $O(\log^2 n)$ which refers to access to and location of new equivalence classes in the hierarchical structure $\mathcal{H}$. This results in the total complexity $O(|S|n \log^2 n)$.

Entropy Function Calculation. The entropy function is computed dynamically on the basis of a directed acyclic graph $\vec{G}$ (see Figure 4) gradually expanded during consecutive rounds of Algorithm 2. We keep at each node in $\vec{G}$ the name and the size of the equivalence class it represents. At the end of round i, for $0 \leq i \leq n-1$, all values of the entropy function $H_{\mathcal{T}(i) \cup \{T_j\}}$, for all $T_j \in \mathcal{S} \setminus \mathcal{T}(i)$, are calculated. The set T_j which minimises $H_{\mathcal{T}(i) \cup \{T_j\}}$ is selected as $T(i)$. The use of $\vec{G}$ allows to reduce the overall cost (on the top of handling the hierarchical structure $\mathcal{H}$) of computation of the entropy function to $O(|\mathcal{S}|n)$. We prove later that this cost is linear in the total number of *splits* of equivalence classes $E(\mathcal{T}(i) \cup \{T_j\})$ represented by nodes in $\mathcal{H}$ through out consecutive rounds of Algorithm 2.

Recall that in a directed graph, nodes without successors are called *sinks*, and nodes with no predecessors are called *source* nodes. The acyclic graph $\vec{G}$ is created and maintained as follows. At the top level of $\overrightarrow{G}$, see Figure 4, we place $|\mathcal{S}|$ nodes labelled by R_j, for $1 \leq j \leq |\mathcal{S}|$, where each node represents the whole range $[0..n - 1]$ before any of $T_j \in \mathcal{S}$ is introduced. These nodes will be the only sinks in $\overrightarrow{G}$ throughout the duration of the algorithm. In round 0, each set R_j is partitioned into T_j and $\overline{T_j}$ (the complement of T_j). The nodes labelled by the names of T_j and $\overline{T_j}$ become temporary sources. They are inserted into $\overrightarrow{G}$ as predecessors of the sink labelled by R_j. The entropy function $H_{\{R_j\} \cup \{T_j\}} = H_{\{T_j\}}$ is calculated directly on the basis of information available in newly generated nodes (the sizes of T_j and $\overline{T_j}$) and its value $H_{\{T_j\}}$ is stored at the sink labelled by R_j, for $1 \leq j \leq |\mathcal{S}|$. A set T_j which minimises $H_{\{T_j\}}$ is selected as $T(0)$.

Later, at the beginning of round i, each source node in $\overrightarrow{G}$ is labelled by the name of some equivalence class $\mathcal{E} \in E(\mathcal{T}(i - 1) \cup \{T_j\})$ represented by some node in $\mathcal{H}$. We also have $T(i - 1)$ which is calculated during round $i - 1$. Note that if $\mathcal{E} \not\subseteq T(i - 1)$ and $\mathcal{E} \cap T(i - 1) \neq \emptyset$ (intersection of $\mathcal{E}$ and $T(i - 1)$ is non-trivial), the source node labelled by the name of $\mathcal{E}$ becomes a successor of two new nodes. We also say that $\mathcal{E}$ is split. The two new nodes are labelled by the names of two new equivalence classes $\mathcal{E} \cap T(i-1), \mathcal{E} \cap \overline{T(i-1)} \in E(\mathcal{T}(i)) \otimes E(\{T_j\})$. If any two newly obtained equivalence classes $\mathcal{E}_j \in E(\mathcal{T}(i) \cup \{T_j\})$ and $\mathcal{E}_{j'} \in E(\mathcal{T}(i) \cup \{T_{j'}\})$, for $j \neq j'$, have the same content, they are represented by the same node in $\overrightarrow{G}$ called a *branching node*. We colour all branching nodes as well as all sinks to black, see Figure 4. All other nodes in $\overrightarrow{G}$ remain white. Moreover, we create a collection of *express links* such that every (black or white) node v is connected via express link to the first black nodes w on a directed path leading to any R_j reachable from v. The following lemma holds.

Lemma 3. *A structure of all nodes connected via express links from any node v in $\overrightarrow{G}$ forms a tree rooted in v with all R_js reachable from v as leaves where the number of leaves subsumes of the number of internal nodes.*

Proof. By the construction, every node v is connected via some express link to the first black node w which can either be a branching node or a sink. Moreover, the directed express path rooted from v will finally reach some sink labelled by R_j. This holds due to the fact that the set represented by v must be a subset of some range $[0..n - 1]$ which is a sink in $\overrightarrow{G}$. Therefore, the structure of all nodes connected via express links from any node v forms a tree where v is the root and all R_js reachable from v are the leaves. Assume that a node v in $\overrightarrow{G}$ is connected to x sinks which are the leaves in the spanning tree of v. There are at most $x - 1$ branching nodes in the spanning tree of v.

The value $H_{\mathcal{T}(i) \cup \{T_j\}}$ computed in round i only needs to be updated when a temporary source v connected to R_j is split into two new nodes. Let the size of v be s and the sizes of the two new nodes be s_1 and s_2, respectively. Recall the definition of the entropy function, when a split happens, $H_{\mathcal{T}(i) \cup \{T_j\}} = \frac{s_1! s_2!}{s!} H_{\mathcal{T}(i)}$. The fraction $\frac{s_1! s_2!}{s!}$ can be delivered to the sink via the spanning tree of v. In such a way, $H_{\mathcal{T}(i) \cup \{T_j\}}$ can be calculated efficiently and the set T_j which minimises $H_{\mathcal{T}(i) \cup \{T_j\}}$ is selected as $T(i)$.

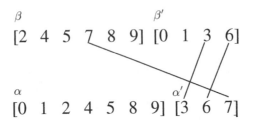

Fig. 5. Two different pairs of equivalence classes

Lemma 4. *Let $C(i)$ be the number of splits of equivalence classes in round i. The maintenance cost (on the top of manipulation of H) of $\overrightarrow{G}$ in round i is $O(C(i))$.*

Proof. Assume that a node v (corresponding to some equivalence class) in $\overrightarrow{G}$ is connected to x (the number of equivalence relations containing v as a class) sinks. When v gets split, x sinks have to be informed and updated. By Lemma 3, this is done in time $O(x)$ due to the presence of express links in the spanning tree spanned on at most $x-1$ branching nodes. Therefore, the maintenance cost of $\overrightarrow{G}$ in round i is $O(C(i))$.

Corollary 1. *Since the total number of splits in all sets R_j is no more than $|\mathcal{S}|(n-1)$ the on-line maintenance of all current values of the entropy function is done at cost $O(|\mathcal{S}|n)$.*

2.5 Amortised Analysis

We show here that the total cost of our string barcoding procedure is $O(n|\mathcal{S}|\log^3 n)$.

Assume that during round i two different pairs of equivalence classes (α, α') and (β, β'), where $\alpha = E_x^L$ and $\alpha' = E_x^R$ form a split of a class $E_x = E(\mathcal{T}(i-1), x)$ (in $E(\mathcal{T}(i-1))$) caused by $\mathcal{T}(i-1)$ and $\beta = E_x^{L(j)}$ and $\beta = E_x^{R(j)}$ form a split of the same class E_x caused by $\mathcal{T}(j)$, are available in the list L_x (see Figure 5). As the result of cross-examination we obtain new four sets $\alpha\beta = \alpha \cap \beta$, $\alpha\beta' = \alpha \cap \beta'$ $\alpha'\beta = \alpha' \cap \beta$, and $\alpha'\beta' = \alpha' \cap \beta'$ that form two new pairs of equivalence classes $(\alpha\beta, \alpha\beta')$ and $(\alpha'\beta, \alpha'\beta')$ to be considered during the next round. Assume also that $|\alpha'| \le |\alpha|$ (note that information about the sizes of α and α' can be either kept in the hierarchical structure $\mathcal{H}$ or it could be computed on line from the size of a superclass formerly split into α and α'). Our algorithm takes all elements (one by one) from α' and searches for their occurrences in β and β'. When an element is located in β it is moved to the set $\alpha'\beta$ otherwise it is moved from β' to $\alpha'\beta'$. When this process is finished whatever is left in β becomes $\alpha\beta$ and whatever remains in β' becomes $\alpha\beta'$. The split operation is completed.

In our amortised analysis argument we would like to trade in tested elements from α' for the total cost of our string barcoding procedure. Thus in round i the equivalence classes in all $E(\mathcal{T}(i) \cup T_j)$ overlapping with E_x will be updated at the uniform cost $|\alpha'|$, for each T_j outside of $\mathcal{T}(i)$. And this is happening only when α' is non-empty, otherwise no cost is charged (there will be no pair (α, α') in L_x since there will be no immediate split of classes β and β' in $E(\mathcal{T}(i-1) \cup T_j)$. Also when $(\alpha, \alpha') = (\beta, \beta')$ no split is required, and indeed in this case both pairs appear as one in L_x. Since we always charge the cost of a split to a smaller set α' (i.e., $|\alpha'| \le |\alpha|$) every element in each $E(\mathcal{T}(i-1) \cup T_j)$ will be charged at most $\log n$ times during the whole execution of Algorithm 2. Note also that the search in β and β' for each charged element takes time $O(\log^2 n)$. This is done with a help of the hierarchical structure $\mathcal{H}$ and the procedures $Delete()$ and $Insert()$, see section 2.4. This means that the total charge across all $(|\mathcal{S}|)$ equivalence relations $E(\mathcal{T}(i-1) \cup T_j)$ can be limited to $O(|\mathcal{S}|n \log^3 n)$.

Theorem 1. *Algorithm 2 is $O(|\mathcal{S}|n \log^3 n)-$ time string barcoding approximation procedure with the approximation ratio $O(1 + \log n)$.*

Proof. The time complexity follows from the amortised argument presented above. The $O(1 + \log n)$ approximation ratio is ensured by the algorithm presented in [1], i.e., our string barcoding procedure does not change sets selected to the final solution. Our primary focus was on improved performance of the selection process.

3   Conclusion

In this paper, we improve on the time complexity $O(n^2|\mathcal{S}|)$ of the approximation algorithm for the set variant of the string barcoding problem proposed by Berman *et al.* in [1]. Our algorithm works in almost optimal time $O(n|\mathcal{S}| \log^3 n)$ in view of the fact that the size of the input to the studied problem is of size $\Omega(|S|n)$. Note also that the time improvement presented here applies also to other test set problems considered in [1].

Among problems to be still addressed is efficient design of fault-tolerant (more robust) barcodes in which every pair of strings must separated by two (or more) probes available in the pool of precomputed probes.

References

1. Berman, P., DasGupta, B., Kao, M.Y.: Tight approximability results for test set problems in bioinformatics. Journal of Computer and System Sciences 71(2), 145–162 (2005)
2. Borneman, J., Chrobak, M., Vedova, G.D., Figueroa, A., Jiang, T.: Probe selection algorithms with applications in the analysis of microbial communities. Bioinformatics 17, 39–48 (2001)
3. DasGupta, B., Konwar, K.M., Mandoiu, I.I., Shvartsman, A.A.: Dna-bar: distinguisher selection for dna barcoding. Bioinformatics 21(16), 3424–3426 (2005)
4. DasGupta, B., Konwar, K.M., Mandoiu, I.I., Shvartsman, A.A.: Highly scalable algorithms for robust string barcoding. International Journal of Bioinformatics Research and Applications 1(2), 145–161 (2005)
5. Gerhold, D., Rushmore, T., Caskey, C.T.: DNA chips: promising toys have become powerful tools. Trends Biochem. Sci. 24(5), 168–173 (1999)
6. Karp, R.M., Miller, R.E., Rosenberg, A.L.: Rapid identification of repeated patterns in strings, trees and arrays. In: Proc. 4th Symposium on Theory of Computing (STOC), pp. 125–136 (1972)

7. Klau, G.W., Rahmann, S., Schliep, A., Vingron, M., Reinert, K.: Optimal robust non-unique probe selection using Integer Linear Programming. Bioinformatics 20, 186–193 (2004)
8. Lancia, G., Rizzi, R.: The approximability of the string barcoding problem. Algorithms for Molecular Biology 1(12), 1–7 (2006)
9. Rash, S., Gusfield, D.: String Barcoding: Uncovering Optimal Virus Signatures. In: Proc. 6th Annual International Conference on Research in Computational Molecular Biology (RE-COMB), pp. 254–261 (2002)

Probabilistic Arithmetic Automata and Their Application to Pattern Matching Statistics

Tobias Marschall and Sven Rahmann

Bioinformatics for High-Throughput Technologies
at the Chair of Algorithm Engineering,
Computer Science Department, TU Dortmund,
D-44221 Dortmund, Germany
{tobias.marschall,sven.rahmann}@tu-dortmund.de

Abstract. We present *probabilistic arithmetic automata (PAAs)*, which can be used to model chains of operations whose operands depend on chance. We provide two different algorithms to exactly calculate the distribution of the results obtained by such probabilistic calculations. Although we introduce PAAs and the corresponding algorithm in a generic manner, our main concern is their application to pattern matching statistics, i.e. we study the distributions of the number of occurrences of a pattern under a given text model. Such calculations play an important role in computational biology as they give access to the significance of pattern occurrences. To assess the practicability of our method, we apply it to the Prosite database of amino acid motifs and to the Jaspar database of transcription factor binding sites. Regarding the latter, we additionally show that our framework permits to take binding affinities predicted from a physical model into account.

1 Introduction

Biological sequence analysis is often concerned with the search for structure in long strings like DNA, RNA or amino acid sequences. Frequently, "search for structure" means to look for patterns that occur very often. An important point in this process is to define sensibly a notion of "very often". One option is to consult the statistical significance of an event: Suppose we have found a certain pattern n times in a given sequence. What is the probability of observing n or more matches just by chance? To answer this question we have to specify the meaning of "by chance" and define an appropriate null model. In the most simple case of independent, uniformly distributed characters, all strings of the same length m have equal probabilities of occurrence. Then, the posed question turns out to be of purely combinatorial nature and can be rephrased as follows: How many strings of length m exist that contain n or more instances of the given pattern? For many applications, however, this simple model is not sufficient. In DNA, for example, the GC-content often differs considerably from 50%, making a uniform model inappropriate. Then, one has to employ at least an i. i. d.[1] model

---

[1] independent, identically distributed.

P. Ferragina and G. Landau (Eds.): CPM 2008, LNCS 5029, pp. 95–106, 2008.
© Springer-Verlag Berlin Heidelberg 2008

(also known as Bernoullian model), which is defined by a distribution over an alphabet. Should the application require an even more elaborate model, the most common choice is a Markov model that allows the probability of each character to depend on a finite history.

Given a suitable null model, a procedure to compute the significance of a pattern is a powerful tool in the context of motif discovery, as it allows the comparison of different patterns regardless of their structure and length.

There are many different types of patterns that are relevant in computational biology, such as single strings, sets of strings, Prosite patterns[2], consensus strings together with a distance measure and a distance threshold, abelian patterns, position weight matrices in connection with a threshold, etc. All these pattern types may be seen as a way to describe a finite set of strings. Thus, all these patterns may be expressed in the form of deterministic finite automata (DFAs) that recognize the respective string set. As our method is based on this DFA representation, it is very general and flexible regarding the pattern type.

Besides specifying a pattern, one has to decide how overlaps are to be handled. We refer to the used strategy as *counting scheme*. The easiest case is to disallow overlaps at all; we call this scheme *non-overlapping count*. Consequently, we define the *overlapping count* to be the number of substrings that match the given pattern. In the case of a set of strings without any restrictions, this scheme makes counting more complicated as some words may be substrings of others. Many authors neglect the problem—at least partly—by simply counting the positions where at least one pattern ends, which we name *match position count*.

Related Work and New Results. The topic of statistics of words on random texts has been studied extensively. An overview is provided in the book by Lothaire [2]; Chapter 6 ("Statistics on Words with Applications to Biological Sequences"), which is particularly interesting to us, is based on the overview article by Reinert et al. [3].

In most approaches developed until now, a generating function is derived for the sought quantity. Then, typically using symbolic Taylor expansion, the concrete values can be computed. This procedure is, for instance, described by Régnier [4], who gives formulas for mean, variance and higher statistical moments of the exact occurrence count distribution. Her framework is general enough to admit Markovian sources as well as finite sets of patterns to be treated in the overlapping as well as in the non-overlapping case. Closely related is the approach of Nicodème et al. [5], who present an algorithmic chain to compute the distribution of the match position count for regular expressions.

Lladser et al. [6] recently reviewed the field. Their main concern is to bring together involved concepts in a consistent and rigorous manner. They make the connection to the classical field of automata theory and pattern matching explicit and therefore speak of *probabilistic pattern matching*. Furthermore, they describe the relation between finite automata and Markov chains in terms of the *Markov chain embedding* technique.

---

[2] Like used in the Prosite database (see Hulo et al. [1]). A syntax description can be found under http://www.expasy.org/tools/scanprosite/scanprosite-doc.html.

In this article, we introduce the concept of *probabilistic arithmetic automata* and demonstrate how it paves the way for a dynamic programming approach to exact pattern matching statistics. The notion of probabilistic arithmetic automata can be seen as a generalization of *Markov additive chains* which were used by Kaltenbach et al. [7] for fragment statistics of cleavage reactions. Another dynamic programming approach was recently presented by Zhang et al. [8]. They use it to compute exact p-values for position weight matrices describing transcription factor binding sites (TFBS).

For finite string sets, our framework is able to handle overlaps of words of different lengths accurately. We are not aware of any previous work permitting that; all those mentioned above use the match position count instead. The presented approach is fast and flexible, as we show by examples from computational biology. Especially in the application to statistics of TFBS, the generality of our framework proves itself advantageous; effortlessly we can take binding affinities obtained from a physical model into account.

2   Probabilistic Arithmetic Automata

In this section, we define *probabilistic arithmetic automata (PAAs)*. They later allow us to establish a well-grounded connection between a pattern set and its statistics on random texts.

Definition 1 (Probabilistic Arithmetic Automaton). *A probabilistic arithmetic automaton is a tuple $\left(Q, T, q_0, N, n_0, E, \theta = (\theta_q)_{q \in Q}, \pi = (\pi_q)_{q \in Q}\right)$, where*

- Q *is a finite set of states,*
- $T : Q \times Q \to [0, 1]$ *is a transition function with $\sum_{q \in Q} T(p, q) = 1$ for all $p \in Q$, i.e. $\big(T(p, q)\big)_{p, q \in Q}$ is a stochastic matrix,*
- $q_0 \in Q$ *is called* start state,
- N *is a finite set called* value set,
- $n_0 \in N$ *is called* start value,
- E *is a finite set called* emission set,
- *each $\theta_q : N \times E \to N$ is an operation associated with the state q,*
- *each $\pi_q : E \to [0, 1]$ is a distribution associated with the state q.*

At first, the automaton is in its start state q_0, as for a classical deterministic finite automaton (DFA). In a DFA, the transitions are triggered by input symbols. In a PAA, however, the transitions are purely probabilistic; $T(p, q)$ gives the chance of going from state p to state q. Note that the tuple (Q, T, δ_{q_0}) defines a Markov chain on state set Q with transition matrix T, where δ_{q_0} is the Dirac distribution assigning probability 1 to $\{q_0\}$ as the initial distribution. Besides the similarity to Markov chains, this part of the definition may also be seen as a special case of probabilistic automata[3] (see [9] for an introduction to probabilistic automata).

---

[3] Probabilistic automata have two features that we do not need here. Firstly, they allow non-deterministic choices, and, secondly, each transition is associated with an action.

While going from state to state, a PAA performs a chain of calculations on a set of values N. In the beginning, it starts with the value n_0. Whenever a state transition is made, the entered state, say state q, generates an emission according to the distribution π_q. The current value and this emission are then subject to the operation θ_q, resulting in the next value.

Notice that the Markov chain (Q, T, δ_{q_0}), together with the emission distributions $\pi = (\pi_q)_{q \in Q}$, defines a hidden Markov model (HMM). In the context of HMMs, however, the focus usually rests on the sequence of emissions, whereas we are interested in the value resulting from a chain of operations on these emissions.

Let us introduce some notation. Let $(Y_k)_{k \in \mathbb{N}_0}$ denote the automaton's state process, i.e. $\mathbb{P}(Y_k = q)$ is the probability of being in state q after k steps. Analogously, we write $(Z_k)_{k \in \mathbb{N}_0}$ and $(V_k)_{k \in \mathbb{N}_0}$ to denote the sequence of emissions and the sequence of values resulting from the performed operations, respectively. Using this terminology, we can describe the value process formally:

$$V_0 \equiv n_0 \,, \tag{1}$$

$$V_k = \theta_{Y_k}(V_{k-1}, Z_k) \,. \tag{2}$$

Example 1 (Dice). To illustrate the definition, let us examine a simple dice experiment. Suppose you have a bag (or urn) containing three dice, a 6-faced, a 12-faced, and a 20-faced die. Now a die is drawn from the bag, rolled, and put back. This procedure is repeated m times. In the end one may, for example, be interested in the distribution of the maximum number observed. Using a PAA, we can model each die as a state. Then, all transition probabilities would be $1/3$ and the emissions would be uniform distributions over the number of faces of the respective dice (the dice are assumed to be fair). As we are interested in the maximum, each state's operation would be to choose the maximum. Note that it would also be possible to associate individual operations with each state (each dice), for instance: "sum up all numbers from the 6- and 12-faced dice and subtract the numbers seen on the 20-faced die".

3   Computing the State-Value Distributions of PAAs

Having introduced the definition a PAA, we now take the next step and present two algorithms to compute the distribution of results. In other words, we seek to calculate the distribution $\mathcal{L}(V_m)$ of the random variable V_m for a given m. The idea is to compute the joint distribution $\mathcal{L}(Y_m, V_m)$ and then to derive the sought distribution:

$$\mathbb{P}(V_m = v) = \sum_{q \in Q} \mathbb{P}(Y_m = q, V_m = v) \,. \tag{3}$$

For the sake of a shorter notation, we define $p_k(q, v) := \mathbb{P}(Y_k = q, V_k = v)$.

3.1 Basic Algorithm

We now briefly discuss a simple algorithm to compute the distribution $\mathcal{L}(Y_m, V_m)$. Following the semantics introduced in Section 2, we can form the recurrence equation

$$p_{k+1}(q, v) = \sum_{q' \in Q} \sum_{(v', e) \in \theta_q^{-1}(v)} p_k(q', v') \cdot T(q', q) \cdot \pi_q(e), \tag{4}$$

where $\theta_q^{-1}(v)$ denotes the inverse image set of v under θ_q.

We start with the distribution p_0 and calculate the subsequent distributions by applying Equation (4) until we obtain the desired p_m. A straightforward implementation of Equation (4) results in a *pull-strategy*; that means each entry in the table representing p_{k+1} is calculated by "pulling over" the required probabilities from table p_k. Note that this approach makes it necessary to calculate θ^{-1} in a preprocessing step. In order to avoid this, we may implement a *push-strategy*, meaning that we iterate over all entries in p_k rather than p_{k+1} and "push" the encountered summands over to the appropriate places in table p_{k+1}; in effect, we just change the order of summation.

In the course of the computation, we have to store two distributions, p_k and p_{k+1}, at a time. Once p_{k+1} is calculated, p_k can be discarded. Since each table has a size of $|Q| \times |N|$, the total space consumption is $\mathcal{O}(|Q| \cdot |N|)$. To perform a transition from p_k to p_{k+1}, we have to evaluate Equation (4) exactly $|Q| \cdot |N|$ times. We sum over $\theta_q^{-1}(v)$ for all $v \in N$, but $\theta_q^{-1}(v_1)$ and $\theta_q^{-1}(v_2)$ are disjoint for $v_1 \neq v_2$ and $\bigcup_{v \in N} \theta_q^{-1}(v) = N \times E$. Therefore, we get a runtime of $\mathcal{O}(|Q|^2 \cdot |N| \cdot |E|)$ for one transition and, hence, a total runtime of $\mathcal{O}(m \cdot |Q|^2 \cdot |N| \cdot |E|)$ to calculate p_m.

3.2 Doubling Technique

In case of a large m, executing the above algorithm is cumbersome. In this section, we present an alternative algorithm that is favorable for large m. To derive this algorithm, we consider the conditional probability

$$U^{(k)}(q_1, q_2, v_1, v_2) = \mathbb{P}\big(Y_{i+k} = q_2, V_{i+k} = v_2 | Y_i = q_1, V_i = v_1\big). \tag{5}$$

Note that this definition does not depend on i, because transition as well as emission probabilities do not change over "time" (a property called *homogeneity*). Once $U^{(m)}$ is known, we can simply read off the desired distribution $\mathcal{L}(Y_m, V_m)$:

$$\mathbb{P}(Y_m = q, V_m = v) = U^{(m)}(q_0, q, n_0, v). \tag{6}$$

In the following, we show how $U^{(k)}$ can be computed. For $k = 1$, we get

$$U^{(1)}(q_1, q_2, v_1, v_2) = T(q_1, q_2) \cdot \sum_{\substack{e \in E: \\ \theta_{q_2}(v_1, e) = v_2}} \pi_{q_2}(e). \tag{7}$$

From this starting point, we can calculate $U^{(k_1+k_2)}$ from $U^{(k_1)}$ and $U^{(k_2)}$ for any $k_1, k_2 \in \mathbb{N}$ by summing over all possible intermediate states and values:

$$U^{(k_1+k_2)}(q_1, q_2, v_1, v_2) = \sum_{q' \in Q} \sum_{v' \in N} U^{(k_1)}(q_1, q', v_1, v') \cdot U^{(k_2)}(q', q_2, v', v_2). \quad (8)$$

Equation (8) is a generalization of the Chapman-Kolmogorov Equation for homogeneous Markov chains.

The transition from $U^{(k_1)}$ and $U^{(k_2)}$ to $U^{(k_1+k_2)}$ takes $\mathcal{O}(|Q|^3 \cdot |N|^3)$ time, as follows immediately from Equation (8). On the other hand, one transition suffices to get $U^{(2k)}$ out of $U^{(k)}$. Thus, we can compute all $U^{(2^b)}$ for $0 \leq b \leq \lceil \log(m) \rceil$ in $\lceil \log(m) \rceil$ steps, which in turn can be combined to $U^{(m)}$ in at most $\lceil \log(m) \rceil$ steps. Hence, we get a total runtime of $\mathcal{O}(\log m \cdot |Q|^3 \cdot |N|^3)$.

4   Pattern Matching Statistics

In this section, we discuss the application of probabilistic arithmetic automata to pattern matching statistics. We see how a deterministic finite automaton together with a text model, either i. i. d. or Markovian, can be transformed into a PAA for the overlapping count.

4.1   Sets of Generalized Strings

Generalized strings are finite sequences of sets of characters over an alphabet Σ, for example [abc][ac][ab] (which matches aaa, ccb but not aba). We now explain the construction of PAAs from finite sets of generalized strings for i. i. d. text models.

The first step is to construct a non-deterministic finite automaton (NFA) for the given set of generalized strings. To be more precise, we need to construct a NFA that recognizes all strings ending with an instance of a generalized string from the given set. The NFA corresponding to one generalized string is just a linear chain of states with a start state plus one state for each position, where the start state is additionally equipped with a self-transition. The NFA for the set of generalized strings can be constructed by simply merging all individual start states into one common start state.

The next step towards a PAA is to build a DFA. In order to obtain it, we employ the subset construction, a classical procedure that is for example explained in the book by Navarro and Raffinot [10]. Although, in the worst case, it results in an exponential increase in the number of states, this method is feasible in many practical cases, as we demonstrate shortly.

Before we come to that point, let us complete the construction of a PAA. We define it to operate on the same state set Q as the DFA. The transition function T can then be derived from the text model and the DFA's transition function by "replacing" all characters with their probability. Let us state this transformation precisely. Let $\delta : Q \times \Sigma \to Q$ denote the transition function of the DFA and p_σ the occurrence probability of each character $\sigma \in \Sigma$, then T is defined by

$$T : (q, q') \mapsto \sum_{\sigma \in \{\sigma' \in \Sigma : \delta(q, \sigma') = q'\}} p_\sigma .\qquad (9)$$

This technique is called *Markov chain embedding* by Lladser et al. [6].

Let us specify the emission distribution π_q of each state q. We use it to model the number of matches to count upon entering state q. Since this number does not depend on chance, the emissions are deterministic. By the subset construction, there corresponds a set B_q of NFA states to each state q. Due to the construction of the NFA, the number of final states in B_q equals the number of matching substrings that end when q is entered; so q emits this number with probability 1.

Assume we have observed the given pattern P (a set of generalized strings) n times in a text of length m. Therefore, we wish to compute the probability of finding n or more occurrences of P in a random text of length m. To achieve this, we choose the value set $N = \{0, \dots, n\}$ and the operation

$$\theta_q(v, e) := \begin{cases} v + e & \text{if } v + e \leq n, \\ n & \text{otherwise}, \end{cases}\qquad (10)$$

for all $q \in Q$. Thus, the value n has the meaning "n or more matches observed".

Before turning the DFA into a PAA, one may wish to minimize it. Using an algorithm by Hopcroft [11], a classical DFA can be minimized in $\mathcal{O}(|Q| \log |Q|)$ time, where Q is the set of states (see Knuutila [12] for a tutorial-like introduction and an in-depth analysis of this algorithm). Hopcroft's algorithm can be adapted by using the partition induced by the different emissions as an initial partition, i.e. states with the same emission are grouped together.

Runtime. The runtime bounds given in the general analysis in Section 3 can be tightened for this concrete application. Let us first examine the basic algorithm from Section 3.1. Firstly, observe that now the emissions are deterministic, that means we only have to consider one possible emission per state, reducing the runtime by a factor of $\mathcal{O}(|E|)$. Secondly, note that by Equation (9) the transition matrix contains at most $|Q| \cdot |\Sigma|$ non-zero entries, allowing us to further speed up the evaluation of Equation (4). In total, we get a runtime of $\mathcal{O}(m \cdot |Q| \cdot |\Sigma| \cdot |N|)$.

The doubling algorithm presented in Section 3.2 can also be simplified. This time, we exploit a property of the operations θ_q, which are simple additions in our case. Thus, $U^{(k)}(q_1, q_2, v_1, v_2) = U^{(k)}(q_1, q_2, v_3, v_4)$ if $v_2 - v_1 = v_4 - v_3$, which means that we can fix $v_1 = 0$ and thereby save a factor of $|N|$. This results in a total runtime of $\mathcal{O}(\log m \cdot |Q|^3 \cdot |N|^2)$.

Prosite Patterns. Prosite is a database of biologically meaningful amino acid motifs (see Hulo et al. [1]). Release 20.17 contains 1319 patterns, 16 of which refer to the start or ending of a sequence. Those entries were ignored, leaving a database of 1303 patterns.

Prosite patterns can be seen as generalized strings with the extension that, for each position, a "multiplicity range" can be specified. In the pattern A-x(2,3)-C, for example, an A is followed by either two or three arbitrary characters followed by a C. We translate every Prosite pattern into a set of generalized strings. The above example would result in the two patterns A-x-x-C and A-x-x-x-C. This set can then be dealt with as explained above.

We implemented the algorithms in Java and ran them on a customary PC[4]. Unfortunately, for 42 patterns (3.2%) the computation did not succeed due to memory limitations. This can happen if either the Prosite pattern translates into too many generalized strings or if the DFA resulting from the subset construction grows too large.

For 1236 of the 1261 remaining patterns, the subset construction was completed within 2 seconds while the computation took 69.9 seconds for the "worst pattern". The resulting automata were then minimized using Hopcroft's algorithm. Many automata, however, already were minimal or close to minimal; for 1209 automata the minimized automaton was larger than half the size of the original automaton. The majority of resulting minimal automata were of reasonable size. We obtained 1198 automata with less than 10000 states, among which 1036 had less than 500 states.

To give an impression of the runtimes to be expected, consider the pattern C-x-H-R-[GAR]-x(7,8)-[GEKVI]-[NERAQ]-x(4,5)-C-x-[FY]-H from the Prosite database. It results in an automaton with 462 states. Assuming $n = 50$ (number of occurrences) and $m = 1000$ (text length), computing the distribution of the occurrence count took 1 second.

4.2 Finite String Sets

Assume that the pattern is given in the form of an enumerated set of strings. Obviously, this is a special case of a set of generalized strings. In this situation, however, the intermediate step of constructing a NFA is unnecessary. Instead, an Aho-Corasick automaton (Aho and Corasick [13]), which essentially is a DFA, can be built directly. It can be constructed in linear time by either using the algorithm given in the original paper or by employing a recent elegant algorithm based on the suffix tree of the reverse strings (Dori and Landau [14]). The latter has the advantage that the runtime does not depend on the alphabet size. The emissions (number of matches) of the states can directly be read off the Aho-Corasick automaton's output function.

Transcription Factor Binding Site Statistics. Transcription factors are proteins that play an important role in gene regulation. By binding to special DNA regions, they influence the transcription of DNA to RNA and, thereby, the expression of genes. Due to their significance for gene regulation, representations of the corresponding DNA binding sites have been studied extensively; an overview is provided by Stormo [15]. These transcription factor binding sites (TFBSs) are now commonly represented by position weight matrices (PWMs). Frequently, one wishes to compute the significance of a high number of occurrences of a PWM, for instance in a given promoter region. This raises the question

---

[4] Intel Core 2 Duo 2.66GHz, 4GB RAM, running Linux.

of when to consider a substring to be an occurrence of the PWM. In this section, we base our considerations on a threshold, while in Section 4.3, we present a threshold-free approach.

Recently, new approaches—which are also based on a threshold—have been proposed. On the one hand, Pape et al. [16] introduce a method to approximate the significance of PWM occurrences. On the other hand, Zhang et al. [8] describe a dynamic programming algorithm to solve the problem exactly.

In our framework, we can calculate the distribution of the occurrence count by enumerating all patterns above a threshold and using them to build an Aho-Corasick automaton. As explained above, this automaton can then be transformed into a PAA.

In order to assess the practicability, we consulted Jaspar (see Sandelin et al. [17]), a database containing 138 position frequency matrices (PFMs). All PFMs were converted into PWMs using the method of Rahmann et al. [18]. We controlled the threshold by fixing the probability of false positives on a random text of length 500 at $\alpha = 0.01$. In other words, we set the threshold such that the probability to get one or more matches just by chance is (at least approximately) α. Using this threshold and assuming a uniform distribution on the alphabet of nucleotides, we computed the distribution of the occurrence count (up to 100 occurrences) for the binding sites found in the Jaspar database. It can be computationally demanding to enumerate all strings above the threshold. In some cases, the memory requirement could not be met and the calculation was aborted. For 126 of all 138 PFMs, however, the computation was completed successfully and took 10.0 seconds on average and 9 minutes in the worst case.

4.3   Probabilistic String Sets

Above, we discussed the matching statistics of string sets given in one form or another. Now we generalize the notion of a string set and define a *probabilistic string set* by associating a weight between 0 and 1 with each string in the set. This mechanism is useful in computational biology as it allows to model the chance that a protein binds to a specific sequence.

TFBS Statistics Accounting for Binding Affinities. Recently, Roider et al. [19] presented a procedure to predict a transcription factor's affinity to a sequence based on a physical model. Based on their implementation[5], we can estimate the probability that a TF binds to a particular sequence. Then, we can modify the emission distributions of the PAA accordingly. That means, a state corresponding to a binding site instance emits the match count 1 with the probability given by the affinity and 0 with the remaining probability.

Again, we assessed the practicability on the Jaspar database. For every PWM, we generated the 1000 best-scoring strings, calculated their binding probability, constructed a PAA and computed the distribution of the occurrence count (again up to 100 occurrences and on a random text of length 500). The calculations took 11.6 seconds on average and 63 seconds in the worst case.

---

[5] See http://www.molgen.mpg.de/~manke/papers/TFaffinities.

It arises the question, if the choice of 1000 strings is appropriate. The binding probability of the string with the lowest binding probability p_{min} may give us a hint. In our case, p_{min} is below 0.001 for 60 of the 138 matrices and below 0.01 for 86 matrices. These numbers can be improved at the cost of longer runtimes. We propose to study the influence of the number of strings on the obtained distributions in future work.

4.4 Further Generalizations

Markovian Text Models. So far, we only considered i. i. d. text models, but the generalization to Markovian models can be done without much effort.

For a first-order model, the distribution of a character depends on the last character. Equation (9) could be modified accordingly if all incoming edges of each DFA state were labeled with the same character, i.e. if from $\delta(q_1, \sigma_1) = \delta(q_2, \sigma_2)$ it followed that $\sigma_1 = \sigma_2$. This property, however, can easily be established by the following procedure: If a state has incoming edges labeled with k different characters, duplicate the state $k - 1$ times (along with the outgoing edges) and reroute the incoming edges such that those labeled with different character end in different clones of the original state. For n-th-order models, repeat this procedure until for each state the n last characters are known.

Note that the computational expense is especially low for Aho-Corasick automata. By construction, each state corresponds to a prefix of a string from the string set. Therefore, for a k-th-order model, only the states corresponding to states whose distance to the root node is lower than k potentially have to be duplicated. For all other states, a sufficiently long history is already known.

Different Counting Schemes. Regarding the counting scheme, we already dealt with the most complicated case of the overlapping count. If one wishes to disallow overlaps, the automaton can be modified accordingly. We just have to change the outgoing transitions of each accept state q_a to act like the start state, i.e. $\delta'(q_a, \sigma) = \delta(q_s, \sigma)$ for all $\sigma \in \Sigma$. To get the match position count, all emissions of values larger than 1 just have to be changed to emit 1.

Inhomogeneous PAAs. We defined the transition function T to be constant over "time", which is reflected in Equation (4), where T does not depend on k. In the basic algorithm of Section 3.1, however, such a dependency can be incorporated straightforwardly.

One application of an inhomogeneous PAA lies in the field of motif discovery. Assume that the motif with the lowest (or at least a low) p-value has been found, let us call it "best motif". When seeking the "second best motif", one is likely to obtain a variant of the best motif that matches essentially at the same positions. Now it is an option to judge the second motif according to a modified text model. In this model, the character distribution is changed at those positions where the first motif matches. There, Dirac distributions are used such that the chance of finding the former motif at this position is 1. This text model only gives small p-values for new motifs, rather than for variants of the best motif.

Another application is the calculation of the binding count distribution for a TF on a particular sequence. Here, we remove all randomness from the text model and assign probability 1 to the given sequence. The only random choices are done by the emission distributions, which are chosen according to the binding affinity as explained in Section 4.3.

5 Discussion

We have introduced the abstract concept of probabilistic arithmetic automata and have presented two generic algorithms to compute the joint distribution of states and values. The algorithms constitute different trade-offs between the factors governing the runtime. While the basic algorithm is applicable in most cases, the doubling technique is favorable for long texts and relatively small state and value spaces.

The notion of PAAs blends into the landscape of existing concepts like probabilistic automata, Markov chains, hidden Markov models and, last but not least, Markov additive chains (Kaltenbach et al. [7]). A strength of PAAs lies in their flexibility. In this paper, we have discussed various applications. As we showed, our method is applicable to the majority of motifs from the Prosite database. In contrast to existing methods, the proposed one is able to handle overlaps accurately.

Another possible application is the calculation of TFBS statistics. Concerning this matter, we examined two approaches. On the one hand, we did statistics based on the enumeration of all words with a PWM score above a threshold. This approach is similar to that of Zhang et al. [8], who also developed a dynamic programming algorithm. Their approach is comparable in space and time consumption, but lacks the flexibility of our method. It is unclear if different counting schemes or probabilistic emissions could be incorporated into their method. Another advantage of our concept lies in its roots in theoretical computer science; we can take advantage of well-studied methods like DFA minimization or the subset construction. The latter allows us to handle pattern sets, for example based on generalized strings, whose enumeration would not be feasible.

Besides the threshold-based approach, we demonstrated that our model allows for more advanced TFBS statistics. As we showed, it is possible to take binding affinities derived from a physical model into account. To the best of our knowledge, this has not been done before.

A promising direction of future research seems to be the application of PAAs to field of motif discovery. Fast and exact significance calculations seem to be a helpful tool as they allow the comparison of sets of motifs with different structure and length.

References

1. Hulo, N., Bairoch, A., Bulliard, V., Cerutti, L., De Castro, E., Langendijk-Genevaux, P., Pagni, M., Sigrist, C.: The PROSITE database. Nucleic Acids Research 34(S1), D227–230 (2006)

2. Lothaire, M.: Applied Combinatorics on Words (Encyclopedia of Mathematics and its Applications). Cambridge University Press, Cambridge (2005)
3. Reinert, G., Schbath, S., Waterman, M.S.: Probabilistic and statistical properties of words: An overview. Journal of Computational Biology 7(1-2), 1–46 (2000)
4. Régnier, M.: A unifed approach to word occurrence probabilities. Discrete Applied Mathematics 104, 259–280 (2000)
5. Nicodème, P., Salvy, B., Flajolet, P.: Motif statistics. Theoretical Computer Science 287, 593–617 (2002)
6. Lladser, M., Betterton, M.D., Knight, R.: Multiple pattern matching: A Markov chain approach. Journal of Mathematical Biology 56(1-2), 51–92 (2008)
7. Kaltenbach, H.M., Böcker, S., Rahmann, S.: Markov additive chains and applications to fragment statistics for peptide mass fingerprinting. In: Ideker, T., Bafna, V. (eds.) Joint RECOMB 2006 Satellite Workshops on Systems Biology and on Computational Proteomics. LNCS (LNBI), vol. 4532, pp. 29–41. Springer, Heidelberg (2007)
8. Zhang, J., Jiang, B., Li, M., Tromp, J., Zhang, X., Zhang, M.Q.: Computing exact p-values for DNA motifs. Bioinformatics 23(5), 531–537 (2007)
9. Stoelinga, M.: An introduction to probabilistic automata. In: Rozenberg, G. (ed.) EATCS bulletin, vol. 78 (2002)
10. Navarro, G., Raffinot, M.: Flexible pattern matching in strings. Cambridge University Press, Cambridge (2002)
11. Hopcroft, J.: An $n \log n$ algorithm for minimizing the states in a finite automaton. In: Kohavi, Z., Paz, A. (eds.) The theory of machines and computations, pp. 189–196. Academic Press, New York (1971)
12. Knuutila, T.: Re-describing an algorithm by Hopcroft. Theoretical Computer Science 250, 333–363 (2001)
13. Aho, A.V., Corasick, M.J.: Efficient string matching: an aid to bibliographic search. Communications of the ACM 18(6), 333–340 (1975)
14. Dori, S., Landau, G.M.: Construction of Aho Corasick automaton in linear time for integer alphabets. Information Processing Letters 98(2), 66–72 (2006)
15. Stormo, G.D.: DNA binding sites: representation and discovery. Bioinformatics 16(1), 16–23 (2000)
16. Pape, U.J., Grossmann, S., Hammer, S., Sperling, S., Vingron, M.: A new statistical model to select target sequences bound by transcription factors. Genome Informatics 17(1), 134–140 (2006)
17. Sandelin, A., Alkema, W., Engström, P.G., Wasserman, W.W., Lenhard, B.: JASPAR: an open access database for eukaryotic transcription factor binding profiles. Nucleic Acids Research 32(1) (2004) (Database Issue)
18. Rahmann, S., Müller, T., Vingron, M.: On the power of profiles for transcription factor binding site detection. Statistical Applications in Genetics and Molecular Biology (Article 7), 2(1) (2003)
19. Roider, H., Kanhere, A., Manke, T., Vingron, M.: Predicting transcription factor affinities to DNA from a biophysical model. Bioinformatics 23(2), 134–141 (2007)

Analysis of the Size of Antidictionary in DCA

Julien Fayolle

LRI; Univ. Paris-Sud, CNRS, F-91405 Orsay, France
julien.fayolle@lri.fr
http://www.lri.fr/~fayolle

Abstract. We analyze the lossless data compression scheme using antidictionary. Its principle is to build the dictionary of a set of words
that do **not** occur in the text (*minimal forbidden words*). We prove here
that the number of words in the antidictionary, i.e., minimum forbidden
words, behaves asymptotically linearly in the length of the text under a
memoryless model on the generation of texts. The linearity constant is
explicited. We use methods from analytic combinatorics.

1 Introduction

Data compression using anti-dictionaries (DCA) was introduced by Crochemore,
Mignosi, Restivo, and Salemi in 1999 [3,4]. It focuses on words that do not occur
in the text T. There is an infinite number of words that do not occur in a finite
text, so the authors use *minimal forbidden words* (MFW for short) [2] to obtain
a finite set of words. A word w is forbidden for a text T if it does not occur in
the text. A word w is a minimal forbidden word for T if it does not occur in
T and all its factors do occur. The MFWs are also called antifactors. The DCA
algorithm relies on the fact that text are written on a binary alphabet. DCA has
been used by Ota and Morita [12] to compress an electrocardiogram. The output
is 10% smaller than that of a Lempel-Ziv-like algorithm. They also proved in [11]
that the number of MFWs is bounded by the size of the dictionary of the text.

In this paper, texts are generated on a binary alphabet $\mathcal{A} = \{0, 1\}$ by a
memoryless source model, meaning the letter emitted at a given time does not
depend on the preceding letters. The letter 0 is emitted with probability p and
1 with probability q. We suppose that $p \geq q$. For a pattern w, the occurrence
probability p_w is the probability that the source emits the pattern w. A source
is said to be *periodic* if $\log p / \log q$ is rational, otherwise it is *aperiodic*.

For a pattern w of length k its prefix of length $k-1$ is noted w_L and its suffix
of length $k-1$, w_R. The letter $\bar{\alpha}$ is the letter that is not α *e.g.*, $\bar{0} = 1$.

The parameter $\mathcal{S}$ is defined as the size of the antidictionary (or the number of
MFWs). For a text T, $\mathcal{S}(T)$ is the number of MFWs in the text T. Our goal is to
obtain the asymptotic behavior of $\mathbb{E}_n(\mathcal{S})$, the mean of $\mathcal{S}$ over texts of length n:

$$\mathbb{E}_n(\mathcal{S}) = \mathbb{E}_n \left(\sum_{w \in \mathcal{A}^\star} [\![w \in \mathrm{MFW}]\!] \right) = \sum_{w \in \mathcal{A}^\star} \mathbb{P}_n(w \in \mathrm{MFW}), \qquad (1)$$

P. Ferragina and G. Landau (Eds.): CPM 2008, LNCS 5029, pp. 107–117, 2008.

where $[\![\,.\,]\!]$ is Iverson's notation for the indicator function and $\mathbb{P}_n(w \in \text{MFW})$ is the probability of texts of length n for which w is an MFW.

In Sect. 2 the DCA algorithm is presented. We are interested by the asymptotic behavior of $\mathbb{E}_n(\mathcal{S})$. The sum $\sum_{w \in \mathcal{A}^*} \mathbb{P}_n(w \in \text{MFW})$ is split in three sums depending on the length of the patterns. The sum over the patterns of small (resp. intermediate, long) lengths is denoted by $S(n)$ (resp. $I(n)$, $L(n)$). For a text T of length n, patterns of small length (hereafter small patterns) are defined as those of length smaller than $k_s(n) := aC_q \log n$ and patterns of long length (hereafter long patterns) are those of length larger than $k_l(n) := bC_p \log n$ for $a < 1$, $b > 1$, and $C_r = -\frac{1}{\log r}$. The sums $S(n)$ and $L(n)$ are asymptotically sublinear as shown in Sect. 3. For intermediate pattern, the asymptotic behavior of $I(n)$ is harder to derive. We introduce an approximate model in Sect. 4 consisting of two hypotheses. The asymptotic behavior of the contribution of intermediate patterns under the approximate model is obtained in Section 5 for a symmetric memoryless model. The result is stated for a biased model. In Sect. 6, we show that the asymptotic behavior under the approximate model matches the asymptotic behavior of $I(n)$ under the exact model up to a sublinear term. Full details of the computations are available in Chap. 4 of Fayolle [7].

Theorem. *Under a memoryless biased model on the generation of texts, the mean size of the antidictionary i.e., the number of minimal forbidden words over all texts of length n behaves asymptotically as*

$$\mathbb{E}_n(\mathcal{S}) = K\frac{n}{h} + \frac{n}{h}\epsilon(n) + o(n)$$

for a periodic source, and as

$$\mathbb{E}_n(\mathcal{S}) = K\frac{n}{h} + o(n)$$

for an aperiodic source, where $\epsilon(x)$ is a function fluctuating around zero of very small amplitude (roughly 10^{-5}),

$h = -p \log p - q \log q$ *is the entropy of the source, and,*

$K := 2h + (1 - p^2) \log(1 - p^2) + (1 - q^2) \log(1 - q^2) + 2(1 - pq) \log(1 - pq).$

2   Description of DCA

In this section, we describe the compression and decompression in DCA. Three phases are distinguished: construction of the trie containing the antidictionary for the text T, compression, and decompression.

The construction of the trie containing the MFWs for the binary text T is performed in linear time. First the suffix tree of the text T is built in linear time (with the use of suffix links). Let $u = \alpha v$ be a node with no left child (resp. no right child) in the suffix tree of Fig. 1 then $u0$ (resp. $u1$) does not occur in the text T. If the node v (this node is reached following a suffix link from u) has a

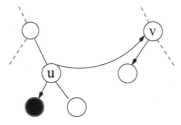

Fig. 1. The left child of u is empty and the left child of v exists, hence $u0$ is an MFW

left child (resp. right) then $v0$ (resp. $v1$) occurs in the text and thus $u0$ (resp. $u1$) is an MFW for T (an MFW is marked by a leaf in the tree). The tree structure is modified accordingly and branches that do not lead to a leaf are cut off.

Morita and Ota described in [11] an algorithm AD2D to go from the trie built on MFWs (the antidictionary) to the suffix tree of the text.

The compression phase is based on a simple idea: if at a given position in the text the next letter can not be 0 then it must be 1 (since the alphabet is binary). Suppose the text $T[0 \ldots i]$ has been compressed, if there is a suffix v of $T[0 \ldots i]$ such that $v0$ (resp. $v1$) is an MFW then the letter $T[i+1]$ is 1 (resp. 0). The letter $T[i+1]$ carries redundant information with the antidictionary and is not included in the compressed text. (Otherwise the letter is included.)

Decompression uses the antidictionary, the compressed text, and the length of the initial text. Suppose the text has been decompressed up to position i, if there is a suffix v of the text $T[0 \ldots i]$ such that $v0$ (resp. $v1$) is an MFW, then the letter $T[i+1]$ is forced to be 1 (resp. 0). The decompression is linear in time. The length is used to indicate the end of the decompression.

3   Patterns of Small and Long Lengths

In this section, we show that the asymptotic contributions of patterns of small $S(n)$ and long $L(n)$ lengths to the mean size of antidictionary are sublinear.

A *small pattern* occurs often in a text, so the probability that w does not occur is small and decreases exponentially with n. Furthermore the number of small patterns is polynomial, hence the contribution of small patterns to $\mathbb{E}_n(\mathcal{S})$ is

$$S(n) := \sum_{k=1}^{k_s(n)} \sum_{w \in \mathcal{A}^k} \mathbb{P}_n(w \in \mathrm{MFW}) \leq M 2^{k_s(n)} \exp((p-1)n^{1-a}), \qquad (2)$$

where M is a positive constant. The decay of the exponential implies $S(n)$ is $o(1)$.

A pattern w of length k is an MFW for a text if the pattern $u(w) = u := w_2 \cdots w_{k-1}$ occurs at least twice in the text (once with an occurrence of w_L, and once with an occurrence of w_R). For a *long pattern* w, the pattern u is roughly of the same length as w hence it also qualifies as long pattern. A long pattern occurs rarely in a text hence the probability that u occurs at least twice in the text is small. Yet there are an infinite number of long patterns. The probability to have at least two occurrences of u in a text of length n is bounded by

$$\mathbb{P}_n(N_u \geq 2) \leq np_u \left(\frac{1}{c_u(1)} - 1 + \frac{np_u}{c_u(1)} \right), \tag{3}$$

where N_u is the parameter counting the number of occurrences of the pattern u in the text (with overlaps) and $c_u(1)$ is the value in 1 of the autocorrelation polynomial of u. The autocorrelation polynomial of a pattern w is defined as

$$c_w(z) = \sum_{j=0}^{|w|-1} c_j z^j,$$

where c_j is 1 if the suffix of length $|w| - j$ and the prefix of length $|w| - j$ of w match, and zero otherwise. For long patterns, the quantity np_u tends to zero and the difference between 1 and $\frac{1}{c(1)}$ is controled. For $k_l(n) = bC_p \log n$, $L(n)$ is $O(n^{1-b/2})$.

4   Approximate Model

In this section, we introduce an approximate model under which the contribution of patterns of intermediate lengths *i.e.*, neither small nor long, to $\mathbb{E}_n(\mathcal{S})$ is computed. The approximate model is defined for any pattern, not only those of intermediate lengths. The model consists in two hypotheses H_1 and H_2. A pattern w is of intermediate length if its length k ranges within $k_s(n)$ and $k_l(n)$.

If a pattern w of length k is an MFW for a text T then w_L and w_R each occur at least once in T. Hence there are at least two occurrences of the pattern $u(w) = u = w_2 \cdots w_{k-1}$.[1] We note $\alpha = w_1$ and $\beta = w_k$ the end letters of w. The *extensions* of u are the four patterns built by adding a letter at each end of u.

We count texts with the following constraints on extensions of u: no occurrence of $\alpha u \beta$ (no occurrence of w), at least one occurrence of $\alpha u \bar{\beta}$ (occurrences of w_L without w occurring), and at least one occurrence of $\bar{\alpha} u \beta$ (occurrences of w_R without w occurring). The correlation of patterns is hard to deal with and usually has a small impact on means (see for instance the impact of autocorrelation on mean size and path length [10,6,8]). To ensure that none of the extensions of u overlap, the approximate model states that the occurrences of u are at least two letters apart from one another. Let $\mathfrak{N}_u$ be the set of texts with at least two occurrences of u and for which the occurrences of u are separated by at least two letters from one another. Hence the first hypothesis (H_1):

Hypothesis 1. *The probability of texts of length n for which w is an MFW is approximated by the probability of texts of length n for which w is an MFW* **and** *in which any two occurrences of $u(w)$ are separated by at least two letters.*

---

[1] If $w_L = w_R$ then $w = \alpha^k$, and the pattern $u = \alpha^{k-2}$ also occurs at least twice.

Definition 1. *Let u be a pattern, the parameter $\widetilde{N}_u$ is defined for texts as: If any two occurrences of u are separated by at least two letters in the text T then $\widetilde{N}_u$ counts the number of occurrences of the pattern u in the text. Otherwise, i.e., if two occurrences of u overlap in T, the parameter is zero.*

For a text T, a pattern u, and an integer $j \geq 2$, $\widetilde{N}_u(T) = j$ means there are j occurrences of u in the text T and each occurrence of u is at least two letters apart from any other.

The second hypothesis is common to model number of occurrences of events:

Hypothesis 2. *On texts of length n, $\widetilde{N}_u$ is Poisson of parameter np_u:*

$$\forall j \in \mathbb{N}, \ \forall u \in \mathcal{A}^\star, \ \mathbb{P}_n(\widetilde{N}_u = j) = \frac{(np_u)^j}{j!} \exp(-np_u). \tag{4}$$

Under these hypotheses the probability that a pattern w is an MFW is:

$$
\begin{aligned}
\mathbb{P}_n(w \in \mathrm{MFW}) &\overset{H_1}{=} \mathbb{P}_n(\{w \in \mathrm{MFW}\} \cap \mathfrak{N}_u) = \sum_{j \geq 2} \mathbb{P}_n(w \in \mathrm{MFW} | \widetilde{N}_u = j)\mathbb{P}_n(\widetilde{N}_u = j) \\
&\overset{H_2}{=} \sum_{j \geq 2} \mathbb{P}_n(w \in \mathrm{MFW} \mid \widetilde{N}_u = j)\frac{(np_u)^j}{j!} \exp(-np_u).
\end{aligned}
\tag{5}
$$

The sum is on j's greater than 2 since u must occur at least twice for w to be an MFW. Under the approximate model, the mean size of the antidictionary is:

$$\mathcal{E}(n) := \sum_{k \geq 2} \sum_{w \in \mathcal{A}^k} \sum_{j \geq 2} \mathbb{P}_n(w \in \mathrm{MFW} \mid \widetilde{N}_u = j)\frac{(np_u)^j}{j!} \exp(-np_u). \tag{6}$$

5 Asymptotic Contribution under the Approximate Model

In this section, the asymptotic behavior of the contribution of intermediate patterns to $\mathcal{E}(n)$ is obtained under the approximate model presented in Sect. 4.

The sum $\mathcal{E}(n)$ is the contribution for pattern of all lengths ($k \geq 2$). The contribution for intermediate patterns is the sum over lengths between $k_s(n)$ and $k_l(n)$. Both the summand and the summation indices depend on n therefore we can not obtain the asymptotic behavior of this sum directly. We circumvent this dependency problem with a threefold approach: first an explicit expression for $\mathcal{E}(n)$ is obtained, then the asymptotic behavior of $\mathcal{E}(n)$ is computed and thirdly the contributions of small and long patterns to $\mathcal{E}(n)$ is shown to be asymptotically sublinear. By a subtraction we get the asymptotic behavior of $\mathcal{E}$.

The asymptotic behavior is derived under a symmetric memoryless model. The same method is used to derive the asymptotic behavior of the contribution of intermediate patterns under a biased memoryless model but details are omitted.

5.1 Combinatorics

We use a combinatorial approach to obtain an explicit expression for the probability $C_{j,w} := \mathbb{P}_n(w \in \text{MFW} \mid \widetilde{N}_u = j)$, the probability on texts of length n that w is an MFW knowing there are j occurrences of $u = u(w)$ each separated one from another by at least two letters. In the memoryless model, $C_{j,w}$ depends only on the end letters α and β of w.

Let us suppose that u occurs twice, there are 16 possibilities for the four letters adjacent to each of the two occurrence of u. In a memoryless symmetric model, each 4-uple of letters is equiprobable. The word w is an MFW if the letters adjacent to the first (resp. second) occurrence of u need create an occurrence of w_L (resp. w_R) without any occurrence of w **or** an occurrence of w_R (resp. w_L) without any occurrence of w. Hence the two possibilities are the pairs of adjacent letters $((\alpha, \bar\beta), (\bar\alpha, \beta))$ and $((\bar\alpha, \beta), (\alpha, \bar\beta))$ therefore $C_{2,w} = \frac{1}{8}$.

In the general case where u occurs j times ($j \geq 2$), there are $2j$ letters adjacent to the occurrences of u, hence 2^{2j} choices of adjacent letters, each with the same probability. The word w is an MFW if the letters adjacent to each occurrence of u are such that w_L and w_R each occur at least once and w does not occur in the text. The probability $C_{j,w}$ is a coefficient of the exponential generating function counting the pairs of adjacent letters for each occurrence of u with the constraint to have at least one pair $(\alpha, \bar\beta)$, at least one pair $(\bar\alpha, \beta)$, and no pair (α, β):

$$\sum_{j \geq 2} \frac{C_{j,w}}{j!} z^j = (\exp(z p_\alpha p_{\bar\beta}) - 1)(\exp(z p_{\bar\alpha} p_\beta) - 1) \exp(z p_{\bar\alpha} p_{\bar\beta}).$$

For a symmetric source, one has

$$C_{j,w} = \frac{1}{4^j}\left(3^j - 2^{j+1} + 1\right).$$

In the symmetric case, one has $p_u = 2^{-(k-2)}$ and it leads to

$$\begin{aligned}\mathcal{E}(n) &:= \sum_{k \geq 2} \sum_{w \in \mathcal{A}^k} \sum_{j \geq 2} C_{j,w} \frac{(n p_u)^j}{j!} \exp(-n p_u)\\ &= \sum_{k \geq 2} 2^k \left(\exp\left(-\frac{n}{2^k}\right) - 2\exp\left(-2\frac{n}{2^k}\right) + \exp\left(-3\frac{n}{2^k}\right) \right).\end{aligned}$$

5.2 Computation

In this section the asymptotic behavior of the sum $\mathcal{E}(n)$ is obtained with the use of Mellin transform. The Mellin transform is an analytic tool relating the asymptotic behavior of a function f and complex information of its transform $f^\star$ (see Flajolet, Gourdon, and Dumas [9] for an overview of the Mellin transform). The Mellin transform $\mathcal{E}^\star$ of $\mathcal{E}$ is defined within the strip $\langle -2, -1 \rangle$ and is

$$\mathcal{E}^\star(s) = \sum_{k \geq 2} 2^k \Gamma(s) 2^{ks}(1 - 2.2^{-s} + 3^{-s}) = \Gamma(s) \frac{2^{2(s+1)}}{1 - 2^{s+1}}(1 - 2.2^{-s} + 3^{-s}). \quad (7)$$

An analysis of the poles and the residues of $\mathcal{E}^\star(s)$ shows that the asymptotic behavior of the sum $\mathcal{E}$ is

$$(3\log_2(3) - 4)n + \frac{n}{\log 2}\epsilon(n) + o(1), \tag{8}$$

where

$$\epsilon(n) := \sum_{m \in \mathbb{Z}^*} \Gamma\left(-1 + \frac{2im\pi}{\log 2}\right) 3(3^{-\frac{2im\pi}{\log 2}} - 1)n^{\frac{2im\pi}{\log 2}}$$

is a function of small modulus (about 10^{-5}) oscillating around zero.

5.3   Bounding the Contribution of Small and Long Patterns

In this section, the asymptotic behavior of the contribution of small and long patterns to the sum $\mathcal{E}(n)$ is derived. The contribution of the small patterns under the approximate model behaves asymptotically as $o(1)$ and the long patterns as $O(\sqrt{n})$ for a bound $k_l(n) = 1.5C_p \log n$. For other reals $b > 1$ in the definition of $k_l(n)$, the contribution remains sublinear.

The contribution of small patterns is

$$\sum_{k=2}^{k_s(n)} 2^k f\left(\frac{n}{2^k}\right),$$

where $f(z) = \exp(-z) - 2\exp(-2z) + \exp(-3z)$. Since the lengths of these patterns are smaller than $k_s(n) = aC_q \log n$, the quantity $n/2^k$ tends to infinity and is lower-bounded by n^{1-a}.

For long patterns, the quantity $n/2^k$ tends to zero. The dominant term in the Taylor expansion of $f(x)$ when x tends to zero is x^2. Hence the contribution of long patterns behaves as

$$\sum_{k \geq k_l(n)} 2^k f\left(\frac{n}{2^k}\right) \simeq \sum_{k \geq k_l(n)} 2^k \left(\frac{n}{2^k}\right)^2 = n^2 \sum_{k \geq k_l(n)} 2^{-k} = O(\sqrt{n}) \tag{9}$$

The asymptotic contribution of the small and long patterns to the sum $\mathcal{E}(n)$ are $O(\sqrt{n})$, thus

Proposition 2. *The contribution of intermediate patterns to $\mathcal{E}(n)$, i.e., the contribution of intermediate patterns to the size of the antidictionary under the approximate model and for a symmetric memoryless source is asymptotically*

$$(3\log_2 3 - 4)n + \frac{n}{\log 2}\epsilon(n) + O(\sqrt{n}), \tag{10}$$

where

$$\epsilon(n) := \sum_{m \in \mathbb{Z}^*} \Gamma\left(-1 + \frac{2im\pi}{\log 2}\right) 3\left(3^{-\frac{2im\pi}{\log 2}} - 1\right) n^{\frac{2im\pi}{\log 2}}$$

is a function oscillating around zero of very small amplitude (about 10^{-5}).

5.4 Result under a Memoryless Biased Model

The asymptotic behavior of $\mathcal{E}(n)$ for a memoryless biased (p, q) source is linear. It is obtained using the same method as for a symmetric source (see preceding sections). Nevertheless, under a biased model, different letters have different probabilities so there are four different sums (instead of one) depending on the end letters of w. Furthermore, in the symmetric model the probability p_u depends only on the length fo the pattern u; under a biased model, the letters composing the pattern u do matter. Lastly Mellin analysis is more technical.

Once the asymptotic behavior of $\mathcal{E}(n)$ is obtained, we show the contribution of small and long patterns is $O(\sqrt{n})$ for a bound $k_l(n) := 1.5 C_p \log n$. Hence

Proposition 3. *The contribution of intermediate patterns to $\mathcal{E}(n)$ under a memoryless biased model behaves asymptotically as*

$$\frac{n}{h}[2h + (1 - p^2)\log(1 - p^2) + (1 - q^2)\log(1 - q^2) + 2(1 - pq)\log(1 - pq)]$$

$$+ \frac{n}{h} \sum_{k \in \mathbb{Z}^*} n^{-s_k - 1} \Gamma(s_k) \Big[[p^{-2s_k} - 2p^{-s_k} + (1 - q^2)^{-s_k}] + [q^{-2s_k} - 2q^{-s_k} + (1 - p^2)^{-s_k}]$$

$$+ 2\big[(pq)^{-s_k} - q^{-2s_k} - p^{-2s_k} + (1 - pq)^{-s_k}\big] \Big] + O(\sqrt{n}),$$

for a periodic memoryless biased source and

$$\frac{n}{h}[2h + (1 - p^2)\log(1 - p^2) + (1 - q^2)\log(1 - q^2) + 2(1 - pq)\log(1 - pq)] + O(\sqrt{n}),$$

for an aperiodic memoryless biased sources. The constant $h := -p\log p - q\log q$ is the entropy of the source and $s_k = -1 + \frac{2ik\pi}{\log p - \log q}$ for $k \in \mathbb{Z}^\star$.

6 Validation of the Hypotheses

In this section we prove that the contribution of intermediate patterns to $\mathcal{E}(n)$ under the approximate model (*i.e.*, with Hypotheses H_1 and H_2) differs from $I(n)$ by a $O(n^{1-\delta})$ term for a $\delta > 0$ depending on $k_s(n)$.

The impact of Hypothesis H_1 on the contribution of intermediate patterns is

$$\sum_{k=k_s(n)}^{k_l(n)} \sum_{w \in \mathcal{A}^k} \mathbb{P}_n(w \in \text{MFW}) - \mathbb{P}_n(\{w \in \text{MFW}\} \cap \mathfrak{N}_u). \tag{11}$$

The two probabilities differ only on texts with at least two occurrences of u that are separated by less than two letters *i.e.*, either overlapping occurrences, or adjacent, or separated by one letter. This set of texts is denoted by $\mathfrak{X}_u$. The sum over intermediate patterns of the probability of texts from $\mathfrak{X}_u$ is $O(n^{1-\delta})$.

Hypothesis H_2 ($\widetilde{N}_u$ is Poisson of parameter np_u) modifies $I(n)$ by a $O(n^{1-\delta})$ term. The difference between the contributions under Hypothesis H_1 (sum of $\mathbb{P}_n(\{w \in \text{MFW}\} \cap \mathfrak{N}_u)$) and under the approximate model (H_1 and H_2) is

$$\sum_{k \geq k_s(n)} \sum_{w \in \mathcal{A}^k} \sum_{j \geq 2} C_{j,w} \left[\underbrace{\mathbb{P}_n(\widetilde{N}_u = j) - \mathbb{P}_n(N_u = j)}_{S_1(j,w)} + \underbrace{\mathbb{P}_n(N_u = j) - \frac{(np_u)^j}{j!} e^{-np_u}}_{S_2(j,w)} \right],$$

where N_u is a parameter counting the number of occurrences of u in a text. This sum is split in two sums of general terms $C_{j,w} S_1(j,w)$ and $C_{j,w} S_2(j,w)$. In the first sum, the probabilities differ only on texts from the set $\mathfrak{X}_u$. The contribution of those texts for intermediate patterns has already been dealt with. In the second sum, the Stein-Chen method (see Barbour, Holst, and Janson [1] for details) is used to estimate the distance between the distributions of N_u and a Poisson distribution.

7   Conclusion

In Sect. 3 we showed that the contribution of *small* and *long* patterns to $\mathbb{E}_n(\mathcal{S})$ is asymptotically sublinear. Then in Sect. 5 we proved that, under an approximate model, the contribution of *intermediate* patterns is asymptotically linear. Finally in Sect. 6 we showed that the contribution of intermediate patterns under the approximate model does not differ from the contribution of intermediate patterns to $\mathbb{E}_n(\mathcal{S})$ by more than a $O(n^{1-\beta})$ term (for $\beta > 0$), hence

Theorem 4. *Under a memoryless biased model for the generation of texts, the mean size of the antidictionary $\mathbb{E}_n(\mathcal{S})$, i.e., the number of minimal forbidden words, over all texts of length n behaves asymptotically as*

$$K \frac{n}{h} + \frac{n}{h} \epsilon(n) + o(n)$$

for a periodic source, and as

$$K \frac{n}{h} + o(n)$$

for an aperiodic source, where $\epsilon(x)$ is a function fluctuating around zero of amplitude 10^{-5},

$$K := 2h + (1 - p^2) \log(1 - p^2) + (1 - q^2) \log(1 - q^2) + 2(1 - pq) \log(1 - pq),$$

$$and\ h = -p \log p - q \log q.$$

Takahiro Ota ran experiments on the size of the antidictionary on texts of length 3000. His results match the formula of Theorem 4 with a margin of error of 0.6% for a p $(p \geq q)$ between 0.5 and 0.9. If p is closer to 1, the error increases but remains satisfactory: 5% for $p = 0.9$.

The study of the performances of DCA is far from complete. Several research directions arise: first obtaining the variance and the distribution of the number of words in an antidictionary. The length of MFWs is of interest, for instance to estimate the mean time to find a suffix in the compression step of DCA. In the original paper Crochemore *et alii* consider an antidictionary with patterns of bounded length; informations on the length would help set an optimal threshold. Questions like what is the length of the compressed text, *i.e.*, how many letters are removed in the compression step, and what is the size of the data structure storing the MFWs provide additional informations on the behavior of the algorithm. Crochemore *et alii* have shown that DCA attains a compression rate close to the entropic ratio for balanced sources. We are interested in a generalization of this result with tools from analytic combinatorics.

The analysis should be extended from a memoryless probabilistic model for the generation of texts to more general sources (Markovian, dynamical source).

Crochemore and Navarro [5] developed the notion of *almost antifactors*, words with very little occurrence in the text. A version of DCA encodes the occurrences of each almost antifactor in a file of exceptions, but in the compression phase these almost antifactors are considered as *true* antifactors. Experiments show this improves the compression ratio. A complete analysis will help characterize the threshold on the number of occurrences between almost antifactors that make the compression more efficient and those that do not.

Acknowledgements

The author thanks Hiroyoshi Morita for introducing him to the problem and Takahiro Ota for providing simulations. This work was done during my Ph.D. and i also thank my advisor Philippe Flajolet for discussing the issue and providing helpful advices.

References

1. Barbour, A.D., Holst, L., Janson, S.: Poisson approximation. The Clarendon Press Oxford University Press, New York (1992) (Oxford Science Publications)
2. Béal, M.-P., Mignosi, F., Restivo, A.: Minimal forbidden words and symbolic dynamics. In: Puech, C., Reischuk, R. (eds.) STACS 1996. LNCS, vol. 1046. Springer, Heidelberg (1996)
3. Crochemore, M., Mignosi, F., Restivo, A., Salemi, S.: Text compression using antidictonaries. In: Wiedermann, J., Van Emde Boas, P., Nielsen, M. (eds.) ICALP 1999. LNCS, vol. 1644, pp. 261–270. Springer, Heidelberg (1999)
4. Crochemore, M., Mignosi, F., Restivo, A., Salemi, S.: Data compression using antidictonaries. In: Storer, J. (ed.) Proceedings of the I.E.E.E., Lossless Data Compression, pp. 1756–1768 (2000)
5. Crochemore, M., Navarro, G.: Improved antidictionary based compression. In: SCCC 2002, Chilean Computer Science Society, pp. 7–13. I.E.E.E. CS Press (November 2002)
6. Fayolle, J.: An average-case analysis of basic parameters of the suffix tree. In: Drmota, M., Flajolet, P., Gardy, D., Gittenberger, B. (eds.) Mathematics and Computer Science. Proceedings of a colloquium organized by TU, Wien, Vienna, Austria, pp. 217–227. Birkhäuser, Basel (2004)

7. Fayolle, J.: Compression de données sans perte et combinatoire analytique. PhD thesis, Université Paris VI (2006)
8. Fayolle, J., Ward, M.D.: Analysis of the average depth in a suffix tree under a Markov model. In: Proceedings of the 2005 International Conference on the Analysis of Algorithms (2005), DMTCS. Proceedings of a colloquium organized by Universitat Politècnica de Catalunya, Barcelona, Catalunya, June 2005, pp. 95–104 (2005)
9. Flajolet, P., Gourdon, X., Dumas, P.: Mellin transforms and asymptotics: Harmonic sums. Theoretical Computer Science 144, (1–2), 3–58 (1995)
10. Jacquet, P., Szpankowski, W.: Autocorrelation on words and its applications: analysis of suffix trees by string-ruler approach. Journal of Combinatorial Theory. Series A 66(2), 237–269 (1994)
11. Morita, H., Ota, T.: An upper bound on size of antidictionary. In: Proceedings of SITA 2004 (2004)
12. Ota, T., Morita, H.: One-path ECG lossless compression using antidictionaries. IEICE Trans. Fundamentals (Japanese Edition) J87-A 9, 1187–1195 (2004)

Approximate String Matching with Address Bit Errors

Amihood Amir[1,2,*], Yonatan Aumann[1], Oren Kapah[1],
Avivit Levy[3,**], and Ely Porat[1]

[1] Department of Computer Science, Bar Ilan University,
Ramat Gan 52900, Israel
{amir,aumann,kapaho,porately}@cs.biu.ac.il
[2] Department of Computer Science, Johns Hopkins University,
Baltimore, MD 21218
[3] CRI, Haifa University, Mount Carmel, Haifa 31905, Israel
avivitlevy@gmail.com

Abstract. A string $S \in \Sigma^m$ can be viewed as a set of pairs $S = \{(\sigma_i, i) : i \in \{0, \ldots, m-1\}\}$. We consider approximate pattern matching problems arising from the setting where errors are introduced to the location component (i), rather than the more traditional setting, where errors are introduced to the content itself (σ_i). In this paper, we consider the case where bits of i may be erroneously flipped, either in a consistent or transient manner. We formally define the corresponding approximate pattern matching problems, and provide efficient algorithms for their resolution, while introducing some novel techniques.

1 Introduction

1.1 Background

Consider a text $T = t_0 \cdots t_{n-1}$ and pattern $P = p_0 \cdots p_{m-1}$, both over an alphabet Σ. Traditional pattern matching regards T and P as *sequential* strings, provided and stored in sequence (e.g. from left to right). Therefore, an implicit in the conventional approximate pattern matching is the assumption that there may indeed be errors in the **content** of the data, but the **order** of the data is inviolate. However, some non-conforming problems have been gnawing at the walls of this assumption. Some examples are:

Text Editing: The *swap* error, motivated by the common typing error where two adjacent symbols are exchanged [11,2], does not assume error in the content of the data, but rather, in the order. The data content is, in fact, assumed to be correct. The swap error seemed initially to be akin to the other Levenshtein errors, in that it could be added to the other edit operations and solved with the same dynamic programming [11]. However, when isolated, it

---

* Partly supported by ISF grant 35/05.
** This work is part of A. Levy's Ph.D. thesis.

P. Ferragina and G. Landau (Eds.): CPM 2008, LNCS 5029, pp. 118–129, 2008.

turned out to be surprisingly simple to handle [3]. This scarcely seems to be the case for indels or mismatch errors. We stress that the main importance of this work is in the theoretical understanding of the combinatorics involved since spell-checking is an easier practical solution to the problem.

Computational Biology: During the course of evolution areas of genome may be shifted from one location to another. Considering the genome as a string over the alphabet of genes, these cases represent a situation where the difference between the original string and resulting one is in the locations rather than contents of the different elements. Several works have considered specific versions of this biological setting, primarily focusing on the sorting problem (*sorting by reversals* [5,6], *sorting by transpositions* [4], and *sorting by block interchanges* [7]).

Bit Torrent and Video on Demand: The inherently distributed nature of the web is already causing the phenomenon of transmission of a stream of data in tiny pieces from different sources. This creates the problem of putting scrambled data back together again.

Computer Architecture: In computer architecture, it is by no means taken for granted that when seeking a word from a given address, no errors will occur in the address bits [9]. This problem is relevant even when reading a buffer of consecutive words since these words are not necessarily consecutive in the disk or in an interleaved cache[1].

Therefore, the traditional view of strings is becoming, at times, less natural. In such cases, it is more natural to view the string as a set of pairs (σ, i), where i denotes a location in the string, and σ is the value appearing at this location. Given this view of strings, we reconsider the problem of *approximate pattern matching*. Practically, the content Hamming error and the address error are both solved via error correcting codes. However, from a theoretical point of view, it would be interesting to consider searching where address errors are not corrected at all. What are the types of uncorrected address errors that can still be reasonably handled by a search application? Is the address error similar to content error from a pattern matching point of view? Can it be solved by the same means?

Motivated by these questions a new pattern matching paradigm – *pattern matching with address errors* – was proposed by [1]. In this model, the pattern *content* remains intact, but the relative positions (addresses) may change. Efficient algorithms for several different natural types of rearrangement errors are presented in [1]. These types of address errors are inspired by biology, i.e. pattern elements exchanging their locations due to some external process.

In this paper we suggest another broad class of address errors inspired by computer architecture. Specifically, we consider errors which arise from a process of flipping some or all of the bits in the binary representation of $[0..m-1]$. Such address errors may arise in situations where the text and the pattern are generated by two different systems, which may use different naming conventions.

---

[1] Practically, these problems are solved by means of redundancy bits, checksum bits, error detection and correction codes, and communication protocols.

Alternatively, address errors may result from failures in the wires of the address bus (the wires connecting the CPU and the memory which are used to transmit the address of operands), or failure in the transmitted address bits. Finally, address errors may actually not constitute an error, but rather represent different legitimate ways to order the given set of elements. Following [1], in this paper we do not consider content errors at all, since our aim to to analyze whether there is novelty in the address error scheme.

1.2   The Problem Definition

Consider a string $S \in \Sigma^m$. Using the alternate view of strings described above we write $S = \{(\sigma, i) : i \in \{0, 1\}^{\log m}\}$. We consider two types of errors in the bits of the i entries:

Flipped bits: there exists a subset of bit positions $F \subseteq \{0, \ldots, \log m - 1\}$, such that in each i, all bits in positions $f \in F$ are flipped (i.e. 1 is turned into a 0 and visa versa).

For example, for the string $S = 1234 = \{(1, 00), (2, 01), (3, 10), (4, 11)\}$ and $F = \{1\}$, the resulting string is $S' = 3412 = \{(1, 10), (2, 11), (3, 00), (4, 01)\}$.

Faulty bits: there exists a subset of bit positions $F \subseteq \{0, \ldots, \log m - 1\}$, such that in each i, the bits in positions $f \in F$ *may* be flipped, and may not.

For example, for the string $S = 1234 = \{(1, 00), (2, 01), (3, 10), (4, 11)\}$ and $F = \{1\}$, the resulting string may be $S' = \{(1, 10), (2, 01), (3, 10), (4, 01)\}$ (the bit was flipped for 1 and 4 but not for 2 and 3).

Note that in this case the resulting set is actually a multi-set, and may not represent a valid string, as some locations may appear multiple times, while others not at all.

We consider approximate pattern matching problems associated with each of the above types of errors. Specifically, given a pattern P and text T, we wish to find:

- the smallest set F such that if the bits of F are consistently flipped, then P has a match in T. We call this problem the *flipped bits* problem.
- the smallest set F such that if the bits of F may be transiently flipped, then P has a match in T. We call this problem the *faulty bits* problem.

We prove (omitted proofs will be given in the full paper):

- For pattern and text of size m, the flipped bits problem can be solved in $O(m \log m)$ steps.
- For pattern and text of size m, the faulty bits problem can be solved deterministically in $O(m^{\log_2 3} |\Sigma|)$ steps and randomly in $O(m \log m)$ steps.
- For pattern and text of size m, the faulty bits problem can be deterministically approximated to a constant $c > 1$ in $O(|\Sigma| \frac{m^{\log 3}}{\log^{c-1} m})$.
- For text and pattern of sizes n and m, respectively, m power of 2, the faulty bits problem can be solved deterministically in $O(|\Sigma| nm \log m)$ steps.

2   Flipped Bits Errors

In this section we consider the flipped bits problem. In this setting, one or more of the bit positions may exhibit a faulty behavior whereby the bit at this position is consistently flipped. Given two strings $P, T \in \Sigma^m$, the distance between the two is the least number of *flipped bits* positions that can explain the differences between the two, and ∞ if no such set of position can explain the difference. Formally,

Definition 1. *For an index $k \in [0..m-1]$,[2] we view k as a binary string, i.e. $k = k[0] \cdots k[\log m - 1] \in \{0,1\}^{\log m}$ (w.l.o.g. m is a power of 2). Consider $F \subseteq [0..\log m - 1]$. The bit flip transformation induced by F, denoted f_F, is a function $f_F : \{0,1\}^{\log m} \to \{0,1\}^{\log m}$, such that for any k and i*

$$f_F(k)[i] = \begin{cases} 1 - k[i] & i \in F \\ k[i] & i \notin F \end{cases}$$

i.e. the value of $f_F(k)$ is flipped at bits of F and identical on other bits.

For strings $P, T \in \Sigma^m$ we say that T is a F-flip-bits match of P if for all $k \in \{0,1\}^{\log m}$, $T[k] = P[f_F(k)]$. The flip-bit distance between P and T is the cardinality of the smallest F such that T is an F-flip-bits match of P. If no such F exists, then the distance is ∞.

Note that there are $2^{\log m}$ possible faulty sets F. Checking each possibility separately takes $O(m)$, so a naive algorithm takes time $O(m^2)$ per position. We show how to reduce this to $O(m \log m \log |\Sigma|)$. We begin with an efficient solution for the case $\Sigma = \{0,1\}$, and then use it to obtain an efficient solution for general alphabets.

Let $k, j \in \{0,1\}^{\log m}$, denote $k \oplus j$ to be the result of the bitwise XOR of the two, i.e. for each i, $(k \oplus j)[i] = k[i] \oplus j[i]$ (where $\oplus$ is the XOR operation, i.e. addition over Z_2). For strings $T, P \in \mathbb{Z}^m$, define the *binary convolution* of the two to be a vector, also of size m, $T \otimes P \in \mathbb{Z}^m$, such that for all $k \in \{0,1\}^{\log m}$: $(T \otimes P)[k] = \sum_{j \in \{0,1\}^{\log m}} T[j] \cdot P[k \oplus j]$.

Lemma 1. *For a set $F \subseteq 0..\log m - 1$, let $\chi_F \in \{0,1\}^{\log m}$ be the characteristic vector of F. Consider binary P and T, both of size m, and let α_P be the number of ones in P and α_T be the number of ones in T. Then, T is an F-flip-bits match of P iff $\alpha_T = \alpha_P$ and $(T \otimes P)[\chi_F] = \alpha_T$.*

Proof. For any index j, $T[j] \cdot P[\chi_F \oplus j] = 1$ iff both $T[j] = 1$ and $P[\chi_F \oplus j] = P[f_F(j)] = 1$. Thus, $(T \otimes P)[\chi_F]$ counts the number of ones in T that are mapped to ones in P under the transformation f_F. Since, $(T \otimes P)[\chi_F] = \alpha_T$, then all ones in T are mapped to ones in P. But, $\alpha_T = \alpha_P$, so also all zeros in T are mapped to zeros in P.

---

[2] For integers i, j, we denote by $[i..j]$ the set of integers from i to j. Thus, $[0..m-1]$ is the set $\{0, 1, \ldots, m-1\}$.

Thus, in order to find the *flip-bit distance between P and T* we compute the entire vectors $T \otimes P$. We then seek all locations k for which $(T \otimes P)[k] = \alpha_T$, and among these k's, find the one with the minimum weight (i.e. least number of 1's).

It thus remains to explain how to efficiently compute the binary convolution. The convolution can easily be computed in $O(m^2)$ time. We explain how to compute it in $O(m \log m)$ time.

For a vector $v \in \mathbb{Z}^t$ (t power of 2), define two vectors $v^+, v^- \in \mathbb{Z}^{t/2}$, as follows. For each $k \in \{0,1\}^{\log t - 1}$, $v^+[k] = v[0k] + v[1k]$ and $v^-[k] = v[0k] - v[1k]$. The key lemma for the computation is:

Lemma 2. *For any $v, w \in \{0,1\}^t$, and $k \in \{0,1\}^{\log t - 1}$:*

$$(v \otimes w)[0k] = \frac{(v^+ \otimes w^+)[k] + (v^- \otimes w^-)[k]}{2}, \quad (v \otimes w)[1k] = \frac{(v^+ \otimes w^+)[k] - (v^- \otimes w^-)[k]}{2}.$$

Thus, in order to compute $T \otimes P$, our algorithm recursively computes $T^+ \otimes P^+$ and $T^- \otimes P^-$, and then uses lemma 2 in order to compute the convolution $T \otimes P$. In each recursion level we need to compute $O(m)$ values, each taking $O(1)$ time. Thus, we get a recursive recurrence $time(m) = 2 \cdot time(m/2) + cm$, for a total $time(m) = O(m \log m)$. We obtain:

Theorem 1. *The flipped bit problem can be solved in $O(m \log m)$ time for binary text and pattern of size m.*

For a general alphabet, the same techniques as in [8] can be used to handle with only one convolution. Hence,

Theorem 2. *The flipped bit problem can be solved in $O(m \log m)$ time for text and pattern of size m and alphabet Σ.*

Remark. The above algorithm can also be viewed as a form of Fast Fourier Transform over $\mathbb{Z}_2$ (rather than over the complexes). We omit the details.

3   The Faulty Bits Problem

This section studies the faulty bits problem. In this model a faulty position inconsistently produces errors. It may sometimes hold the correct value and sometimes the wrong one. Given two strings, the objective is to find the least number of faulty positions that explain the differences between the two. We begin by formally defining the faulty bits distance problem.

3.1   Problem Definition

Let Σ be a finite alphabet. Let $P, T \in \Sigma^m$ be two strings of length m, such that P is the *query string* and T is the *stored string*. Denote $P = p[0]p[1] \cdots p[m-1]$ and similarly for T. Consider $F \subseteq \{0, \ldots, \log m - 1\}$, and suppose that the address bits carrying bits in the set F are faulty. We now formulate the criterion that determines if the stored string T matches the query string P, assuming that the bits of F are faulty.

Consider an address k, and let $k = k[0]k[1] \cdots k[\log m - 1]$ be the binary representation of k. Let $[k]_F$ be the set of all the addresses ℓ such $k[i] = \ell[i]$ for all $i \notin F$, i.e. k and ℓ agree on all bits not in F. Note that $[k]_F$ is an equivalence class, so $[\ell]_F = [k]_F$ if $\ell \in [k]_F$. Then, if the address bits in F are faulty, a value intended to location k can end up in any location $\ell \in [k]_F$. Thus, we obtain the following criterion for a match of T to the query string P while using the faulty bits of F:

Definition 2. *For strings P and T and set $F \subseteq \{0, \ldots, \log m - 1\}$ we say that T is an F-faulty-bit match of P if for each equivalence class $[k]_F$: for each $\sigma \in \Sigma$*

$$|\{\ell : \ell \in [k]_F, P[\ell] = \sigma\}| = |\{\ell : \ell \in [k]_F, T[\ell] = \sigma\}|$$

The Optimization Problem. Given any of the above match conditions and strings P and T, we wish to find the set F of minimal cardinality such that T is an F-faulty-bit match of P. We call this the *faulty-bits problem*.

3.2   A Deterministic Algorithm

For each equivalence class $[k]_F$ and $\sigma \in \Sigma$ let

$$\text{BUCKET}(P, [k]_F, \sigma) = \{\ell : \ell \in [k]_F, P[k] = \sigma\}$$

the elements of P with locations in $[k]_F$ that have value σ. Similarly,

$$\text{BUCKET}(T, [k]_F, \sigma) = \{\ell : \ell \in [k]_F, T[\ell] = \sigma\}$$

the elements of T with locations in $[k]_F$ that have value σ. The criteria for an F-faulty-bit match is that for all k : $|\text{BUCKET}(P, [k]_F, \sigma)| = |\text{BUCKET}(T, [k]_F, \sigma)|$ for all σ. Thus, it remains to explain how to compute the sizes of the buckets.

For any fixed F, all buckets can be computed in a total of $O(m)$ steps, with a single pass over the strings T and P. Thus, for a given F, the condition can be tested in $O(m)$ steps. There are $2^{\log m} = m$ different possible sets F, which provides a naive $O(m^2)$ algorithm. We now show how to reduce this to $O(m^{\log 3}|\Sigma|)$.

For an address k and index $i \in \{0, \ldots, \log m - 1\}$, let $k^{(i)}$ be the address which has the same representation as k except for the i-th bit which is flipped. Then, it is easy to see that for any i, σ and $X \in \{T, P\}$, $\text{BUCKET}(X, [k]_F, \sigma) = \text{BUCKET}(X, [k]_{F-\{i\}}, \sigma) \cup \text{BUCKET}(X, [k^{(i)}]_{F-\{i\}}, \sigma)$. That is, the bucket with faults at F can be obtained as the union of buckets with one less fault, and fixing the two possible values for this bits. In particular,

$$|\text{BUCKET}(X, [k]_F, \sigma)| = |\text{BUCKET}(X, [k]_{F-\{i\}}, \sigma)| + |\text{BUCKET}(X, [k^{(i)}]_{F-\{i\}}, \sigma)| \tag{1}$$

Note that for $F = \emptyset$,

$$|\text{BUCKET}(X, [k]_F, \sigma)| = \begin{cases} 1 & X[k] = \sigma \\ 0 & X[k] \neq \sigma \end{cases} \tag{2}$$

Thus, combining (2) and (1), we obtain that for any σ all sizes of all buckets can be computed in an inductive fashion, with $O(1)$ steps per bucket.

For a given σ, the overall total number of buckets – for all fault patterns F, is the overall total number of equivalence classes $[k]_F$ for all F. Each equivalence class can be identified with a string $w \in \{0, 1, *\}^{\log m}$ such that $w[i] = *$ denotes a bit in F and the other $w[i]$'s are fixed as in k. Thus, the number of equivalence classes is: $|\{w \in \{0, 1, *\}^{\log m}\}| = 3^{\log m} = m^{\log 3}$. We thus obtain:

Theorem 3. *The faulty-bits problem can be solved in* $O(|\Sigma|m^{\log 3})$ *time.*

3.3 A Randomized Algorithm

We now describe how to solve the faulty-bits problem in $O(m \log m)$ time for unbounded alphabet using formal polynomials and the Schwartz-Zippel Lemma [13,14]. A pseudo-code of the algorithm is given in Fig. 1. Every fault pattern F can be regarded as a $\log m$-length mask with values in $\{*, +\}$, where $*$ specifies that this bit is faulty therefore its specific value is insignificant, and $+$ specifies that this bit is non-faulty and its specific value is significant. Given an m-length array A (T or P) and a mask M, we inductively define a formal polynomial $P_M(A)$ as follows (Lemma 3 describes the explicit form of the polynomial).

The Definition of the Formal Polynomial . For $i = 0$, let j be any index of the array A, then $P_M(A_j) = X_{A[j]}$, where A_j is a sub-array of A of length 1.

FAULTY-BIT RANDOMIZED ALGORITHM
Input: $P \in \Sigma^m, T \in \Sigma^m$
1   Assign a random value $r(\sigma) \in \{1, \ldots, m^3\}$ to every $\sigma \in \Sigma$.
2   for $j = 0$ to $m - 1$ do
3          $P^{(0)}[j] \leftarrow r(P[j])$
4          $T^{(0)}[j] \leftarrow r(T[j])$
5   for $i = 0$ to $\log m - 1$ do
6   Assign a random value $r_i \in \{1, \ldots, m^3\}$.
7   for $j = 0$ to $\frac{m-1}{2}$ do
8              $P^{(i+1)}[j] \leftarrow P^{(i)}[j] + P^{(i)}[j + 2^i]$
9              $T^{(i+1)}[j] \leftarrow T^{(i)}[j] + T^{(i)}[j + 2^i]$
10            $P^{(i+1)}[j + 2^i] \leftarrow P^{(i)}[j] + r_i \cdot P^{(i)}[j + 2^i]$
11           $T^{(i+1)}[j + 2^i] \leftarrow T^{(i)}[j] + r_i \cdot T^{(i)}[j + 2^i]$
Output:
12 for $j = 0$ to $m - 1$ do
13   if $P^{(\log m)}[j] = T^{(\log m)}[j]$ then
14   Report "there is an F match of P and T,
                 where F is the set of indices of 0-bits in the binary
                 representation of j"

Fig. 1. A butterfly design of the faulty bit randomized algorithm

Let $P_M(A_j)$ be the polynomials for sub-arrays of size 2^i (the indices j have length $\log m - i$) and define the polynomials $P_M(A_{j'})$ for sub-arrays of size 2^{i+1} (the index j' has length $\log m - i - 1$, and the indices j are $j'0$ and $j'1$). Consider $M[i+1]$, the $i+1$ bit of the mask M, then:

$$P_M(A_{j'}) = \begin{cases} P_M(A_{j'0}) + X_{i+1} \cdot P_M(A_{j'1}), & \text{if M[i+1]=+} \\ P_M(A_{j'0}) + P_M(A_{j'1}), & \text{if M[i+1]=*} \end{cases}$$

Lemma 3. *Given an array A and a $\log m$-bit mask M, define*

$$f_M(i) = \begin{cases} 0, & \text{if } M[i]=* \\ 1, & \text{if } M[i]=+ \end{cases}$$

Then, $P_M(A) = \sum_{j=0}^{m-1} X_{A[j]} \cdot \prod_{i=1}^{\log m} X_i^{j[i-1] \cdot f_M(i)}$

Lemma 4. *Given two arrays A and B and a $\log m$-bit mask M. Let $F_M = \{k \in \{0, \ldots, \log m - 1\} | M[k+1] = *\}$ (i.e., the set of assumed faulty-bits according to the mask M), then $P_M(A) \equiv P_M(B)$ iff A is an F_M-faulty-bit match of B.*

Theorem 4. *The faulty-bits problem can be solved in $O(m \log m)$ time with high probability.*

4   Approximate Faulty Bits Problem

In this section we describe a scheme to reduce the cost of the deterministic algorithm while paying in some loss of precision. The general idea is to avoid checking all possible subsets F with a guarantee that the deviation of the resulting F from the minimal F is bounded. The following is the crucial observation enabling such a scheme.

Observation 1. *Let $F \subseteq \{0, \ldots, \log m - 1\}$ such that T is an F-faulty-bit match of P and let $F' \subseteq \{0, \ldots, \log m - 1\}$ such that $F \subset F'$, then T is an F'-faulty-bit match of P.*

Let $F \subseteq \{0, \ldots, \log m - 1\}$, and $F' \subseteq \{0, \ldots, \log m - 1\}$ such that $F \subset F'$, we call F' an *ancestor* of F, and F a *descendent* of F'. Our algorithm then, for any $k \in \{1, \ldots, \log m\}$, checks only a part (that depends on k) of the subsets F of size k. By the observation, if we missed a subset F of size k such that T is an F-faulty-bit match of P (these are the only interesting subsets that we care about missing) then we can still detect an F'-faulty-bit match between T and P, but with the loss of minimality, since $|F'| > |F|$. The important property is that the greater the precision we are willing to lose is, the greater is the number of ancestors F' that by Observation 1 are an F'-faulty-bit match between T and P. Since the number of such F' is exactly the number of choices to fix k of the i members of F' to be the k members of F, the next lemma follows.

Lemma 5. *Given $F \subseteq \{0, \ldots, \log m - 1\}$ of size k, $0 \le k < i \le \log m$, the number of ancestors F' of size i of F is $\binom{i}{k}$.*

Denote by $|L_k|$ the number of subsets of size k. Given a function $f : \mathbb{N} \mapsto \mathbb{N}$, we define the following scheme for subsets to be checked by the algorithm of section 3.2:

---

f-**scheme**

    For $k = 1$ to $\frac{\log m}{2} - 1$

        Randomly choose $\frac{|L_k|}{f(k)}$ subsets from all subsets of size k.

Lemma 6. *Given a function f, and a subset F of size k, for any $i \geq k$ denote $\alpha = \frac{\binom{i}{k}}{f(i)}$. Then, under the f-scheme the probability that there is no ancestor of F with size i is at most $\exp^{-\alpha}$.*

By taking $f(k) = k^c$ in the f-scheme we get theorem 5.

Theorem 5. *There exists a deterministic algorithm for the faulty-bits problem that, for any constant $c \in \mathbb{N}$, $c > 1$, runs in $O(|\Sigma|\frac{m^{\log 3}}{\log^{c-1} m})$ time and returns a subset F that is greater than the minimum by at most c.*

Remark. Using standard de-randomization techniques the explicit structure of the deterministic scheme can be found (see [12]).

5   The Faulty Bits Problem with Text Longer Than Pattern

In this section we show how to solve a variant of the minimum faulty-bits match problem where, the stored string T is of length n and the query string P is of length m, where $m < n$. Denote by $T^{(i)}$ the m-long string starting at position i in T. We wish to find for each position i in the stored string T, the set F of minimal cardinality such that $T^{(i)}$ is an F-faulty-bit match of P. Using the algorithms from section 3 for each position in T separately give an $O(|\Sigma|nm^{\log 3})$ deterministic algorithm or an $O(nm \log m)$ randomized algorithm. For the case that m is a power of 2, we show how to construct an $O(|\Sigma|nm \log m)$ deterministic algorithm. The algorithm is based on a core algorithm which, given a *specific* set $F \subseteq [0.. \log m - 1]$ and binary pattern and text, finds all locations i, such that $T^{(i)}$ is an F-faulty-bit match of P in $O(n \log m)$ steps. Since there are $2^{\log m} = m$ possible sets F, we obtain a solution for the binary case in $O(nm \log m)$ steps. This translates into an $O(|\Sigma|nm \log m)$ algorithm for a general alphabet Σ, by counting each symbol separately.

General Structure. Consider a binary alphabet and a set F. We find all locations i, such that $T^{(i)}$ is an F-faulty-bit match of P using a variant of the Karp-Miller-Rosenberg [10] string matching algorithm. The KMR-algorithm solves the exact matching problem by a process of parallel renaming of pairs, quadruplets, etc. for the pattern and for each text location, until each text location gets a name representing the m-length string that starts at this location. Locations with name equal to the pattern name are matches. The key observations that allow to use the KMR algorithm are:

- The pairing process used in the KMR algorithm need not be done in the standard order of bits (from right to left), but can rather be done in any order.
- The KMR algorithm can be adapted to the faulty-bit case as follows. In the renaming process, give the names based on the *number* of occurrences of each symbol, rather than the exact order. For the binary case the name given to a sub-string is simply the number of ones therein. We call this *count renaming*. This provides that a sub-string can be converted to another by faulty-bits iff they have the same count. Clearly, this renaming only works for the case that *all* bits are faulty.

Our algorithm employs a KMR-like structure in two phases. In the first phase we use the count-renaming convention according to the bits of F. In the second phase, using the names from the first phase as the starting point, we rename using the standard KMR process according to the remaining bits. The details follows.

The Algorithm. A pseudo-code of the algorithm is provided in Fig. 2. Let $P \in \{0,1\}^m$, $T \in \{0,1\}^n$, $F = (f_1, f_2, \ldots, f_k) \subseteq [0..\log m - 1]$ and $G = [0..\log m - 1] - F$. Denote $G = (g_1, \ldots, g_d)$. The algorithm processes the pattern and then the text in the same way. The processing has two phases. In the first phase there are $|F|$ steps, where in each step we consider another faulty bit from F. When the bit f_i is considered to be faulty, we add to each position in the string from the previous step the entry with a shift of 2^{f_i} (lines 4-6,15-17). The guarantee of this phase is given in Lemma 7. In the second phase there are $\log m - |F|$ steps, where in each step we consider another non-faulty bit from G. For the bit $g_{i'}$, where $i' = i - |F|$ we rename every pair of positions (in the string from the previous step) with shift $2^{g_{i'}}$ (lines 7-12,18-25). The guarantee of this phase is given in Lemma 8. Theorem 6 follows.

Lemma 7. *For $i = 1, \ldots, |F|$, and $j = 0, \ldots, n - 1$, $t^{(i)}[j] = |\{T[j'] = 1 : \forall k \notin \{f_1, \ldots, f_i\}, (j' - j)[k] = 0\}|$ (i.e. $t^{(i)}[j]$ is the number of elements $T[j']$ which are equal to 1, and such that in the binary representation of $j' - j$, all bits not of F are 0), and for $j = 0, \ldots, m - 1$, $p^{(i)}[j] = |\{P[j'] = 1 : \forall k \notin \{f_1, \ldots, f_i\}, (j' - j)[k] = 0\}|$.*

Denote by $l = (l_1, \ldots, l_i)$ the list of bit indices in $F \cup \{g_1, \ldots, g_{i'}\}$ (where $i' = i - |F|$) ordered from the least to most significant. Let $l(j)$ be the number resulting from the assignment of the binary representation of j to the bits of l and 0 to the bits of $G - l$. Denote by $P^{(j)}|_{2^i}$ the 2^i-long substring of P starting at location j having the values $P[j]P[j + l(1)], P[j + l(2)], \ldots, P[j + l(2^i - 1)]$, and similarly $T^{(j)}|_{2^i}$.

Lemma 8. *For $i = |F| + 1, \ldots, \log m$ and all j_1, j_2, $\hat{p}^{(i)}[j_1] = \hat{t}^{(i)}[j_2]$ iff $T^{(j_2)}|_{2^i}$ is an F-faulty-bit match of $P^{(j_1)}|_{2^i}$.*

Theorem 6. *The faulty-bits problem where $m < n$, m is a power of 2, can be solved in $O(|\Sigma|nm \log m)$ time by a deterministic algorithm.*

F-FAULTY-BIT MATCH ALGORITHM
Input: $P \in \{0,1\}^m, T \in \{0,1\}^n, F = (f_1, f_2, \ldots, f_k) \subseteq [0.. \log m - 1]$

Pattern Processing
1   $G \leftarrow [0.. \log m - 1] - F$. Denote $G = (g_1, \ldots, g_d)$.
2   for $j = 0$ to $m - 1$ do
3         $p^{(0)}[j] \leftarrow P[j]$
4   for $i = 1$ to $|F|$ do
5   for $j = 0$ to $m - 1$ do
6                $p^{(i)}[j] \leftarrow p^{(i-1)}[j] + p^{(i-1)}[j + 2^{f_i}]$ (whenever both are defined)
7   for $i = |F| + 1$ to $\log m$ do
8         $i' \leftarrow i - |F|$
9   for $j = 0$ to $m - 1$ do
10         $\hat{p}^{(i)}[j] \leftarrow \langle p^{(i-1)}[j], p^{(i-1)}[j + 2^{g_{i'}}] \rangle$ (whenever both are defined)
11   Let $h^{(i)}$ be any function $h^{(i)} : \{\hat{p}^{(i)}[j]\} \rightarrow [1..m]$
           ($h^{(i)}$ is the *renaming function*).
12   For all j, $p^{(i)}[j] \leftarrow h^{(i)}(\hat{p}^{(i)}[j])$

Text Processing
13 for $j = 0$ to $n - 1$ do
14       $t^{(0)}[j] \leftarrow$ the number of text elements with address j which are 1
15 for $i = 1$ to $|F|$ do
16   for $j = 0$ to $n - 1$ do
17           $t^{(i)}[j] \leftarrow t^{(i-1)}[j] + t^{(i-1)}[j + 2^{f_i}]$ (whenever both are defined)
18 for $i = |F| + 1$ to $\log m$ do
19       $i' \leftarrow i - |F|$
20   for $j = 0$ to $n - 1$ do
21          $\hat{t}^{(i)}[j] \leftarrow \langle t^{(i-1)}[j], t^{(i-1)}[j + 2^{g_{i'}}] \rangle$ (whenever both are defined)
22   if the $\hat{t}^{(i)}[j]$ appeared as one of the values $\hat{p}^{(i)}[j']$ then
23            $t^{(i)}[j] \leftarrow h^{(i)}(\hat{p}^{(i)}[j'])$ (use the same renaming as for p)
24   else
25            $t^{(i)}[j] \leftarrow \perp$

Output
26 for $j = 0$ to $n - m$ do
27   if $t^{(\log m)}[j] = p^{(\log m)}[0]$ then
28   Report "there is an F match at location j"

Fig. 2. The faulty bit algorithm for text longer than pattern

6   Conclusions

The main contributions of this paper are:

1. A new and flexible model that encompasses the growing number of *address errors* in pattern matching problems.

2. Evidence that address errors are indeed conceptually and algorithmically different from the traditional content errors.

3. Some novel techniques, such as FFT over $\mathbb{Z}_2$, that have never been used in pattern matching and rarely in the theoretical algorithms community, and a non-conventional form of the KMR algorithm.

We believe that both [1] and this paper are just the tip of the iceberg in *pattern matching with address errors*. Other reasonable types of address errors, rearrangements or extensions to the proposed address bit errors could and should be considered. As the current set of problems necessitates techniques some of which are new to the classical string matching, it also gives hope for new research directions in the field of pattern matching.

References

1. Amir, A., Aumann, Y., Benson, G., Levy, A., Lipsky, O., Porat, E., Skiena, S., Vishne, U.: Pattern matching with address errors: Rearrangement distances. In: Proc. 17th ACM-SIAM Symp. on Discrete Algorithms (SODA) (2006)

2. Amir, A., Cole, R., Hariharan, R., Lewenstein, M., Porat, E.: Overlap matching. Information and Computation 181(1), 57–74 (2003)

3. Amir, A., Lewenstein, M., Porat, E.: Approximate swapped matching. Information Processing Letters 83(1), 33–39 (2002)

4. Bafna, V., Pevzner, P.A.: Sorting by transpositions. SIAM J. on Discrete Mathematics 11, 221–240 (1998)

5. Berman, P., Hannenhalli, S.: Fast sorting by reversal. In: Hirschberg, D.S., Meyers, G. (eds.) CPM 1996. LNCS, vol. 1075, pp. 168–185. Springer, Heidelberg (1996)

6. Carpara, A.: Sorting by reversals is difficult. In: Proc. 1st Annual Intl. Conf. on Research in Computational Biology (RECOMB), pp. 75–83. ACM Press, New York (1997)

7. Christie, D.A.: Sorting by block-interchanges. Information Processing Letters 60, 165–169 (1996)

8. Cole, R., Hariharan, R.: Verifying candidate matches in sparse and wildcard matching. In: Proc. 34st Annual Symposium on the Theory of Computing (STOC), pp. 592–601 (2002)

9. Hennessy, J.L., Patterson, D.A.: Computer architecture: A quantitative approach, 3rd edn. Morgan Kaufmann, San Francisco (2002)

10. Karp, R., Miller, R., Rosenberg, A.: Rapid identification of repeated patterns in strings, arrays and trees. In: Symposium on the Theory of Computing, vol. 4, pp. 125–136 (1972)

11. Lowrance, R., Wagner, R.A.: An extension of the string-to-string correction problem. J. of the ACM, 177–183 (1975)

12. Motwani, R., Raghavan, P.: Randomized algorithms. Cambridge University Press, Cambridge (1995)

13. Schwartz, J.T.: Fast probabilistic algorithms for verification of polynomial identities. J. of the ACM 27, 701–717 (1980)

14. Zippel, R.: Probabilistic algorithms for sparse polynomials. In: Ng, K.W. (ed.) EUROSAM 1979 and ISSAC 1979. LNCS, vol. 72, pp. 216–226. Springer, Heidelberg (1979)

On-Line Approximate String Matching with Bounded Errors

Marcos Kiwi[1,*], Gonzalo Navarro[2,**], and Claudio Telha[3,***]

[1] Departamento de Ingeniería Matemática, Centro de Modelamiento Matemático
UMI 2807 CNRS-UChile
www.dim.uchile.cl/∼mkiwi
[2] Department of Computer Science, University of Chile
gnavarro@dcc.uchile.cl
[3] Operations Research Center, MIT
ctelha@mit.edu

Abstract. We introduce a new dimension to the widely studied on-line approximate string matching problem, by introducing an *error threshold* parameter ϵ so that the algorithm is allowed to miss occurrences with probability ϵ. This is particularly appropriate for this problem, as approximate searching is used to model many cases where exact answers are not mandatory. We show that the relaxed version of the problem allows us breaking the average-case optimal lower bound of the classical problem, achieving average case $O(n \log_\sigma m/m)$ time with any $\epsilon = \mathrm{poly}(k/m)$, where n is the text size, m the pattern length, k the number of errors for edit distance, and σ the alphabet size. Our experimental results show the practicality of this novel and promising research direction.

1 Introduction

In string matching one is interested in determining the positions (sometimes just deciding the occurrence) of a given pattern P on a text T, where both pattern and text are strings over some fixed finite alphabet Σ of size σ. The lengths of P and T are typically denoted by m and n respectively. In approximate string matching there is also a notion of distance between strings, given say by $d : \Sigma^* \times \Sigma^* \to \mathbf{R}$. One is given an additional non-negative input parameter k and is interested in listing all positions (or just deciding the occurrence) of substrings S of T such that S and P are at distance at most k. In the "on-line" or "sequential" version of the problem, one is not allowed to preprocess the text.

Since the 60's several approaches were proposed for addressing the approximate matching problem, see for example the survey by Navarro [5]. Most of the work focused on the *edit* or *Levenshtein* distance d, which counts the number

---

* Gratefully acknowledges the support of CONICYT via FONDAP in Applied Mathematics and Anillo en Redes ACT08.
** Funded in part by Fondecyt Grant 1-050493, Chile.
*** Gratefully acknowledges the support of CONICYT via Anillo en Redes ACT08 and Yahoo! Research Grant "Compact Data Structures".

P. Ferragina and G. Landau (Eds.): CPM 2008, LNCS 5029, pp. 130–142, 2008.
© Springer-Verlag Berlin Heidelberg 2008

of character insertions, deletions, and substitutions needed to make two strings equal. This distance turns out to be sufficiently powerful to model many relevant applications (e.g., text searching, information retrieval, computational biology, transmission over noisy channels, etc.), and at the same time sufficiently simple to admit efficient solutions (e.g., $O(mn)$ and even $O(kn)$ time).

A lower bound to the (worst-case) problem complexity is obviously $\Omega(n)$ for the meaningful cases, $k < m$. This bound can be reached by using automata, which introduce an extra additive term in the time complexity which is exponential in m or k. If one is restricted to polynomially-bounded time complexities on m and k, however, the worst-case problem complexity is unknown.

Interestingly, the average-case complexity of the problem is well understood. If the characters in P and T are chosen uniformly and independently, the average problem complexity is $\Theta(n(k + \log_\sigma m)/m)$. This was proved in 1994 by Chang and Marr [2], who gave an algorithm reaching the lower bound for $k/m < 1/3 - O(\sigma^{-1/2})$. In 2004, Fredriksson and Navarro [3] gave an improved algorithm achieving the lower bound for $k/m < 1/2 - O(\sigma^{-1/2})$. In addition to covering the range of interesting k values for virtually all applications, the algorithm was shown to be highly practical.

It would seem that, except for determining the worst-case problem complexity (which is mainly of theoretical interest), the on-line approximate string matching problem is closed. In this paper, however, we reopen the problem under a relaxed scenario that is still useful for most applications and admits solutions that beat the lower bound. More precisely, we relax the goal of listing *all* positions where pattern P occurs in the text T to that of listing each such position with probability $1 - \epsilon$, where ϵ is a new input parameter.

There are several relevant scenarios where fast algorithms that make errors (with a user-controlled probability) are appropriate. Obvious cases are those where approximate string matching is used to increase recall when searching data that is intrinsically error-prone. Consider for example an optical character recognition application, where errors will inevitably arise from inaccurate scanning or printing imperfections, or a handwriting recognition application, or a search on a text with typos and misspells. In those cases, there is no hope to find exactly all the correct occurrences of a word. Here, uncertainty of the input translates into approximate pattern matching and approximate searching is used to increase the chance of finding relevant occurrences, hopefully without introducing too many false matches. As the output of the system, even using a correct approximate string matching technique, is an approximation to the ideal answer, a second approximation might be perfectly tolerable, and even welcome if allows for faster searches.

A less obvious application arises in contexts where we might have a priori knowledge that some pattern is either approximately present in the text many times, or does not even approximately occur. Some examples are genetic markers that might often appear or not at all, some modisms that might appear in several forms in certain type of texts, some typical pattern variants that might appear in the denomination of certain drugs, people names, or places, of which typically several instances occur in the same text. Further, we might only be interested in determining whether the pattern occurs or not. (A feature actually available in the well known grep string searching utility as options -1 and -L, and also the approximate string searching utilities agrep and ngrep.) In this context, a text with N approximate pattern occurrences will be misclassified by the inexact algorithm with very low probability, ϵ^N.

Another interesting scenario is that of processing data streams which flow so fast that there is no hope for scanning them exhaustively (e.g. radar derived meteorological data, browser clicks, user queries, IP traffic logs, peer-to-peer downloads, financial data, etc.). Hence even an exact approximate search over part of the data would give only partial results. A faster inexact algorithm could even give better quality answers as it could scan a larger portion of the data, even if making mistakes on it with controlled probability.

The new framework proposed in this work comes from the so-called testing and property testing literature where the aim is to devise sublinear time algorithms obtained by avoiding having to read all of the input of a problem instance. These procedures typically read a very small fraction of the input. For most natural problems the algorithm must use randomization and provide answers which in some sense are approximate, or wrong with some probability. See the many surveys on the topic, e.g. [7,8].

1.1   Main Contributions

We focus in particular on the so-called filtering algorithms [5, § 8]. These algorithms quickly discard areas of the text that cannot approximately match the pattern, and then verify the remaining areas with a classical algorithm. In practice, filtering algorithms are also the fastest approximate string matching algorithms. They also turn out to be natural candidates to design probabilistic variants in this paper.

In Section 3.1 we describe a procedure based on sampling q-grams motivated by the filtering algorithm of Ukkonen [9]. For a fixed constant $t > 0$ and $k < m/\log_\sigma m$, the derived algorithm has an average case complexity of $O(tn \log_\sigma m/m)$ and misses pattern occurrences with probability $\epsilon = O((k \log_\sigma m/m)^t)$. Note that the time equals Yao's lower bound for *exact* string matching ($k = 0$). In contrast, Ukkonen's original algorithm takes $O(n)$ time. In Section 3.2 we describe an algorithm based on Chang and Marr's [2] average-optimal algorithm. For fixed $t > 0$, we derive an $O(tn \log_\sigma m/m)$ average-time approximate matching algorithm with error $\epsilon = O((k/m)^t)$. Note that the latter achieves the same time complexity for a smaller error, and that it works for any $k < m$, whereas the former needs $k < m/\log_\sigma m$.

The discrepancy between both algorithms inherits from that of the original classical algorithms they derive from, where the original differences in time complexities has now translated into their error probabilities. It is important to stress that both algorithms beat the average-complexity lower bound of the problem when errors are not allowed, $\Omega(n(k + \log_\sigma m)/m)$, as they remove the $\Omega(kn/m)$ term in the complexity (the k/m term now shows up in the error probability).

The aforementioned average case complexity results are for random text, but hold even for fixed patterns. Our analyzes focus exclusively on Levenshtein distance d, but should be easily adapted to other metrics.

In Section 4 we present some experimental results that corroborate the theoretical results of Section 3 and give supporting evidence for the practicality of our proposals. In particular, the experiments favor the technique of Section 3.1 over that of Section 3.2, despite the theoretical superiority of the latter.

2 Model for Approximate Searching Allowing Errors

In this section we formalize the main concepts concerning the notion of approximate matching algorithms with errors. We adopt the standard convention of denoting the substring $S_i \ldots S_j$ of $S = S_1 \ldots S_n$ by $S_{i..j}$ and refer to the number of characters of S by the *length of S* which we also denote by $|S|$. We start by recalling the formal definition of the approximate string matching problem when the underlying distance function is d. Henceforth, we abbreviate d-APPROXIMATE STRING MATCHING as d-ASM.

PROBLEM d-APPROXIMATE STRING MATCHING

INPUT Text $T \in \Sigma^*$, pattern $P \in \Sigma^*$ and parameter $k \in \mathbf{N}$.

OUTPUT $S = S(T, P, k) \subseteq \{1, \ldots, n\}$ such that $j \in S$ if and only if there is an i such that $d(T_{i..j}, P) \leq k$.

When the text T and pattern P are both in Σ^*, and the parameter k is in $\mathbf{N}$ we say that (T, P, k) is *an instance of the d-ASM problem*, or simply *an instance* for short. We henceforth refer to $S(T, P, k)$ as the *solution set* of instance (T, P, k). We say that algorithm $\mathcal{A}$ solves the d-ASM problem if on instance (T, P, k) it outputs the solution set $S(T, P, k)$. Note that $\mathcal{A}$ might be a probabilistic algorithm, however its output is fully determined by (T, P, k).

For a randomized algorithm $\mathcal{A}$ that takes as input an instance (T, P, k), let $\mathcal{A}(T, P, k)$ be the distribution over sets $S \subseteq \{1, \ldots, n\}$ that it returns.

Henceforth we denote by $X \leftarrow \mathcal{D}$ the fact that the random variable X is chosen according to distribution $\mathcal{D}$. For a set C we denote the probability that $X \in C$ when X is chosen according to the distribution $\mathcal{D}$ by $\mathbf{Pr}\,[X \in C; X \leftarrow \mathcal{D}]$ or $\mathbf{Pr}_{X \leftarrow \mathcal{D}}\,[X \in C]$. Also, we might simply write $\mathbf{Pr}_X\,[X \in C]$ or $\mathbf{Pr}\,[X \in C]$ when it is clear from context that $X \leftarrow \mathcal{D}$. The notation generalizes in the obvious way to the case where X is a random vector, and/or when instead of a probability one is interested in taking expectation.

We say that randomized algorithm $\mathcal{A}$ *solves the d-ASM problem with (ϵ, ϵ')-error* provided that on any instance (T, P, k) the following holds:

Completeness: if $i \in S(T, P, k)$, then $\mathbf{Pr}\,[i \in S'; S' \leftarrow \mathcal{A}(T, P, k)] \geq 1 - \epsilon$,
Soundness: if $i \notin S(T, P, k)$, then $\mathbf{Pr}\,[i \in S'; S' \leftarrow \mathcal{A}(T, P, k)] \leq \epsilon'$,

where the two probabilities above are taken only over the source of randomness of $\mathcal{A}$.

When $\epsilon' = 0$ we say that $\mathcal{A}$ has *one–sided* ϵ-*error* or that it *is one–sided* for short. When $\epsilon = \epsilon' = 0$ we say that $\mathcal{A}$ is an *errorless* or *exact* algorithm.

We say that randomized algorithm $\mathcal{F}$ is a d-ASM *probabilistic filter with α-error* or simply *is an α-filter* for short, provided that on any instance (W, P, k) the following holds: if $d(P_{i..j}, W) \leq k$ for some pattern substring $P_{i..j}$, then $\mathbf{Pr}\left[\mathcal{F}(W, P, k) = \texttt{Check}\right] \geq 1 - \alpha$, where the probability is taken over the source of randomness of $\mathcal{F}$. If a filter does not return $\texttt{Check}$ we assume without loss of generality that it returns $\texttt{Discard}$.

The notion of an α-filter is crucial to the ensuing discussion. Roughly said, a filter $\mathcal{F}$ will allow us to process a text T by considering non-overlapping consecutive substrings W of T, running the filter on instance (W, P, k) and either: (1) in case the filter returns $\texttt{Check}$, perform a costly approximate string matching procedure to determine whether P approximately occurs in T in the surroundings of window W, or (2) in case the filter does not return $\texttt{Check}$, discard the current window from further consideration and move forward in the text and process the next text window. The previously outlined general mechanism is the basis of the generic algorithm we illustrate in Fig. 1 and describe below. The

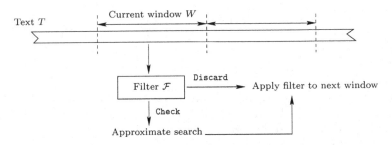

Fig. 1. Generic algorithm d-Approximate String Matching algorithm

attentive reader would have noticed that when defining probabilistic filters we substituted the notation T for texts by W. This is done in order to stress that the probabilistic filters that we will talk about access the text T by sequentially examining substrings of T which we will refer to as *windows*. These windows will typically have a length which is independent of n, more precisely they will be of length $O(m)$.

We now precisely describe the central role played by probabilistic filters in the design of d-ASM algorithms with errors. First, from now on, let w denote $\lfloor (m - k)/2 \rfloor$. Henceforth let $W_1, \ldots, W_s$ be such that $T = W_1 \ldots W_s$ and $|W_p| = w$ (pad T with an additional character not in Σ as necessary). Note that $s = \lceil n/w \rceil$ and $W_p = T_{(p-1)w+1..pw}$. Given any probabilistic filter $\mathcal{F}$ and an exact algorithm $\mathcal{E}$ we can devise a generic d-ASM algorithm with errors such as the one specified in Algorithm 1.[1]

We will shortly show that the generic algorithm $\mathcal{G}$ is correct. We also would like to analyze its complexity in terms of the efficiencies of both the probabilistic filter

---

[1] For $A \subseteq \mathbf{Z}$ we use the standard convention of denoting $\{a + x : x \in A\}$ by $a + A$.

Algorithm 1. Generic d-Approximate String Matching with Errors

```
1: procedure G(T, P, k)                              ▷ T ∈ Σⁿ, P ∈ Σᵐ, k ∈ N
2:     S ← ∅
3:     w ← ⌊(m − k)/2⌋
4:     s ← ⌈n/w⌉
5:     for p ∈ {1, . . . , s} do
6:         if F(Wₚ, P, k) = Check then              ▷ Where Wₚ = T₍ₚ₋₁₎w+1..pw
7:             S ← S ∪ ((pw − m − k + 1) + E(T_{pw−m−k+1..(p−1)w+m+k−1}, P, k))
8:     return S
```

$\mathcal{F}$ and the exact algorithm $\mathcal{E}$. However, we first need to introduce the complexity measures that we will be looking at. Let $\mathbf{Time}_{\mathcal{A}}(T, P, k) \in \mathbf{N} \cup \{+\infty\}$ be the expected time complexity of $\mathcal{A}$ on the instance (T, P, k), where the expectation is taken over the random choices of $\mathcal{A}$. We also associate to $\mathcal{A}$ the following average time complexity measures:

$$\mathbf{Avg}_{\mathcal{A}}(n, P, k) = \mathbf{Ex}_T \left[\mathbf{Time}_{\mathcal{A}}(T, P, k) \right],$$
$$\mathbf{Avg}_{\mathcal{A}}(n, m, k) = \mathbf{Ex}_{T,P} \left[\mathbf{Time}_{\mathcal{A}}(T, P, k) \right].$$

Let $\mathbf{Mem}_{\mathcal{A}}(T, P, k) \in \mathbf{N} \cup \{+\infty\}$ be the maximum amount of memory required by $\mathcal{A}$ on instance (T, P, k), where the maximum is taken over all possible sequences of random bits on which $\mathcal{A}$ may act, and let

$$\mathbf{Mem}_{\mathcal{A}}(n, P, k) = \max_{T \in \Sigma^n} \mathbf{Mem}_{\mathcal{A}}(T, P, k),$$
$$\mathbf{Mem}_{\mathcal{A}}(n, m, k) = \max_{T \in \Sigma^n, P \in \Sigma^m} \mathbf{Mem}_{\mathcal{A}}(T, P, k).$$

We similarly define $\mathbf{Rnd}_{\mathcal{A}}(T, P, k)$, $\mathbf{Rnd}_{\mathcal{A}}(n, P, k)$, and $\mathbf{Rnd}_{\mathcal{A}}(n, m, k)$, but with respect to the maximum number of random bits used by $\mathcal{A}$. Also, the same complexity measures can be defined for probabilistic filters and exact algorithms.

Theorem 1. *Suppose $m > k$. Let $\mathcal{F}$ be an α-filter and let $\mathcal{E}$ be the standard deterministic $O(kn)$ dynamic programming algorithm for the d-ASM problem. Let $w = \lfloor (m − k)/2 \rfloor$, $s = \lceil n/w \rceil$, and $\mathcal{W} \subseteq \Sigma^w$. Then, the generic algorithm $\mathcal{G}$ is a d-ASM algorithm with one-sided α-error such that*

$\mathbf{Avg}_{\mathcal{G}}(n, P, k) \leq s \cdot \mathbf{Avg}_{\mathcal{F}}(w, P, k)$
$+ s \cdot O(mk) \cdot (\mathbf{Pr}_{W \leftarrow \Sigma^w} [W \in \mathcal{W}] + \max_{W \notin \mathcal{W}} \mathbf{Pr} [\mathcal{F}(W, P, k) = \textit{Check}]) + O(s).$

Also, $\mathbf{Mem}_{\mathcal{G}}(n, P, k) = \mathbf{Mem}_{\mathcal{E}}(3w + 4k + 2, P, k)$ (ignoring the space required to output the result), and $\mathbf{Rnd}_{\mathcal{G}}(n, P, k) = O(\frac{n}{m-k}) \cdot \mathbf{Rnd}_{\mathcal{F}}(w, P, k)$.

Proof. First, let us establish completeness of $\mathcal{G}$. Assume $i \in S(T, P, k)$. Let $p + 1$ be the index of the window to which the character T_i belongs. As any occurrence has length at least $m − k$, W_p is completely contained in the occurrence finishing at i, and thus W_p must be at distance at most k of a substring of P. It follows that $\mathcal{F}(W_p, P, k) = \textit{Check}$ with probability at least $1 − \alpha$, in which case line 7 of the algorithm will run an exact verification with $\mathcal{E}$ over a text area comprising any substring of length $m + k$ that contains W_p. Since $m + k$ is the maximum length of an occurrence, it follows that i will be included in the output returned by $\mathcal{G}$. Hence, with probability at least $1 - \alpha$ we have that i is in the output of $\mathcal{G}$.

To establish soundness, assume $i \notin S(T, P, k)$. In this case, i will never be included in the output of $\mathcal{G}$ in line 7 of the algorithm.

We now determine $\mathcal{G}$'s complexity. By linearity of expectation and since $\mathbf{Time}_{\mathcal{E}}(O(m), m, k) = O(mk)$, we have

$$\mathbf{Avg}_{\mathcal{G}}(n, P, k) =$$

$$\sum_{p=1}^{s} \left(\mathbf{Ex}_T \left[\mathbf{Time}_{\mathcal{F}}(W_p, P, k) \right] + O(mk) \cdot \mathbf{Pr}_T \left[\mathcal{F}(W_p, P, k) = \mathtt{Check} \right] + O(1) \right)$$

$$= s \cdot \mathbf{Avg}_{\mathcal{F}}(w, P, k) + O(mk) \cdot \sum_{p=1}^{s} \mathbf{Pr}_T \left[\mathcal{F}(W_p, P, k) = \mathtt{Check} \right] + O(s) .$$

Conditioning according to whether W_p belongs to $\mathcal{W}$, we get for any $\mathcal{W}$ that

$$\mathbf{Pr}_T \left[\mathcal{F}(W_p, P, k) = \mathtt{Check} \right] \leq \mathbf{Pr}_{W \leftarrow \Sigma^w} [W \in \mathcal{W}] + \max_{W \notin \mathcal{W}} \mathbf{Pr} \left[\mathcal{F}(W, P, k) = \mathtt{Check} \right] .$$

The stated bound on $\mathbf{Avg}_{\mathcal{G}}(n, P, k)$ follows immediately. The memory and randomized complexity bounds are obvious.   □

The intuition behind the preceding theorem is that, given any class $\mathcal{W}$ of "interesting" windows, if we have a filter that discards the uninteresting windows with high probability, then the probability that the algorithm has to verify a given text window can be bounded by the sum of two probabilities: (i) that of the window being interesting, (ii) the maximum probability that the filter fails to discard a noninteresting window. As such, the theorem gives a general framework to analyze probabilistic filtration algorithms. An immediate consequence of the result is the following:

Corollary 1. *Under the same conditions as in Theorem 1, if in addition*

$$\mathbf{Pr}_{W \leftarrow \Sigma^w} [W \in \mathcal{W}] = \max_{W \notin \mathcal{W}} \mathbf{Pr} \left[\mathcal{F}(W, P, k) = \mathit{Check} \right] = O \left(1/m^2 \right) ,$$

then $\mathbf{Avg}_{\mathcal{G}}(n, P, k) = O(s \cdot \mathbf{Avg}_{\mathcal{F}}(w, P, k))$. *This also holds if* $\mathcal{E}$ *is the classical* $O(m^2)$ *time algorithm.*

The previous results suggests an obvious strategy for the design of d-ASM algorithms with errors. Indeed, it suffices to identify a small subset of windows $\mathcal{W} \subseteq \Sigma^w$ that contain all windows of length w that are at distance at most k of a pattern substring, and then design a filter $\mathcal{F}$ such that: (1) the probability that $\mathcal{F}(W, P, k) = \mathtt{Check}$ is high when $W \in \mathcal{W}$ (in order not to miss pattern occurrences), and (2) the probability that $\mathcal{F}(W, P, k) = \mathtt{Check}$ is low when $W \notin \mathcal{W}$ (in order to avoid running an expensive procedure over regions of the text where there are no pattern occurrences).

The next result is a simple observation whose proof we omit since it follows by standard methods (running $\mathcal{A}$ repeatedly).

Proposition 1. *Let $\mathcal{A}$ be a randomized algorithm that solves the d-ASM problem with (ϵ, ϵ')-error.*

- *Let $\alpha \leq \epsilon = \epsilon' < 1/2$ and $N = O(\log(1/\alpha)/(1-2\epsilon)^2)$. Then, there is a randomized algorithm $\mathcal{A}'$ that solves the d-ASM problem with (α, α)-error such that $\mathbf{Avg}_{\mathcal{A}'}(n, P, k) = N \cdot \mathbf{Avg}_{\mathcal{A}}(n, P, k)$, $\mathbf{Mem}_{\mathcal{A}'}(n, P, k) = \mathbf{Mem}_{\mathcal{A}}(n, P, k) + O(\log N)$, and where $\mathbf{Rnd}_{\mathcal{A}'}(n, P, k) = N \cdot \mathbf{Rnd}_{\mathcal{A}}(n, P, k)$.*
- *If $\mathcal{A}$ is one-sided, then there is a randomized algorithm $\mathcal{A}'$ solving the d-ASM problem with $(\epsilon^N, 0)$-error such that $\mathbf{Avg}_{\mathcal{A}'}(n, P, k) = N \cdot \mathbf{Avg}_{\mathcal{A}}(n, P, k)$, $\mathbf{Mem}_{\mathcal{A}'}(n, P, k) = \mathbf{Mem}_{\mathcal{A}}(n, P, k) + O(\log N)$, and $\mathbf{Rnd}_{\mathcal{A}'}(n, P, k) = N \cdot \mathbf{Rnd}_{\mathcal{A}}(n, P, k)$.*

3   Algorithms for Approximate Searching with Errors

In this section we derive two probabilistic filters inspired on existing (errorless) filtration algorithms. Note that, according to the previous section, we focus on the design of the window filters, and the rest follows from the general framework.

3.1   Algorithm Based on q-Gram Sampling

A q-gram is a substring of length q. Thus, a pattern of length m has $(m - q + 1)$ overlapping q-grams. Each error can alter at most q of the q-grams of the pattern, and therefore $(m - q + 1 - kq)$ pattern q-grams must appear in any approximate occurrence of the pattern in the text. Ukkonen's idea [9] is to sequentially scan the text while keeping count of the last q-grams seen. The counting is done using a suffix tree of P and keeping the relevant information attached to the $m - q + 1$ important nodes at depth q in the suffix tree. The key intuition behind the algorithms design is that in random text it is difficult to find substrings of the pattern of length $q > \log_\sigma m$. The opposite is true in zones of the text where the pattern approximately occurs. Hence, by keeping count of the last q-grams seen one may quickly filter out many bad pattern alignments.

We now show how to adapt the ideas mentioned so far in order to design a probabilistic filter. The filtering procedure randomly chooses several indices $i \in \{1, \ldots, |W| - q + 1\}$ and checks whether the q-gram $W_{i..i+q-1}$ is a pattern substring. Depending on the number of q-grams that are present in the pattern the filter decides whether or not to discard the window. See Algorithm 2 for a formal description of the derived probabilistic filter **Q-PE-**$\mathcal{F}_{c,\rho,q}$, where c and ρ are parameters to be tuned later. Using the filter as a subroutine for the generic algorithm with errors described in Algorithm 1 gives rise to a procedure to which we will henceforth refer to as **Q-PE**.

Let $\mathcal{W}$ be the collection of all windows in Σ^w for which at least β of its q-grams are substrings of the pattern. Let $w' = w - q + 1$ be the number of q-grams (counting repetitions) in a window of length w. Finally, let p denote the probability that a randomly chosen q-gram is a substring of the pattern P, i.e.

$$p = \frac{1}{\sigma^q} \cdot |\{P_{i..i+q-1} : i = 1, \ldots, m - q + 1\}| .$$

Algorithm 2. Probabilistic filter based on q-grams

1: **procedure** Q-PE-$\mathcal{F}_{c,\rho,q}(W,P,k)$ $\triangleright W \in \Sigma^w, P \in \Sigma^m, k \in \mathbf{N}$
2:    $ctr \leftarrow 0$
3: **for** $i \in \{1,\ldots,c\}$ **do**
4:      Choose j_i uniformly at random in $\{1,\ldots,|W|-q+1\}$
5:      **if** $W_{j_i \cdots j_i+q-1}$ is a substring of P **then** $ctr \leftarrow ctr + 1$
6: **if** $ctr > \rho \cdot c$ **then return** Check **else return** Discard

---

The following result shows that a window chosen randomly in Σ^w is unlikely to be in $\mathcal{W}$.

Lemma 1. Let $\beta \geq pw'$. Then, $\mathbf{Pr}_{W \leftarrow \Sigma^w}[W \in \mathcal{W}] \leq \exp\left(-\dfrac{24(\beta - pw')^2}{25q(\beta + 2pw')}\right)$.

Proof. For $i = 1,\ldots,w'$ let Y_i be the indicator variable of the event "$W_{i..i+q-1}$ is a substring of P" when W is randomly chosen in Σ^w. Clearly, $\mathbf{Ex}[Y_i] = p$. Moreover, $W \in \mathcal{W}$ if and only if $\sum_{i=1}^{w'} Y_i \geq \beta$. Unfortunately, a standard Chernoff type bound cannot be directly applied given that the Y_i's are not independent. Nevertheless, the collection $\{Y_1,\ldots,Y_{w'}\}$ can be partitioned into q families according to $i \bmod q$, each one an independent family of variables. The desired result follows applying a Chernoff type bound for so called q-independent families [4, Corollary 2.4]. $\square$

Lemma 2. If $W \notin \mathcal{W}$, then

$$\mathbf{Pr}\left[\text{Q-PE-}\mathcal{F}_{\rho,c,q}(W,P,k) = \text{Check}\right] \leq \exp\left(\rho c - \frac{c\beta}{w'}\right)\left(\frac{\beta}{\rho w'}\right)^{\rho c}.$$

Proof. Let X_{j_i} denote the indicator of whether $W_{j_i..j_i+1-1}$ turns out to be a substring of the pattern P in line 5 of the description of Q-PE-$\mathcal{F}_{\rho,c,q}$. Note that the X_{j_i}'s are independent, each with expectation at most β/w' when $W \notin \mathcal{W}$. The claim follows by a standard Chernoff type bound from the fact that:

$$\mathbf{Pr}_{W \leftarrow \Sigma^w}\left[\text{Q-PE-}\mathcal{F}_{\rho,c,q}(W,P,k) = \text{Check}\right] = \mathbf{Pr}\left[\sum_{i=1}^{c} X_{j_i} \geq \rho \cdot c\right],$$

where the probabilities are taken exclusively over the sequence of random bits of the probabilistic filter. $\square$

Lemma 3. If $kq \leq w'(1-\rho)$, then Q-PE-$\mathcal{F}_{\rho,c,q}$ is an α-filter for

$$\alpha \leq \exp\left((1-\rho)c - \frac{ckq}{w'}\right)\left(\frac{kq}{w'(1-\rho)}\right)^{c(1-\rho)}.$$

Proof. Let $W \in \Sigma^w$. Assume $d(P_{i..j}, W) \leq k$ for some pattern substring $P_{i..j}$. Then, at least $w' - kq$ of W's q-grams are substrings of P. Defining X_{j_i} as in Lemma 2 we still have that the X_{j_i}'s are independent but now their expectation is at least $1 - kq/w'$. The claim follows by a standard Chernoff type bound from the fact that:

$$\mathbf{Pr}\left[\mathbf{Q\text{-}PE\text{-}}\mathcal{F}_{\rho,c,q}(W, P, k) = \texttt{Discard}\right] = \mathbf{Pr}\left[\sum_{i=1}^{c} X_{j_i} \leq \rho \cdot c\right],$$

where the probabilities are taken exclusively over the sequence of random bits of the probabilistic filter. □

Theorem 2. *If $k < (m - 2\log_\sigma m)/(1 + 4\log_\sigma m)$, then $\mathbf{Q\text{-}PE}$ is a d-ASM algorithm with one-sided error $\epsilon = O((k\log_\sigma m/m)^t)$ for any constant $t > 0$, running in average time $\mathbf{Avg_{Q\text{-}PE}}(n, P, k) = O(tn\log_\sigma m/m)$.*

Proof. The result follows from Theorem 1 and Corollary 1.

Choose $q = 2\lceil\log_\sigma m\rceil$, so $p \leq m/\sigma^q \leq 1/m$. Taking $\beta = \Theta(\log^2 m)$ where the hidden constant is sufficiently large, we have by Lemma 1 that $\mathbf{Pr}_{W\leftarrow w}[W \in \mathcal{W}]$ $= O(1/m^2)$. By Lemma 2 and taking $\rho = 1/2$ and c a sufficiently large constant, we get that $\mathbf{Pr}\left[\mathbf{Q\text{-}PE\text{-}}\mathcal{F}_{\rho,c,q}(W, P, k) = \texttt{Check}\right] = O(1/m^2)$ when $W \notin \mathcal{W}$.

Now, let $k^* = w'(1 - \rho)/q$ and observe that $k < k^*$ satisfies the hypothesis of Lemma 3. Choose $c(1 - \rho) \geq t$ and note that $kq/((1-\rho)w') = 4k\log_\sigma m/(m - k - 2\log_\sigma m)$. Lemma 3 thus implies that $\mathbf{Q\text{-}PE\text{-}}\mathcal{F}_{\rho,c,q}$ has $O((k\log_\sigma m/m)^t)$-error.

Clearly $\mathbf{Avg_{Q\text{-}PE\text{-}}}_{\mathcal{F}_{\rho,c,q}}(w, P, k) = \mathbf{Time_{Q\text{-}PE\text{-}}}_{\mathcal{F}_{\rho,c,q}}(W, P, k) = O(cq) = O(t\log_\sigma m)$. □

3.2 Algorithm Based on Covering by Pattern Substrings

In 1994 Chang and Marr [2] proposed a variant of SET [1] with running time $O(n(k + \log_\sigma m)/m)$ for $k/m \leq 1/3 - O(\sigma^{-1/2})$. As in SET, Chang and Marr consider blocks of text of size $(m - k)/2$, and pinpoint occurrences of the pattern by identifying blocks that approximately match a substring of the pattern. This identification is based on splitting the text into contiguous substrings of length $\ell = t\log_\sigma m$ and sequentially searching the text substrings of length ℓ in the pattern allowing errors. The sequential search continues until the total number of errors accumulated exceeds k. If k errors occur before $(m-k)/2$ text characters are covered, then the rest of the window can be safely skipped.

The adaptation of Chang and Marr's approach to the design of probabilistic filters is quite natural. Indeed, instead of looking at ℓ-grams sequentially we just randomly choose sufficiently many non-overlapping ℓ-substrings in each block. We then determine the fraction of them that approximately appear in the pattern. If this fraction is small enough, then the block is discarded. See Algorithm 3 for a formal description of the derived probabilistic filter $\mathbf{CM\text{-}PE\text{-}}\mathcal{F}_{c,\rho,\ell,g}$. Using the filter as a subroutine for the generic algorithm with errors described in Algorithm 1 gives rise to a procedure to which we will henceforth refer to as $\mathbf{CM\text{-}PE}$.

Remark 1. Note that $asm(S, P)$ of Algorithm 3 can be precomputed for all values of $S \in \Sigma^\ell$.

Algorithm 3. Probabilistic filter based on covering by pattern substrings

---

1: **procedure CM-PE-$\mathcal{F}_{c,\rho,\ell,g}(W,P,k)$**   ▷ $W \in \Sigma^w, P \in \Sigma^m, k \in \mathbf{N}$
2:      $ctr \leftarrow 0$
3:      **for** $i \in \{1,\dots,c\}$ **do**
4:          Choose j_i uniformly at random in $\{1,\dots,\lfloor w/\ell \rfloor\}$
5:          **if** $asm(W_{(j_i-1)\ell+1..j_i\ell}, P) \leq g$ ▷ $asm(S,P) = \min_{a \leq b} d(S, P_{a..b})$
6:              **then** $ctr \leftarrow ctr + 1$
7:      **if** $ctr > \rho \cdot c$ **then return** Check **else return** Discard

---

The analysis of Algorithm 3 establishes results such as Lemmas 1-3, but concerning **CM-PE-$\mathcal{F}_{\rho,c,\ell,g}$**. We can derive the following (proof omitted due to lack of space):

Theorem 3. *If $k < m/5$, then* **CM-PE** *is a d-ASM algorithm with one sided error $\epsilon = (4k/(m-k))^t$, for any constant $t > 0$. Its average running time is* **$\text{Avg}_{\textbf{Q-PE}}(n, P, k) = O(tn \log_\sigma m/m)$.**

4   Experimental Results

We implemented the algorithms of Sections 3.1 and 3.2. We extracted three real-life texts of 50MB from *Pizza&Chili* (http://pizzachili.dcc.uchile.cl): English text, DNA, and MIDI pitches. We used patterns of length 50 and 100, randomly extracted from the text, and some meaningful k values. Each data point is the average over 50 such search patterns, repeating each search 15 times in the case

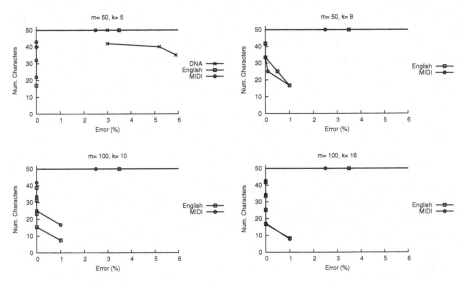

Fig. 2. Experimental results for **Q-PE**. Straight horizontal lines correspond to the errorless version. The y axis represents the number of character inspections times 1024.

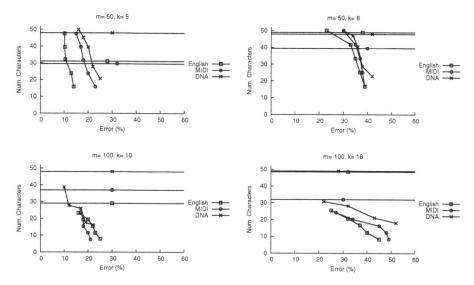

Fig. 3. Experimental results for **CM-PE**. Straight horizontal lines correspond to the errorless version. The y axis represents the number of character inspections times 1024.

of the probabilistic algorithms. We measured the average number of character inspections and the average percentage of missed occurrences.

We used the following setup for the algorithms. For q-gram algorithms (Section 3.1), we used $q = 4$. Our preliminary results show that $\rho = 0.7$ is a good choice. For covering by pattern substrings (Section 3.2), we used $\epsilon = 0.2$ and $\rho = 0.3$. In our algorithms, we only moved parameter c in order to change the accuracy/time trade-off. We compared our algorithms with the corresponding errorless filtering algorithms.

Figure 2 shows the experimental results for the q-gram based procedure, and Fig. 3 for the covering by pattern substrings process. The errorless version of the q-grams algorithm inspects all text characters. In contrast, our q-gram based procedure achieves less than 1% error rate and looks at up to 6 times less characters on English and MIDI corpora. For our second algorithmic proposal, the result of the comparison against the errorless version is not as good. Nevertheless, we emphasize that it beats the average-optimal (errorless) algorithm by a wide margin, specifically it inspects about half the characters with 15% errors on the English corpus.

5   Final Comments

In this paper we have advocated considering a new dimension of the approximate string matching problem, namely the probability of missing an approximate occurrence. This relaxation is particularly natural for a problem that usually arises when modeling processes where errors have to be tolerated, and it opens the door to novel approaches to approximate string matching which break the average-case lower bound of the original problem. In particular, we have shown that much faster text scanning is possible if one allows a small probability of missing occurrences. We achieved $O(n \log_\sigma m/m)$ time (which is the complexity of exact string matching, $k = 0$) with error probability bounded by any polynomial in k/m. Empirically, we have shown that our algorithms inspect a fraction of the text with virtually no mistakes.

We have just scratched the surface of this new area. In particular, we have not considered filtration algorithms that use sliding instead of fixed windows. Sliding-window algorithms have the potential of being more efficient (cf. Fredriksson and Navarro's variant [3] with the original Chang and Marr's average-optimal algorithms [2]). It is not hard to design those variants, yet analyzing them is more challenging. On the other hand, it is rather simple to extend our techniques to multiple ASM. We also applied the techniques to indexed algorithms, where the text can be preprocessed [6]. Several indexes build on sequential filtration algorithms, and thus adapting them is rather natural.

Finally, it is interesting to determine the average complexity of this relaxed problem, considering the error probability ϵ in the formula. This would give an idea of how much can one gain by allowing errors in the outcome of the search. For example, our algorithms break the $\Omega(nk/m)$ term in the problem complexity, yet a term $\text{poly}(k/m)$ appears in the error probability. Which are the best tradeoffs one can achieve?

References

1. Chang, W., Lawler, E.: Sublinear approximate string matching and biological applications. Algorithmica 12(4-5), 327–344 (1994)
2. Chang, W., Marr, T.: Approximate string matching and local similarity. In: Proceedings of the 5th Annual Symposium on Combinatorial Pattern Matching, pp. 259–273. Springer, Heidelberg (1994)
3. Fredriksson, K., Navarro, G.: Average-optimal single and multiple approximate string matching. ACM Journal of Experimental Algorithmics (article 1.4) 9 (2004)
4. Janson, S.: Large deviations for sums of partly dependent random variables. Random Structure & Algorithms 24(3), 234–248 (2004)
5. Navarro, G.: A guided tour to approximate string matching. ACM Computing Surveys 33(1), 31–88 (2001)
6. Navarro, G., Baeza-Yates, R., Sutinen, E., Tarhio, J.: Indexing methods for approximate string matching. IEEE Data Engineering Bulletin 24(4), 19–27 (2001)
7. Ron, D.: Property Testing. In: Handbook of Randomized Computing, volume II of Combinatorial Optimization, vol. 9. Springer, Heidelberg (2001)
8. Rubinfeld, R., Kumar, R.: Algorithms column: Sublinear time algorithms. SIGACT News 34(4), 57–67 (2003)
9. Ukkonen, E.: Approximate string-matching with q-grams and maximal matches. Theoretical Computer Science 92, 191–211 (1992)

A Black Box for Online Approximate Pattern Matching

Raphaël Clifford[1], Klim Efremenko[2], Benny Porat[3], and Ely Porat[3]

[1] University of Bristol, Dept. of Computer Science, Bristol, BS8 1UB, UK
`clifford@cs.bris.ac.uk`
[2] Bar-Ilan University, Dept. of Computer Science, 52900 Ramat-Gan and Weizman
Institute, Dept. of Computer Science and Applied Mathematics, Rehovot, Israel
`klimefrem@gmail.com`
[3] Bar-Ilan University, Dept. of Computer Science, 52900 Ramat-Gan, Israel
`bennyporat@gmail.com, porately@cs.biu.ac.il`

Abstract. We present a deterministic black box solution for online approximate matching. Given a pattern of length m and a streaming text of length n that arrives one character at a time, the task is to report the distance between the pattern and a sliding window of the text as soon as the new character arrives. Our solution requires $O(\Sigma_{j=1}^{\log_2 m} T(n, 2^{j-1})/n)$ time for each input character, where $T(n, m)$ is the total running time of the best offline algorithm. The types of approximation that are supported include exact matching with wildcards, matching under the Hamming norm, approximating the Hamming norm, k-mismatch and numerical measures such as the L_2 and L_1 norms. For these examples, the resulting online algorithms take $O(\log^2 m)$, $O(\sqrt{m \log m})$, $O(\log^2 m/\epsilon^2)$, $O(\sqrt{k \log k} \log m)$, $O(\log^2 m)$ and $O(\sqrt{m \log m})$ time per character respectively. The space overhead is $O(m)$ which we show is optimal.

1 Introduction

Fast approximate string matching is a central problem of modern data intensive applications. Its applications are many and varied, from computational biology and large scale web searching to searching multimedia databases and digital libraries. As a result, string matching has to continuously adapt itself to the problem at hand. Simultaneously, the need for asymptotically fast algorithms grows every year with the explosion of data available in digital form.

A great deal of progress has been made in finding fast algorithms for a variety of important forms of approximate matching. One of the most studied is the Hamming distance which measures the number of mismatches between two strings. Given a text t of length n and a pattern p of length m, the task is to report the Hamming distance at every possible alignment. $O(n\sqrt{m \log m})$ time solutions based on repeated applications of the FFT were given independently by both Abrahamson and Kosaraju in 1987 [1, 17]. Particular interest has been paid to a bounded version of this problem called the k-mismatch problem. Here a bound k is given and we need only report the Hamming distance if it is less

P. Ferragina and G. Landau (Eds.): CPM 2008, LNCS 5029, pp. 143–151, 2008.
© Springer-Verlag Berlin Heidelberg 2008

than or equal to k. If the number of mismatches is greater than the bound, the algorithm need only report that fact and not give the actual Hamming distance. In 1985 Landau and Vishkin gave a beautiful $O(nk)$ algorithm that is not FFT based which uses constant time LCA operations on the suffix tree of p and t [18]. This was subsequently improved to $O(n\sqrt{k \log k})$ time by a method based on filtering and FFTs again [4]. Approximations within a multiplicative factor of $(1 + \epsilon)$ to the Hamming distance can also be found in $O(n/\epsilon^2 \log m)$ time [15].

The problem of determining the time complexity of exact matching with don't cares has also been well studied over many years [13, 15, 16, 12, 6], culminating in two related deterministic $O(n \log m)$ time solutions. This has been accompanied by recent advances for the problem of k-mismatch problem with don't cares [10, 9] as well as a surge in interest in provably fast algorithms for distance calculation and approximate matching between numerical strings. Many different metrics have been considered, with for example the L_1 distance [5, 7, 3] and less-than matching [2] problems both being solvable in $O(n\sqrt{m \log m})$ time and a bounded version of the L_∞ norm which was first discussed in [8] and then improved in [7, 19] requiring $O(\delta n \log m)$ time.

In almost every one of these cases and in many others beside, the algorithms make extensive use of the fast Fourier transform (FFT). The property of the FFT that is required is that in the RAM model, the cross-correlation,

$$(t \otimes p)[i] \stackrel{\text{def}}{=} \sum_{j=1}^{m} p_j t_{i+j-1}, \ \ 0 \le i \le n - m + 1,$$

can be calculated accurately and efficiently in $O(n \log n)$ time (see e.g. [11], Chapter 32). By a standard trick of splitting the text into overlapping substrings of length $2m$, the running time can be further reduced to $O(n \log m)$.

Although the FFT is a very powerful and successful tool, it also brings with it a number of disadvantages. Perhaps most significant of these in this context is that the cross-correlation computation using the FFT is very much an offline algorithm. It requires the entire pattern and text to be available before any search can be performed. Of course, it is not only the FFT that causes this difficulty. For example, the only fast algorithm for the k-mismatch algorithm which does not employ the FFT uses constant time LCA queries [18]. As it is not known how to perform the necessary preprocessing of a suffix tree to compute the LCA online, this algorithm suffers from the same limitations as those that depend on the FFT.

In many situations such as when monitoring Internet traffic or telecommunications networks this model of computation may not be feasible. It is not sufficient simply that a pattern matching algorithm runs fast. It should also require considerably less space than the input and update at least as quickly as the new data are arriving while still maintaining an overall time complexity which is as close as possible to the full offline algorithm. One approach to handle this situation is the data streaming model where it is assumed that it is not possible to ever store all the data seen and in some variants that only one pass over the data is ever allowed. This very successful model has been the source of a great deal of attention in recent years (see [14] and [20] for background on data stream computation), however the techniques developed have largely been randomised whereas our interest is in deterministic solutions.

Our main contribution is a black box for converting offline approximate matching algorithms into efficient online ones. That is, it ensures that the approximate matching algorithm accomplishes its task for the ith input character without requiring the $i + 1$th. The method is deterministic and bounds the worst case running time *per input character* as well as ensuring that the overall running time is within a log factor of the best known offline algorithm. It is an important feature of our method that its running time is not amortised. This is because when processing streaming data it may not be realistic to wait for long periods of time between individual input characters.

Our technique can be applied to a wide class of approximate matching algorithms overcoming one of the main restrictions on their use in data streaming applications. A particularly useful subset that we focus on in this paper includes problems whose distance function Δ is defined so that $\Delta(x, y) = \Sigma_{j=1}^{m} \Delta(x_j, y_j)$, for strings x and y and $|x| = |y| = m$. In other words, distance functions between two strings where the distance is simply measured as the sum of the distances between individual symbols. Many of the most common and widely studied approximate matching problems fall into this category including exact matching with wildcards, matching under the Hamming norm, k-mismatch and matching under the L_2 and L_1 norms. As a result, we provide fast deterministic online algorithms for each one of these problems.

The overall structure of the paper is as follows. In Section 2 we summarise the main results of the paper. In Section 3 we present the main black box solution and in Section 4 we discuss space lower bounds. Finally we conclude with some open problems in Section 5.

2   Our Results

Our black box approach converts an offline approximate matching algorithm into efficient online algorithm. Let $T(n, m)$ be the total running time of the best known offline approximate matching algorithm for the problem being considered. The main results we present are as follows:

– We show how offline approximate matching algorithms can be turned into online algorithms with strict bounds on the computation time per input character. The main idea is to split the pattern into $O(\log m)$ subpatterns of successively halving length and to perform searches in parallel on carefully chosen partitions of the text. The partitions are chosen so that the work needed to compute the distance to a sliding window of the text is started $O(m)$ characters before it is needed.

   Specifically, for each subpattern of length m', we start the approximate matching algorithm $m'/2$ characters before its result is required. This work is carried out in parallel (by time slicing for example) with searches involving a subset of the remaining subpatterns. An auxiliary array of size $O(m)$

is sufficient to keep track of the cumulative counts of the distances found so far. The online algorithm takes $O(\Sigma_{j=1}^{\log_2 m} T(n, 2^{j-1})/n)$ time per input character. As $\Sigma_{j=1}^{\log_2 m} T(n, 2^{j-1}) \leq T(n, m) \log(m)$, this gives a near optimal deterministic solution to a wide class of online approximate matching problems.

– A small adjustment to the algorithm allows us to report the distance to a sliding window in the text in constant time after a new character arrives. Although the computation time per character is unchanged, we are able to move the majority of the work for future symbols until after a new symbol has been processed. This provides a solution for online approximate matching in a model where instant answers are needed once new data arrives.

– Applications of our black box method to exact matching with wildcards, matching under the Hamming norm, approximating the Hamming norm, k-mismatch and matching under the L_2 and L_1 norms result in algorithms that take $O(\log^2 m)$, $O(\sqrt{m \log m})$, $O(\log^2 m/\epsilon^2)$, $O(\sqrt{k \log k} \log m)$, $O(\log^2 m)$ and $O(\sqrt{m \log m})$ time per character respectively.

– Finally we argue that the space requirements for the online approximate matching problem are optimal in the deterministic setting under the assumption that the offline matching algorithm requires $O(m)$ space. This follows immediately from an $\Omega(m)$ communication complexity lower bound for computing "sum-type" functions between strings.

3 The Black Box for Online Approximate Matching

The black box we present will make repeated calls to an offline approximate pattern matching algorithm which we call *offline-pm*. In order to simplify some of the explanation, we assume that the running time $T(n, m)$ of offline-pm can be expressed as $nT'(m)$ with $T'(m) \in O(m)$. This assumption is reasonable as the types of pattern matching problem we consider can all be solved naively in $O(nm)$ time. As a result of this simplification we have $O(n/mT(cm, m)) = O(T(n, m))$ for constant $c > 1$. We will also at times refer to a call to offline-pm as a search for the sake of brevity.

The basic idea of our black box is to split the pattern into $O(\log m)$ consecutive subpatterns each having half the length of the previous one. In this way $P_1 = p[1, \ldots, m/2]$ and subpattern P_j has length $m2^{-j}$ for $1 \leq j \leq \log_2(m)$. $P_{\log_2(m)+1}$ is set to be the last character of the pattern. We then run offline-pm for each subpattern against the whole of the text. The distances found can then be added to an auxiliary array C. Specifically, for any subpattern starting at position j of the pattern, its distance to a substring starting at position i of the text will be added to the count at $C[i - j + 1]$. At the end of this step C will contain $\Delta(p, t[i, \ldots, i + m - 1])$ for every location i in t.

This algorithm will call offline-pm $O(\log m)$ times and requires $O(n)$ extra space for the auxiliary array. The space requirement can be reduced to $O(m)$ by partitioning the text. For any subpattern of length m', we partition the text into $n/(m' - 1)$ overlapping substrings of length $2m'$, each with an overlap of length m' with the previous partition. If we run offline-pm on each partition separately, the total time complexity over the whole text for each subpattern is $O((n/m)T(2m,m)) = O(T(n,m))$. The distances for each subpattern can be added to the auxiliary C in the same way as before. However, now we only need store one auxiliary array of size m at most.

This space reduced algorithm can easily be made online by a lazy execution of the searches performed. For each subpattern P_j there is a list of locations associated with it which marks out the start and end of its associated partitions of the text. These locations do not have to be stored explicilty as they are easily computable as they are needed. For example, for $P_1 = p[1, \ldots, m/2]$, the first partition is $t[1, \ldots, m]$, the second is $t[m/2, 3m/2]$ and so on. When the ith character is read in from the text, offline-pm can now perform the searches for all the subpatterns which have a partition finishing at position i. This online algorithm runs in $O(\Sigma_{j=1}^{\log_2 m} T(n, 2^{j-1}))$ time and $O(m)$ extra space. However the solution is not yet satisfactory as we might potentially have to wait for more than $T(m, m/2)$ time *after* a new character arrives before we are able to compute the distance to the new sliding window. As an example, when the mth character of t is read in, the first search involving P_1 commences for the partition $t[1, \ldots, m]$, thereby delaying the computation of $\Delta(p, t[1, \ldots, m])$ unacceptably. Our aim is to ensure that the maximum computation time per character is limited to $O(\Sigma_{j=1}^{\log_2 m} T(n, 2^{j-1})/n)$.

Bounding the Maximum Time per Character

We can bound the maximum time per character by performing more work earlier on. Instead of splitting the text into partitions of size $2m'$ per subpattern, we change the partition size to $3m'/2$. The overlap between partitions is maintained at m' to ensure no matches are missed.

If the ith character of the text is read in, searches will now be performed for all subpatterns which have a partition of the text ending at position i. The result of a search involving a subpattern will now not be needed until the $i + m'/2$th character is read in. That is $m'/2$ characters after the relevant search has been performed. In this way, whenever a new character is read in, the results for all the subpatterns needed to compute the distance to the new sliding window are already known, except for the last character of the pattern. This last comparison can be carried out in constant time. Figure 1 shows the partitioning of the text for the first subpattern P_1 and the start and finish times of the searches performed.

In order to guarantee the desired upper bound for the amount of work carried out per character, we need to spread the work out evenly over the time it takes to input the text. For each subpattern, the work for a particular search does not have to be completed until $m'/2$ characters after it starts and so we can set this work to be performed over the period between reading in the ith and the $i + m'/2$th character. As an example, the first search for P_1 starts when $t[3m/4]$ is read in and completes at the time that $t[m]$ is. P_2 will on the other hand start to calculate its contribution to $\Delta(p, t[1, \ldots, m])$ when $t[7m/8]$ is read in. The result from these searches and from all the other subpatterns, will be scheduled

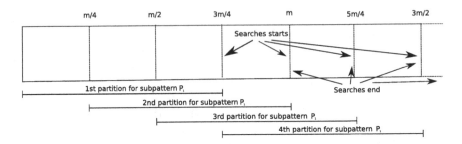

Fig. 1. Partitioning of the text for subpattern $P_1 = p[1, \ldots, m/2]$

to complete by the time $t[m]$ arrives. As there will be a number of searches scheduled to work in this period we will need to perform the work in parallel by time slicing. To guarantee the bound on the work per character without requiring any internal knowledge of offline-pm, we only require an upper bound for the running time of offline-pm for a subpattern of length m' which we then divide by $m'/2$. This gives the amount of work to carry out per input character.

Algorithm 1 gives an overview of the whole process.

Input: Pattern p, a the streaming text t and offline-pm
Output: $\Delta(p, t[i - m + 1, \ldots, i])$ for each $i \geq m$ of streaming text
Initialisation;
    Split p into $\log_2 m$ subpatterns P_j of length $m/2^j$, for $1 \leq j \leq \log_2 m$;
    For each P_j, calculate its partition start and end points;
foreach *symbol $t[i]$ read in* **do**
    Add $\Delta(p[m], t[i])$ to $C[i - m + 1]$;
    Wait for results of offline-pm searches due to end at position i;
    Output $C[i - m + 1]$;
    Start offline-pm searches for each subpattern P_j which has a partition ending at i;
end

Algorithm 1. Black box algorithm for online pattern matching

By making a small adjustment to the scheduling of the work we can also guarantee that all but a constant amount of the work needed to compute $\Delta(p, t[i - m + 1, \ldots, i])$ will have been completed *before* the ith character is read in. Although no change is made to the total work per character, the ability to control at which point work is carried out can have applications when data arrives in bursts. For example, if there is some pause in the data stream before new characters arrive. The following Theorem summarises the main result.

Theorem 1. *Algorithm 1 solves the online approximate problem in $O(\Sigma_{j=1}^{\log_2 m} T(n, 2^{j-1})/n)$ time per input character and $O(m)$ space. Further, with a small*

modification it can report the distance to a new sliding window in constant time after a text symbol arrives.

Proof. The total time taken by Algorithm 1 is $O(T(n, m'))$ per subpattern of length m' making a total of $O(\Sigma_{j=1}^{\log_2 m} T(n, 2^{j-1}))$ time overall for all subpatterns. The work performed by the calls to offline-pm is evenly spread over the whole length of the text. Therefore, the total amount of work per character is $O(\Sigma_{j=1}^{\log_2 m} T(n, 2^{j-1})/n)$.

The space required for the auxiliary array is $O(m)$. We also have to consider the space overhead of offline-pm as there can be $O(\log m)$ searches running simultaneously. Under the assumption that each individual search requires $O(m)$ space, the total space requirement is less than $\Sigma_{j=1}^{\log_2 m}(c2^{j-1}) = O(m)$ overall.

In order to output $\Delta(p, t[i - m + 1, \ldots, i])$ in constant time after $t[i]$ is read in we need only ensure that a search is completed $m'/2 - 1$ (rather than $m'/2$) characters after the end of a partition. In this way, when the ith character is read in only $\Delta(p[m], t[i])$ will remain to be computed. This modification does not affect the time complexity of the algorithm overall.   □

We can now apply our black box to a number of well known matching problems, giving the following time complexities per input character.

Corollary 1. *Algorithm 1 applied to the fastest known offline pattern matching algorithms for the Hamming norm, k-mismatch and matching under the L_2 and L_1 norms gives online algorithms that take $O(\log^2 m)$, $O(\sqrt{m \log m})$, $O(\sqrt{k \log k} \log m)$, $O(\log^2 m)$ and $O(\sqrt{m \log m})$ time per character respectively.*

4   Space Lower Bound for Deterministic Online Approximate Matching

It would seem desirable to reduce the space requirements even further in order to increase the practicality of processing data streams. Unfortunately, there is an $\Omega(m)$ communication complexity lower bound for "sum-type" functions [21] which covers the additive distance functions we have been most interested in. The space lower bound for any deterministic approximate pattern matching algorithm follows directly from the communication complexity lower bound by a standard argument that we briefly summarise. Assuming the communication complexity lower bound, the proof of the space lower bound is by contradiction. If Alice can preprocess the pattern to use $o(m)$ space and then starts her online pattern matching algorithm, she could then transfer a snapshot of the current state to Bob who could then carry on running the algorithm on his string. Bob would then find the distance to his string having received only $o(m)$ items of data, thereby giving the desired contradiction.

5   Discussion

The method we have developed is applicable to a wide class of previously offline approximate matching algorithms. By choosing a black box approach we have not investigated whether particular pattern matching algorithms might be more easily converted to efficient online algorithms without any extra time cost. Also, although we have shown that the space required by our approach is optimal for a wide range of problems, an interesting question is whether randomisation can allow us to solve the same problems with only $o(m)$ space as the communication complexity bounds will no longer hold.

Acknowledgements

The authors would like to thank Inbok Lee and Ashley Montanaro for their helpful comments on a draft of this paper.

References

[1] Abrahamson, K.: Generalized string matching. SIAM journal on Computing 16(6), 1039–1051 (1987)

[2] Amir, A., Farach, M.: Efficient 2-dimensional approximate matching of half-rectangular figures. Information and Computation 118(1), 1–11 (1995)

[3] Amir, A., Lipsky, O., Porat, E., Umanski, J.: Approximate matching in the L_1 metric. In: Apostolico, A., Crochemore, M., Park, K. (eds.) CPM 2005. LNCS, vol. 3537, pp. 91–103. Springer, Heidelberg (2005)

[4] Amir, A., Lewenstein, M., Porat, E.: Faster algorithms for string matching with k mismatches. J. Algorithms 50(2), 257–275 (2004)

[5] Atallah, M.J.: Faster image template matching in the sum of the absolute value of differences measure. IEEE Transactions on Image Processing 10(4), 659–663 (2001)

[6] Clifford, P., Clifford, R.: Simple deterministic wildcard matching. Information Processing Letters 101(2), 53–54 (2007)

[7] Clifford, P., Clifford, R., Iliopoulos, C.S.: Faster algorithms for δ,γ-matching and related problems. In: Apostolico, A., Crochemore, M., Park, K. (eds.) CPM 2005. LNCS, vol. 3537, pp. 68–78. Springer, Heidelberg (2005)

[8] Clifford, R., Iliopoulos, C.: String algorithms in music analysis. Soft Computing 8(9), 597–603 (2004)

[9] Clifford, R., Efremenko, K., Porat, E., Rothschild, A.: k-mismatch with don't cares. In: Arge, L., Hoffmann, M., Welzl, E. (eds.) ESA 2007. LNCS, vol. 4698, pp. 151–162. Springer, Heidelberg (2007)

[10] Clifford, R., Porat, E.: A filtering algorithm for k-mismatch with don't cares. In: Ziviani, N., Baeza-Yates, R. (eds.) SPIRE 2007. LNCS, vol. 4726, pp. 130–136. Springer, Heidelberg (2007)

[11] Cormen, T.H., Leiserson, C.E., Rivest, R.L.: Introduction to Algorithms. MIT Press, Cambridge (1990)

[12] Cole, R., Hariharan, R.: Verifying candidate matches in sparse and wildcard matching. In: Proceedings of the Annual ACM Symposium on Theory of Computing, pp. 592–601 (2002)

[13] Fischer, M., Paterson, M.: String matching and other products. In: Karp, R. (ed.) Proceedings of the 7th SIAM-AMS Complexity of Computation, pp. 113–125 (1974)

[14] Henzinger, M.R., Raghavan, P., Rajagopalan, S.: Computing on data streams. In: External memory algorithms, pp. 107–118. American Mathematical Society, Boston (1999)

[15] Indyk, P.: Faster algorithms for string matching problems: Matching the convolution bound. In: Proceedings of the 38th Annual Symposium on Foundations of Computer Science, pp. 166–173 (1998)

[16] Kalai, A.: Efficient pattern-matching with don't cares. In: Proceedings of the 13th Annual ACM-SIAM Symposium on Discrete Algorithms, pp. 655–656 (2002)

[17] Kosaraju, S.R.: Efficient string matching (1987) (manuscript)

[18] Landau, G.M., Vishkin, U.: Efficient string matching with k mismatches. Theoretical Computer Science 43, 239–249 (1986)

[19] Lipsky, O., Porat, E.: Approximate matching in the L_∞ metric. In: String Processing and Information Retrieval, 12th International Symposium (SPIRE 2005). LNCS, pp. 331–334. Springer, Heidelberg (2005)

[20] Muthukrishnan, S.: Data streams: algorithms and applications. In: SODA 2003: Proceedings of the fourteenth annual ACM-SIAM symposium on Discrete algorithms, p. 413 (2003)

[21] Tamm, U.: Communication complexity of sum-type functions invariant under translation. Inf. Comput. 116(2), 162–173 (1995)

An(other) Entropy-Bounded Compressed Suffix Tree

Johannes Fischer[1], Veli Mäkinen[2,*], and Gonzalo Navarro[1,**]

[1] Dept. of Computer Science, Univ. of Chile
{jfischer|gnavarro}@dcc.uchile.cl
[2] Dept. of Computer Science, Univ. of Helsinki, Finland
vmakinen@cs.helsinki.fi

Abstract. Suffix trees are among the most important data structures in stringology, with myriads of applications. Their main problem is space usage, which has triggered much research striving for compressed representations that are still functional. We present a novel compressed suffix tree. Compared to the existing ones, ours is the first achieving at the same time sublogarithmic complexity for the operations, and space usage which goes to zero as the entropy of the text does. Our development contains several novel ideas, such as compressing the longest common prefix information, and totally getting rid of the suffix tree topology, expressing all the suffix tree operations using range minimum queries and a new primitive called next/previous smaller value in a sequence.

1 Introduction

Suffix trees are probably the most important structure ever invented in stringology. They have been said to have "myriads of virtues" [2], and also have myriads of applications in many areas, most prominently bioinformatics [13]. One of the main drawbacks of suffix trees is their considerable space requirement, which is usually close to $20n$ bytes for a sequence of n symbols, and at the very least $10n$ bytes [17]. For example, the Human genome, containing approximately 3 billion bases, could easily fit in the main memory of a desktop computer (as each DNA symbol needs just 2 bits). However, its suffix tree would require 30GB to 60GB, too large to fit in normal main memories. Although there has been some progress in managing suffix trees in secondary storage [15] and it is an active area of research [16], it will always be faster to operate in main memory.

This situation has stimulated research on compressed representations of suffix trees, which operate in compressed form. Even if many more operations are needed to carry out the operations on the compressed representation, this is clearly advantageous compared to having to manage it on secondary memory. A large body of research focuses on compressed suffix arrays [22], which offer a

---

* Funded by the Academy of Finland under grant 119815.
** Partially funded by Millennium Institute for Cell Dynamics and Biotechnology, Grant ICM P05-001-F, Mideplan, Chile.

P. Ferragina and G. Landau (Eds.): CPM 2008, LNCS 5029, pp. 152–165, 2008.

reduced suffix tree functionality. Especially, they miss the important suffix-link operation. The same restrictions apply to early compressed suffix trees [21,12].

The first fully-functional compressed suffix tree is due to Sadakane [26]. It builds on top of a compressed suffix array [25] that uses $\frac{1}{\epsilon}nH_0 + O(n\log\log\sigma)$ bits of space, where H_0 is the zero-order entropy of the text $T_{1,n}$, σ is the size of the alphabet of T, and $0 < \epsilon < 1$ is any constant. In addition, the compressed suffix tree needs $6n + o(n)$ bits of space. Most of the suffix tree operations can be carried out in constant time, except for knowing the string-depth of a node and the string content of an edge, which take $O(\log^\epsilon n)$ time, and moving to a child, which costs $O(\log^\epsilon n \log\sigma)$. One could replace the compressed suffix array they use by Grossi et al.'s [11], which requires less space: $\frac{1}{\epsilon}nH_k + o(n\log\sigma)$ bits for any $k \le \alpha\log_\sigma n$, where H_k is the k-th empirical entropy of T [19] and $0 < \alpha < 1$ is any constant. However, the $O(\log^\epsilon n)$ time complexities become $O(\log_\sigma^{\frac{\epsilon}{1-\epsilon}} n \log\sigma)$ [11, Thm. 4.1]. In addition, the extra $6n$ bits in the space complexity remain, despite any reduction in the compressed suffix array. This term can be split into $2n$ bits to represent (with a bitmap called Hgt) the longest common prefix (LCP) information, plus $4n$ bits to represent the suffix tree topology with parentheses. Many operations are solved via constant-time range minimum queries (RMQs) over the depths in the parentheses sequence. An RMQ from i to j over a sequence $S[1,n]$ of numbers asks for $\text{RMQ}_S(i,j) := \text{argmin}_{i\le\ell\le j}S[\ell]$.

Russo et al. [24] recently achieved fully-compressed suffix trees, that is, requiring $nH_k + o(n\log\sigma)$ bits of space (with the same limits on k as before), which is essentially the space required by the smallest compressed suffix array, and asymptotically optimal under the k-th entropy model. The main idea is to sample some suffix tree nodes and use the compressed suffix array as a tool to find nearby sampled nodes. The most adequate compressed suffix array for this task is the alphabet-friendly FM-index [6]. The time complexities for most operations are logarithmic at best, more precisely, between $O(\log n)$ and $O(\log n \log\log n)$. Others are slightly costlier, e.g. moving to a child costs an additional $O(\log\log n)$ factor, and some less common operations are as costly as $O((\log n \log\log n)^2)$.

We present a new fully-compressed suffix tree, by removing the $6n$ term in Sadakane's space complexity. The space we achieve is not as good as that of Russo et al., but most of our time complexities are sublogarithmic. More precisely, our index needs $nH_k(2\log\frac{1}{H_k} + \frac{1}{\epsilon} + O(1)) + o(n\log\sigma)$ bits of space. Note that, although this is not the ideal nH_k, it still goes to zero as $H_k \to 0$, unlike the incompressible $6n$ bits in Sadakane's structure. Our solution builds on two novel algorithmic ideas to improve Sadakane's compressed suffix tree.

1. We show that array Hgt, which encodes LCP information in $2n$ bits [26], actually contains $2R$ runs, where R is the number of *runs in* ψ [22]. We show how to run-length compress Hgt into $2R\log\frac{n}{R} + O(R) + o(n)$ bits while retaining constant-time access. In order to relate R with nH_k, we use the result $R \le nH_k + \sigma^k$ for any k [18], although sometimes it is extremely pessimistic (in particular it is useful only for $H_k < 1$, as obviously $R \le n$). This gives the $nH_k(2\log\frac{1}{H_k} + O(1))$ upper bound to store Hgt (and the real space is always $\le 2n$ bits).

2. We get rid of the suffix tree topology and identify suffix tree nodes with suffix array intervals. All the tree traversal operations are simulated with RMQs on LCP (represented with Hgt), plus a new type of queries called "Next/Previous Smaller Value", that is, given a sequence of numbers $S[1, n]$, find the first cell in S following/preceding i whose value is smaller than $S[i]$.[1] We show how to solve these queries in sublogarithmic time while spending only $o(n)$ extra bits of space on top of S. We believe this operation might have independent interest, and the challenge of achieving constant time with sublinear space remains open.

2   Basic Concepts

The *suffix tree* $\mathcal{S}$ of a text $T_{1,n}$ over an alphabet Σ of size σ is a compact trie storing all the suffixes $T_{i,n}$ where the leaves point to the corresponding i values [2,13]. For convenience we assume that T is terminated with a special symbol, so that all lexicographical comparisons are well defined. For a node v in $\mathcal{S}$, $\pi(v)$ denotes the string obtained by reading the edge-labels when walking from the root to v (the *path-label* of v [24]). The *string-depth* of v is the length of $\pi(v)$.

Definition 1. *A suffix tree representation supports the following operations:*

- ROOT(): *the root of the suffix tree.*
- LOCATE(v): *the suffix position i if v is the leaf of suffix $T_{i,n}$, otherwise* NULL.
- ANCESTOR(v, w): *true if v is an ancestor of w.*
- SDEPTH(v)/TDEPTH(v): *the string-depth/tree-depth of v.*
- COUNT(v): *the number of leaves in the subtree rooted at v.*
- PARENT(v): *the parent node of v.*
- FCHILD(v)/NSIBLING(v): *the alphabetically first child/next sibling of v.*
- SLINK(v): *the suffix-link of v; i.e., the node w s.th. $\pi(w) = \beta$ if $\pi(v) = a\beta$ for $a \in \Sigma$.*
- SLINK^i(v): *the iterated suffix-link of v; (node w s.th. $\pi(w) = \beta$ if $\pi(v) = \alpha\beta$ for $\alpha \in \Sigma^i$).*
- LCA(v, w): *the lowest common ancestor of v and w.*
- CHILD(v, a): *the node w s.th. the first letter on edge (v, w) is $a \in \Sigma$.*
- LETTER(v, i): *the ith letter of v's path-label, $\pi(v)[i]$.*
- LAQs(v, d)/LAQt(v, d): *the highest ancestor of v with string-depth/tree-depth $\geq d$.*

Existing compressed suffix tree representations include a *compressed full-text index* [22,25,11,6], which encodes in some form the *suffix array* SA$[1, n]$ of T, with access time t_{SA}. Array SA is a permutation of $[1, n]$ storing the pointers to the suffixes of T (i.e., the LOCATE values of the leaves of $\mathcal{S}$) in lexicographic order. Most full-text indexes also support access to permutation SA^{-1} in time $O(t_{SA})$, as well as the efficient computation of permutation $\psi[1, n]$, where $\psi(i) = $ SA$^{-1}[$SA$[i] + 1]$ for $1 \leq i \leq n$ if SA$[i] \neq n$ and SA$^{-1}[1]$ otherwise. $\psi(i)$ is

---

[1] Computing NSVs/PSVs on the fly has been considered in parallel computing [3], yet not in the static scenario.

computed in time t_ψ, which is at most $O(t_{\sf SA})$, but usually less. Compressed suffix tree representations also include array $\mathsf{LCP}[1, n]$, which stores the length of the longest common prefix (lcp) between consecutive suffixes in lexicographic order, $\mathsf{LCP}[i] = |lcp(T_{\mathsf{SA}[i-1],n}, T_{\mathsf{SA}[i],n})|$ for $i > 1$ and $\mathsf{LCP}[1] = 0$. The access time for LCP is t_{LCP}.

We make heavy use of the following complementary operations on bit arrays: $rank(B, i)$ is the number of bits set in $B[1, i]$, and $select(B, j)$ is the position of the j-th 1 in B. Bit vector $B[1, n]$ can be preprocessed to answer both queries in constant time using $o(n)$ extra bits of space [20]. If B contains only m bits set, then the representation of Raman et al. [23] compresses B to $m \log \frac{n}{m} + O(m + \frac{n \log \log n}{\log n})$ bits of space and retains constant-time $rank$ and $select$ queries.

3   Compressing LCP Information

Sadakane [26] describes an encoding of the LCP array that uses $2n + o(n)$ bits. The encoding is based on the fact that values $i + \mathsf{LCP}[i]$ are nondecreasing when listed in text position order: Sequence $S = s_1, \ldots, s_{n-1}$, where $s_j = j + \mathsf{LCP}[\mathsf{SA}^{-1}[j]]$, is nondecreasing.

To represent S, Sadakane encodes each $\mathtt{diff}(j) = s_j - s_{j-1}$ in unary: $1\, 0^{\mathtt{diff}(j)}$, where $s_0 = 0$ and 0^d denotes repetition of 0-bit d times. This encoding, call it U (similar to Hgt [26]), takes at most $2n$ bits. Thus $\mathsf{LCP}[i] = select(U, j+1) - j - 1$, where $j = \mathsf{SA}[i]$, is computed in time $O(t_{\sf SA})$.

Let us now consider how to represent U in a yet more space-efficient form, i.e., in $nH_k(2 \log \frac{1}{H_k} + O(1)) + o(n)$ bits, for small enough k. The result follows from the observation (to be shown below) that the number of 1-bit runs in U is bounded by the number of runs in ψ. We call a run in ψ a maximal sequence of consecutive i values where $\psi(i) - \psi(i - 1) = 1$ and $T_{\mathsf{SA}[i-1]} = T_{\mathsf{SA}[i]}$, including one preceding i where this does not hold [18]. Note that an area in ψ where the differences are not 1 corresponds to several length-1 runs. Let us call $R \le n$ the overall number of runs.

We will represent U in run-length encoded form, coding each maximal run of both 0 and 1 bits. We show soon that there are at most R 1-runs, and hence at most R 0-runs (as U starts with a 1). If we encode the 1-run lengths $o_1, o_2, \ldots$ and the 0-run lengths $z_1, z_2, \ldots$ separately (cf. Sect. 3.2 in [5]), it is easy to compute $select(U, j)$ by finding the largest r such that $\sum_{i=1}^{r} o_i < j$ and then answering $select(U, j) = j + \sum_{i=1}^{r} z_i$. This so-called searchable partial sums problem is easy to solve. Store bitmap $O[1, n]$ setting the bits at positions $\sum_{i=1}^{r} o_i$, hence $\max\{r, \sum_{i=1}^{r} o_i < j\} = rank(O, j - 1)$. Likewise, bitmap $Z[1, n]$ representing the z_i's solves $\sum_{i=1}^{r} z_i = select(Z, r)$. Since both O and Z have at most R 1's, O plus Z can be represented using $2R \log \frac{n}{R} + O(R + \frac{n \log \log n}{\log n})$ bits [23].

We now show the connection between runs in U and runs in ψ. Let us call position i a stopper if $i = 1$ or $\psi(i) - \psi(i - 1) \ne 1$ or $T_{\mathsf{SA}[i-1]} \ne T_{\mathsf{SA}[i]}$. Hence ψ has exactly R stoppers by the definition of runs in ψ. Say now that a chain in ψ is a maximal sequence $i, \psi(i), \psi(\psi(i)), \ldots$ such that each $\psi^j(i)$ is not a stopper except the last one. As ψ is a permutation with just one cycle, it follows that in the path of $\psi^j[\mathsf{SA}^{-1}[1]]$, $0 \le j < n$, we will find the R stoppers, and hence there are also R chains in ψ [10].

We now show that each chain in ψ induces a run of 1's of the same length in U. Let i, $\psi(i)$, ..., $\psi^\ell(i)$ be a chain. Hence $\psi^j(i) - \psi^j(i-1) = 1$ for $0 \le j < \ell$. Let $x = \mathsf{SA}[i-1]$ and $y = \mathsf{SA}[i]$. Then $\mathsf{SA}[\psi^j(i-1)] = x + j$ and $\mathsf{SA}[\psi^j(i)] = y + j$. Then $\mathsf{LCP}[i] = |lcp(T_{\mathsf{SA}[i-1],n}, T_{\mathsf{SA}[i],n})| = |lcp(T_{x,n}, T_{y,n})|$. Note that $T_{x+\mathsf{LCP}[i]} \ne T_{y+\mathsf{LCP}[i]}$, and hence $\mathsf{SA}^{-1}[y + \mathsf{LCP}[i]] = \psi^{\mathsf{LCP}[i]}(i)$ is a stopper, thus $\ell \le \mathsf{LCP}[i]$. Moreover, $\mathsf{LCP}[\psi^j(i)] = |lcp(T_{x+j,n}, T_{y+j,n})| = \mathsf{LCP}[i] - j \ge 0$ for $0 \le j < \ell$. Now consider $s_{y+j} = y+j+\mathsf{LCP}[\mathsf{SA}^{-1}[y+j]] = y+j+\mathsf{LCP}[\psi^j(i)] = y+j+\mathsf{LCP}[i]-j = y + \mathsf{LCP}[i]$, all equal for $0 \le j < \ell$. This produces $\ell - 1$ diff values equal to 0, that is, a run of ℓ 1-bits in U. By traversing all the chains in the cycle of ψ we sweep S left to right, producing at most R runs of 1's and hence at most R runs of 0's. (Note that even an isolated 1 is a run with $\ell = 1$.) Since $R \le nH_k + \sigma^k$ for any k [22], we obtain the bound $nH_k(2\log\frac{1}{H_k} + O(1)) + O(\frac{n\log\log n}{\log n})$ for any $k \le \alpha \log_\sigma n$ and any constant $0 < \alpha < 1$. Although our somewhat crude upper bounds do not show it, our representation is asymptotically never larger than the original Hgt.

4   Next-Smaller and Prev-Smaller Queries

In this section we consider queries *next smaller value* (*NSV*) and *previous smaller value* (*PSV*), and show that they can be solved in sublogarithmic time using only a sublinear number of extra bits on top of the raw data. We make heavy use of these queries in the design of our new compressed suffix tree, and also believe that they can be of independent interest.

Definition 2. *Let $S[1,n]$ be a sequence of elements drawn from a set with a total order $\preceq$ (where one can also define $a \prec b \Leftrightarrow a \preceq b \wedge b \npreceq a$). We define the query* next smaller value *and* previous smaller value *as follows:* $NSV(S,i) = \min\{j, (i < j \le n \wedge S[j] \prec S[i]) \vee j = n+1\}$ *and* $PSV(S,i) = \max\{j, (1 \le j < i \wedge S[j] \prec S[i]) \vee j = 0\}$, *respectively.*

The key idea to solve these queries reminds that for *findopen* and *findclose* operations in balanced parentheses, in particular the recursive version [9]. However, there are several differences because we have to deal with a sequence of generic values, not parentheses.

We will describe the solution for *NSV*, as that for *PSV* is symmetric. For shortness we will write $NSV(i)$ for $NSV(S,i)$. We split $S[1,n]$ into consecutive *blocks* of b values. A position i will be called *near* if $NSV(i)$ is within the same block of i. The first step when solving a *NSV* query will be to scan the values $S[i+1 \ldots b \cdot \lceil i/b \rceil]$, that is from $i+1$ to the end of the block, looking for an $S[j] \prec S[i]$. This takes $O(b)$ time and solves the query for near positions.

Positions that are not near are called *far*. We note that the far positions within a block, $i_1 < i_2 \ldots < i_s$ form a nondecreasing sequence of values $S[i_1] \preceq S[i_2] \ldots \preceq S[i_s]$. Moreover, their *NSV* values form a nonincreasing sequence $NSV(i_1) \ge NSV(i_2) \ldots \ge NSV(i_s)$.

A far position i will be called a *pioneer* if $NSV(i)$ is *not* in the same block of $NSV(j)$, being j the largest far position preceding i (the first far position is also a pioneer). It follows that, if j is the last pioneer preceding i, then $NSV(i)$ is in the same block of $NSV(j) \geq NSV(i)$. Hence, to solve $NSV(i)$, we find j and then scan (left to right) the block $S[\lceil NSV(j)/b \rceil - b + 1 \dots NSV(j)]$, in time $O(b)$, for the first value $S[j'] \prec S[i]$.

So the problem boils down to efficiently finding the pioneer preceding each position i, and to storing the answers for pioneers. We mark pioneers in a bitmap $P[1, n]$. We note that, since there are $O(n/b)$ pioneers overall [14], P can be represented using $O(\frac{n \log b}{b}) + O(\frac{n \log \log n}{\log n})$ bits of space [23]. With this representation, we can easily find the last pioneer preceding a far position i, as $j = select(P, rank(P, i))$. We could now store the NSV answers for the pioneers in an answer array $A[1, n']$ ($n' = O(n/b)$), so that if j is a pioneer then $NSV(j) = A[rank(P, j)]$. This already gives us a solution requiring $O(\frac{n \log b}{b}) + O(\frac{n \log \log n}{\log n}) + O(\frac{n \log n}{b})$ bits of space and $O(b)$ time. For example, we can have $O(\frac{n}{\log \log n})$ bits of space and $O(\log n \log \log n)$ time.

We can do better by recursing on the idea. Instead of storing the answers explicitly in array A, we will form a (virtual) *reduced* sequence $S'[1, 2n']$ containing all the pioneer values i and their answers $NSV(i)$. Sequence S' is not explicitly stored. Rather, we set up a bitmap $R[1, n]$ where the selected values in S are marked. Hence we can retrieve any value $S'[i] = S[select(R, i)]$. Again, this can be computed in constant time using $O(\frac{n \log b}{b} + \frac{n \log \log n}{\log n})$ bits to represent R [23].

Because S' is a subsequence of S, it holds that the answers to NSV in S' are the same answers mapped from S. That is, if i is a pioneer in S, mapped to $i' = rank(R, i)$ in S', and $NSV(i)$ is mapped to $j' = rank(R, NSV(i))$, then $j' = NSV(S', i')$, because any value in $S'[i' + 1 \dots j' - 1]$ correspond to values within $S[i+1 \dots NSV(i)-1]$, which by definition of NSV are not smaller than $S[i]$. Hence, we can find $NSV(i)$ for pioneers i by the corresponding recursive query on S', $NSV(i) = select(R, NSV(S', rank(R, i)))$. We are left with the problem of solving queries $NSV(S', i)$.

We proceed again by splitting S' into blocks of b values. Near positions in S' are solved in $O(b)$ time by scanning the block. Recall that S' is not explicitly stored, but rather we have to use *select* on R to get its values from S. For far positions we define again pioneers, and solve NSV on far positions in time $O(b)$ using the answer for the preceding pioneer. Queries for pioneers are solved in a third level by forming the virtual sequence $S''[1, 2n'']$, $n'' = O(n'/b) = O(n/b^2)$.

We continue the process recursively for r levels before storing the explicit answers in array $A[1, n^{(r)}]$, $n^{(r)} = O(n/b^r)$. We remark that the P^ℓ and R^ℓ bitmaps at each level ℓ map positions directly to S, not to the reduced sequence of the previous level. This permits accessing the $S^\ell[i]$ values at any level ℓ in constant time, $S^\ell[i] = S[select(R^\ell, i)]$. The pioneer preceding i in S^ℓ is found by first mapping to S with $i' = select(R^\ell, i)$, then finding the preceding pioneer directly in the domain of S, $j' = select(P^\ell, rank(P^\ell, i'))$, and finally mapping the pioneer back to S^ℓ by $j = rank(R^\ell, j')$.

Let us now analyze the time and space of this solution. Because we pay $O(b)$ time at each level and might have to resort to the next level in case our position is far, the total time is $O(rb)$ because the last level is solved in constant time. As for the space, all we store are the P^ℓ and R^ℓ bitmaps, and the final array A. Array A takes $O(\frac{n \log n}{b^r})$ bits. As there are $O(n/b^\ell)$ elements in S^ℓ, both P^ℓ and R^ℓ require $O(\frac{n}{b^\ell} \log(b^\ell) + \frac{n \log \log n}{\log n})$ bits of space (actually P^ℓ is about half the size of R^ℓ). The sum of all the P^ℓ and R^ℓ takes order of $\sum_{1 \le \ell \le r} \left(\frac{n}{b^\ell} \log(b^\ell) + \frac{n \log \log n}{\log n} \right) = O \left(\frac{n \log b}{b} + r \frac{n \log \log n}{\log n} \right)$.

We now state the main result of this section.

Theorem 1. *Let $S[1, n]$ be a sequence of elements drawn from a set with a total order, such that access to any $S[i]$ and any comparison $S[i] \prec S[j]$ can be computed in constant time. Then, for any $1 \le r, b \le n$, it is possible to build a data structure on S taking $O(\frac{n \log b}{b} + r \frac{n \log \log n}{\log n} + \frac{n \log n}{b^r})$ bits, so that queries NSV and PSV can be solved in worst-case time time $O(rb)$. In particular, for any $f(n) = O(\frac{\log n}{\log \log n})$, one can achieve $O(\frac{n}{f(n)})$ bits of extra space and $O(f(n) \log \log n)$ time.*

Proof. The general formula for any r, b has been obtained throughout this section. As for the formulas in terms of $f(n)$, let us set the space limit to $O(\frac{n}{f(n)})$. Then $\frac{n \log b}{b} = O(\frac{n}{f(n)})$ implies $b = \Omega(f(n) \log f(n))$. Also, $\frac{n \log n}{b^r} = O(\frac{n}{f(n)})$ implies $r \ge \frac{\log \log n + \log f(n) - O(1)}{\log b}$. Hence $rb \ge \frac{b}{\log b} (\log \log n + \log f(n) - O(1))$. Thus it is best to minimize b. By setting $b = f(n) \log f(n)$, we get $rb = \frac{f(n) \log f(n)}{\log f(n) + \log \log f(n)} (\log \log n + \log f(n) - O(1)) = \Theta(f(n)(\log \log n + \log f(n)))$. The final constraint is $r \frac{n \log \log n}{\log n} = O(\frac{n}{f(n)})$, which, by substituting $r = \frac{\log \log n + \log f(n)}{\log b}$ and since $b = \Omega(f(n) \log f(n))$, yields the condition $f(n) = O(\frac{\log n}{\log \log n})$. Thus $\log \log n + \log f(n) = O(\log \log n)$. $\square$

Note that, if one is willing to spend $4n + o(n)$ bits of extra space, the operations can be solved in constant time. The idea is to reduce PSV and NSV queries to $O(1)$ *findopen* and *findclose* operations in balanced parentheses [9]. For NSV, for $1 \le i \le n + 1$ in this order, write a '(' and then x ')'s if there are x cells $S[j]$ for which $NSV(j) = i$. The resulting sequence B is balanced if a final ')' is appended, and $NSV(i)$ can be obtained by $rank(B, findclose(B, select(B, i)))$, where a 1 in B represents '('. PSV is symmetric, needing other $2n + o(n)$ bits.

5   An Entropy-Bounded Compressed Suffix Tree

Let v be a node in the (virtual) suffix tree $\mathcal{S}$ for text $T_{1,n}$. As in previous works [1,4,24], we represent v by an interval $[v_l, v_r]$ in SA such that $SA[v_l, v_r]$ are exactly the leaves in $\mathcal{S}$ that are in the subtree rooted at v. Let us first consider internal nodes, so $v_l < v_r$. Because $\mathcal{S}$ does not contain unary nodes, it follows from the definition of LCP that at least one entry in $LCP[v_l + 1, v_r]$ is equal to the string-depth h of v; such a position is called h-*index* of $[v_l, v_r]$. We further have

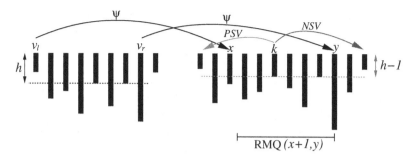

RMQ $(x+1, y)$

Fig. 1. Left: Illustration to the representation of suffix tree nodes. The lengths of the bars indicate the LCP values. All leaves in the subtree rooted at $v = [v_l, v_r]$ share a longest common prefix of length at least h. Right: Schematic view of the SLINK operation. From v, first follow ψ, then perform an RMQ to find an $(h-1)$-index k, and finally locate the defining points of the desired interval by a PSV/NSV query from k.

LCP$[v_l] < h$, LCP$[i] \geq h$ for all $v_l < i \leq v_r$, and LCP$[v_r + 1] < h$. Fig. 1 (left) illustrates. We state the easy yet fundamental

Lemma 1. *Let* $[v_l, v_r]$ *be an interval in* SA *that corresponds to an internal node* v *in* $\mathcal{S}$. *Then the string-depth of* v *is* $h = $ LCP(k), *where* $k = $ RMQ$_{\mathsf{LCP}}(v_l + 1, v_r)$.

For leaves $v = [v_l, v_l]$, the string-depth of v is simply given by $n - $ SA$[v_l] + 1$.

5.1   Range Minimum Queries in Sublinear Space

As Lemma 1 suggests, we wish to preprocess LCP such that RMQ$_{\mathsf{LCP}}$ can be answered in sublogarithmic time, using $o(n)$ bits of additional space. A well-known strategy [7,26] divides LCP iteratively into *blocks* of decreasing size $n > b_1 > b_2 > \cdots > b_r$. On level i, $1 \leq i \leq r$, compute all answers to RMQ$_{\mathsf{LCP}}$ that exactly span over blocks of size b_i, but not over blocks of size b_{i-1} (set $b_0 = n$ for handling the border case). This takes $O(\frac{n}{b_i} \log(\frac{b_{i-1}}{b_i}) \log(b_{i-1}))$ bits of space if the answers are stored relative to the beginning of the blocks on level $i - 1$, and if we only precompute queries that span 2^j blocks for all $j \leq \lfloor \log(\frac{b_{i-1}}{b_i}) \rfloor$ (this is sufficient because each query can be decomposed into at most 2 possibly overlapping sub-queries whose lengths are a power of 2).

A general range minimum query is then decomposed into at most $2r + 1$ non-overlapping sub-queries $q_1, \ldots, q_{2r+1}$ such that q_1 and q_{2r+1} lie completely inside of blocks of size b_r, q_2 and q_{2r} exactly span over blocks of size b_r, and so on. q_1 and q_{2r+1} are solved by scanning in time $O(b_r)$,[2] and all other queries can be answered by table-lookups in total time $O(r)$. The final answer is obtained by comparing at most $2r + 1$ minima.

The next lemma gives a general result for RMQs using $o(n)$ extra space.

---

[2] The constant-time solutions [26,7] also solve q_1 and q_{2r+1} by accessing tables that require $\Theta(n)$ bits.

Lemma 2. *Having constant-time access to elements in an array $A[1, n]$, it is possible to answer range minimum queries on A in time $O(f(n)(\log f(n))^2)$ using $O(\frac{n}{f(n)})$ bits of space, for any $f(n) = \Omega(\log^{[r]} n)$ and any constant r, where $\log^{[r]} n$ denotes r applications of $\log$ to n.*

Proof. We use $r+1 = O(1)$ levels $1 \ldots r+1$, so it is sufficient that $\frac{n}{b_i} \log^2 b_{i-1} = O(\frac{n}{f(n)})$ for all $1 \leq i \leq r + 1$, where $b_0 = n$. From the condition $\frac{n}{b_1} \log^2 b_0 = O(\frac{n}{f(n)})$ we get $b_1 = \Theta(f(n) \log^2 n)$ (the smallest possible b_i values are best). From $\frac{n}{b_2} \log^2 b_1 = O(\frac{n}{f(n)})$ we get $b_2 = \Theta(f(n) \log^2 b_1) = \Theta(f(n)(\log f(n) + \log \log n)^2)$. In turn, from $\frac{n}{b_3} \log^2 b_2 = O(\frac{n}{f(n)})$ we get $b_3 = \Theta(f(n) \log^2 b_2) = \Theta(f(n)(\log f(n) + \log \log \log n)^2)$. This continues until $b_{r+1} = \Theta(f(n) \log^2 b_r) = \Theta(f(n)(\log f(n) + \log^{[r+1]} n))^2) = \Theta(f(n) \log^2 f(n))$. □

5.2   Suffix-Tree Operations

Now we have all the ingredients for navigating in the suffix tree. The operations are described in the following; the intuitive reason why an RMQ is often followed by a PSV/NSV-query is that the RMQ gives us an h-index of the (yet unknown) interval, and the PSV/NSV takes us to the delimiting points of this interval. Apart from t_{SA}, t_{LCP}, and t_ψ, we denote by t_{RMQ} and t_{PNSV} the time to solve, respectively, RMQs or NSV/PSV queries (both on LCP from now on, hence they will be multiplied by t_{LCP}).

ROOT/COUNT/ANCESTOR: ROOT() returns interval $[1, n]$, COUNT(v) is simply $v_r - v_l + 1$, ANCESTOR(w, v) is true iff $w_l \leq v_l \leq v_r \leq w_r$. These take $O(1)$ time.

SDEPTH(v)/LOCATE(v): According to Lemma 1, SDEPTH(v) can be computed in time $O(t_{\mathrm{RMQ}} \cdot t_{\mathsf{LCP}})$ for internal nodes, and both operations need time $O(t_{\mathsf{SA}})$ for leaves. One knows in constant time that $v = [v_l, v_r]$ is a leaf iff $v_l = v_r$.

PARENT(v): If v is the root, return NULL. Else, since the suffix tree is compact, the string-depth of PARENT(v) must be either LCP$[v_l]$ or LCP$[v_r + 1]$, whichever is greater [24]. So, by setting $k =$ **if** LCP$[v_l] >$ LCP$[v_r + 1]$ **then** v_l **else** $v_r + 1$, the parent interval of v is $[PSV(k), NSV(k) - 1]$. Time is $O(t_{\mathrm{PNSV}} \cdot t_{\mathsf{LCP}})$.

FCHILD(v): If v is a leaf, return NULL. Otherwise, because the minima in $[v_l, v_r]$ are v's h-indices [7], the first child of v is given by $[v_l, \mathrm{RMQ}(v_l + 1, v_r) - 1]$, assuming that RMQs always return the leftmost minimum in the case of ties (which is easy to arrange). Time is $O(t_{\mathrm{RMQ}} \cdot t_{\mathsf{LCP}})$.

NSIBLING(v): First move to the parent of v by $w =$ PARENT(v). If $v_r = w_r$, return NULL, since v does not have a next sibling. If $v_r + 1 = w_r$, v's next sibling is a leaf, so return $[w_r, w_r]$. Otherwise, return $[v_r + 1, \mathrm{RMQ}(v_r + 2, w_r) - 1]$. The overall time is $O((t_{\mathrm{RMQ}} + t_{\mathrm{PNSV}}) \cdot t_{\mathsf{LCP}})$.

SLINK(v): If v is the root, return NULL. Otherwise, first follow the suffix links of the leaves v_l and v_r, $x = \psi(v_l)$ and $y = \psi(v_r)$. Then locate an h-index of the target interval by $k = \mathrm{RMQ}(x + 1, y)$; see Lemma 7.5 in [1] (the first character of all strings in $\{T_{\mathsf{SA}[i], n} : v_l \leq i \leq v_r\}$ is the same, so the h-indices in $[v_l, v_r]$

appear also as $(h-1)$-indices in $[\psi(v_l), \psi(v_r)]$. The final result is then given by $[PSV(k), NSV(k) - 1]$. Time is $O(t_\psi + (t_{\mathrm{PNSV}} + t_{\mathrm{RMQ}}) \cdot t_{\mathsf{LCP}})$. See Fig. 1 (right).

SLINK$^i(v)$: Same as above with $x = \psi^i(v_l)$ and $y = \psi^i(v_r)$. If the first LETTER of x and y are different, then the answer is ROOT. Otherwise we go on with k as before. Computing ψ^i can be done in $O(t_{\mathsf{SA}})$ time using $\psi^i(v) = \mathsf{SA}^{-1}[\mathsf{SA}[v] + i]$ [24]. Time is thus $O(t_{\mathsf{SA}} + (t_{\mathrm{PNSV}} + t_{\mathrm{RMQ}}) \cdot t_{\mathsf{LCP}})$.

LCA(v, w): If one of v or w is an ancestor of the other, return this ancestor node. Otherwise, w.l.o.g., assume $v_r < w_l$. The h-index of the target interval is given by an RMQ between v and w [26]: $k = \mathrm{RMQ}(v_r + 1, w_l)$. The final answer is again $[PSV(k), NSV(k) - 1]$. Time is $O((t_{\mathrm{RMQ}} + t_{\mathrm{PNSV}}) \cdot t_{\mathsf{LCP}})$.

CHILD(v, a): If v is a leaf, return NULL. Otherwise, the minima in $\mathsf{LCP}[v_l + 1, v_r]$ define v's child-intervals, so we need to find the position $p \in [v_l + 1, v_r]$ where $\mathsf{LCP}[p] = \min_{i \in [v_l+1, v_r]} \mathsf{LCP}[i]$, and $T_{\mathsf{SA}[p]+\mathsf{LCP}[p]} = \text{LETTER}([p, p], \mathsf{LCP}[p] + 1) = a$. Then the final result is given by $[p, \mathrm{RMQ}(p + 1, v_r) - 1]$, or NULL if there is no such position p. To find this p, split $[v_l, v_r]$ into three sub-intervals $[v_l, x - 1], [x, y - 1], [y, v_r]$, where x (y) is the first (last) position in $[v_l, v_r]$ where a block of size b_r starts (b_r is the smallest block size for precomputed RMQs, recall Sect. 5.1). Intervals $[v_l, x - 1]$ and $[y, v_r]$ can be scanned for p in time $O(t_{\mathrm{RMQ}} \cdot (t_{\mathsf{LCP}} + t_{\mathsf{SA}}))$. The big interval $[x, y - 1]$ can be binary-searched in time $O(\log \sigma \cdot t_{\mathsf{SA}})$, provided that we also store exact median positions of the minima in the precomputed RMQs [26] (within the same space bounds). The only problem is how these precomputations are carried out in $O(n)$ time, as it is not obvious how to compute the exact median of an interval from the medians in its left and right half, respectively. However, a solution to this problem exists [8, Sect. 3.2]. Overall time is $O((t_{\mathsf{LCP}} + t_{\mathsf{SA}}) \cdot t_{\mathrm{RMQ}} + \log \sigma \cdot t_{\mathsf{SA}})$.

LETTER(v, i): If $i = 1$ we can easily solve the query in constant time with very little extra space. Mark in a bitmap $C[1, n]$ the first suffix in SA starting with each different letter, and store in a string $L[1, \sigma]$ the different letters that appear in $T_{1,n}$ in alphabetical order. Hence, if $v = [v_l, v_r]$, LETTER$(v, 1) = L[rank(C, v_l)]$. L requires $O(\sigma \log \sigma)$ bits and C, represented as a compressed bitmap [23], requires $O(\sigma \log \frac{n}{\sigma} + \frac{n \log \log n}{\log n})$ bits of space. Hence both add up to $O(\sigma \log n + \frac{n \log \log n}{\log n})$ bits. Now, for $i > 1$, we just use LETTER$(v, i) = \text{LETTER}(\psi^{i-1}(v_l), 1)$, in time $O(\min(t_{\mathsf{SA}}, i \cdot t_\psi))$. We remark that L and C are already present, in some form, in all compressed text indexes implementing SA [11,25,6].

TDEPTH(v): Tree-depth can be maintained while performing some traversal operations such as FCHILD, CHILD, PARENT, LAQT, but not others.

However, there is also a direct way to support TDEPTH, using $nH_k(2\log\frac{1}{H_k} + O(1)) + o(n)$ further bits of space. The idea is similar to Sadakane's representation of LCP [26]: the key insight is that the tree depth can decrease by at most 1 if we move from suffix $T_{i,n}$ to $T_{i+1,n}$ (i.e., when following ψ). Define TDE$[1, n]$ such that TDE$[i]$ holds the tree-depth of the LCA of leaves $\mathsf{SA}[i]$ and $\mathsf{SA}[i - 1]$ (similar to the definition of LCP). Then the sequence $(\mathsf{TDE}[\psi^k(\mathsf{SA}^{-1}[1])] + k)_{k=0,1,\dots,n-1}$ is nondecreasing and in the range $[1, n]$, and can hence be stored using $2n + o(n)$

bits. Further, the repetitions appear in the same way as in Hgt (Sect. 3), so the resulting sequence can be compressed to $nH_k(2\log\frac{1}{H_k} + O(1)) + o(n)$ bits using the same mechanism as for LCP. The time is thus $O(t_{\mathrm{RMQ}} \cdot t_{\mathrm{LCP}})$. For leaves we can do in $O(t_{\mathrm{SA}})$ time by $\mathrm{TDEPTH}(v) = 1 + \max(\mathrm{TDE}[SA[v]], \mathrm{TDE}[SA[v+1]])$.

$\mathrm{LAQS}(v,d)$: Let $u = [u_l, u_r] = \mathrm{LAQS}(v,d)$ denote the (yet unknown) result. Because u is an ancestor of v, we must have $u_l \leq v_l$ and $v_r \leq u_r$. We further know that $\mathrm{LCP}[i] \geq d$ for all $u_l < i \leq u_r$. Thus, u_l is the largest position in $[1, v_l]$ with $\mathrm{LCP}[u_l] < d$. So the search for u_l can be conducted in a binary manner by means of RMQs: Letting $k = \mathrm{RMQ}(\lfloor v_l/2\rfloor, v_l)$, we check if $\mathrm{LCP}[k] \geq d$. If so, u_l cannot be in $[\lfloor v_l/2\rfloor, v_l]$, so we continue searching in $[1, \lfloor v_l/2\rfloor - 1]$. If not, we know that u_l must be in $[\lfloor v_l/2\rfloor, v_l]$, so we continue searching in there. The search for u_r is handled symmetrically. Total time is $O(\log n \cdot t_{\mathrm{RMQ}} \cdot t_{\mathrm{LCP}})$.

$\mathrm{LAQT}(v,d)$: The same idea as for LAQS can be applied here, using the array TDE instead of LCP, and RMQs on TDE. Time is also $O(\log n \cdot t_{\mathrm{RMQ}} \cdot t_{\mathrm{LCP}})$.

6   Discussion

The final performance of our compressed suffix tree (CST) depends on the compressed full-text index used to implement SA. Among the best choices we have Sadakane's compressed suffix array (SCSA) [25], which is not so attractive for its $O(n\log\log\sigma)$ extra bits of space in a context where we are focusing on using $o(n)$ extra space. The alphabet-friendly FM-index (AFFM) [6] gives the best space, but our CST over AFFM is worse than Russo et al.'s CST (RCST) [24] both in time and space. Instead, we focus on using Grossi et al.'s compressed suffix array (GCSA) [11], which is larger than AFFM but lets our CST achieve better times than RCST. (Interestingly, RCST does not benefit from using the larger GCSA.) Our resulting CST is a space/time tradeoff between Sadakane's CST (SCST) [26] and RCST. Within this context, it makes sense to consider SCST on top of GCSA, to remove the huge $O(n\log\log\sigma)$ extra space of SCSA.

GCSA uses $|GCSA| = (1 + \frac{1}{\epsilon})nH_k + O(\frac{n\log\log n}{\log_\sigma n})$ bits of space for any $k \leq \alpha\log_\sigma n$ and constant $0 < \alpha < 1$, and offers times $t_\psi = O(1)$ and $t_{\mathrm{SA}} = O(\log^\epsilon n\log^{1-\epsilon}\sigma)$. On top of $|GCSA|$, SCST needs $6n + o(n)$ bits, whereas our CST needs $nH_k(2\log\frac{1}{H_k} + O(1)) + o(n)$ extra bits. Our CST times are $t_{\mathrm{LCP}} = t_{\mathrm{SA}}$, whereas t_{RMQ} and t_{PNSV} depend on how large is $o(n)$. Instead, RCST needs $|AFFM| + o(n)$ bits, where $|AFFM| = nH_k + O(\frac{n\log\log n}{\log_\sigma n}) + O(\frac{n\log n}{\gamma})$ bits, for some $\gamma = \omega(\log_\sigma n)$, to maintain the extra space $o(n\log\sigma)$. AFFM offers times $t_\psi = O(1 + \frac{\log\sigma}{\log\log n})$ and $t_{\mathrm{SA}} = O(\gamma(1 + \frac{\log\sigma}{\log\log n}))$. In addition, RCST uses $o(n) = O(\frac{n\log n}{\delta})$ bits for a parameter $\delta = \omega(\log_\sigma n)$.

An exhaustive comparison is complicated, as it depends on ϵ, γ, δ, σ, the nature of the $o(n)$ extra bits in our CST, etc. In general, our CST loses to RCST if they use the same amount of space, yet our CST can achieve sublogarithmic times by using some extra space, whereas RCST cannot. We opt for focusing on a particular setting that exhibits this space/time tradeoff. The reader can easily derive other settings. We focus on the case $\sigma = O(1)$ and all extra spaces not

Table 1. Comparison between ours and alternative compressed suffix trees. The column labeled 'General' assumes $t_\psi \leq t_{SA} = t_{LCP}$. All other columns further assume $\sigma = O(1)$, and that the extra spaces is $O(\frac{n}{\log^{\epsilon'} n})$.

Operation	Our suffix tree		Other suffix trees	
	General	over GCSA [11]	SCST [26]	RCST [24]
ROOT,COUNT, ANCESTOR	1	1	1	1
LOCATE	t_{SA}	$\log^\epsilon n$	$\log^\epsilon n$	$\log^{1+\epsilon'} n$
SDEPTH	$t_{SA} \cdot t_{RMQ}$	$\log^{\epsilon+\epsilon'} n (\log\log n)^2$	$\log^\epsilon n$	$\log^{1+\epsilon'} n$
PARENT	$t_{SA} \cdot t_{PNSV}$	$\log^{\epsilon+\epsilon'} n \log\log n$	1	$\log^{1+\epsilon'} n$
FCHILD	$t_{SA} \cdot t_{RMQ}$	$\log^{\epsilon+\epsilon'} n (\log\log n)^2$	1	$\log^{1+\epsilon'} n$
NSIBLING	$t_{SA}(t_{RMQ} + t_{PNSV})$	$\log^{\epsilon+\epsilon'} n (\log\log n)^2$	1	$\log^{1+\epsilon'} n$
SLINK,LCA	$t_{SA}(t_{RMQ} + t_{PNSV})$	$\log^{\epsilon+\epsilon'} n (\log\log n)^2$	1	$\log^{1+\epsilon'} n$
SLINK^i	$t_{SA}(t_{RMQ} + t_{PNSV})$	$\log^{\epsilon+\epsilon'} n (\log\log n)^2$	$\log^\epsilon n$	$\log^{1+\epsilon'} n$
CHILD	$t_{SA}(t_{RMQ} + \log\sigma)$	$\log^{\epsilon+\epsilon'} n (\log\log n)^2$	$\log^\epsilon n$	$\log^{1+\epsilon'} n \log\log n$
LETTER	t_{SA}	$\log^\epsilon n$	$\log^\epsilon n$	$\log^{1+\epsilon'} n$
TDEPTH	$t_{SA} \cdot t_{RMQ}$ (*)	$\log^{\epsilon+\epsilon'} n (\log\log n)^2$	1	$\log^{2+2\epsilon'} n$
LAQ$_S$	$t_{SA} \cdot t_{RMQ} \cdot \log n$	$\log^{1+\epsilon+\epsilon'} n (\log\log n)^2$	Not supp.	$\log^{1+\epsilon'} n$
LAQ$_T$	$t_{SA} \cdot t_{RMQ} \cdot \log n$ (*)	$\log^{1+\epsilon+\epsilon'} n (\log\log n)^2$	1	$\log^{2+2\epsilon'} n$

(*) Our CST needs other $nH_k(2\log\frac{1}{H_k} + O(1)) + o(n)$ extra bits to implement TDEPTH and LAQ$_T$.

related to entropy limited to $O(\frac{n}{\log^{\epsilon'} n})$ bits, for constant $0 < \epsilon' < 1$ (so $f(n) = \log^{\epsilon'} n$ in Thm. 1 and Lemma 2). Thus, our times are $t_{RMQ} = \log^{\epsilon'} n (\log\log n)^2$ and $t_{PNSV} = \log^{\epsilon'} n \log\log n$. RCST's γ and δ are $O(\log^{1+\epsilon'} n)$. Table 1 shows a comparison under this setting. The first column also summarizes the general complexities of our operations, with no assumptions on σ nor extra space except $t_\psi \leq t_{SA} = t_{LCP}$, as these are intrinsic of our structure.

Clearly SCST is generally faster than the others, but it requires $6n + o(n)$ non-compressible extra bits on top of $|CSA|$. RCST is smaller than the others, but its time is typically $O(\log^{1+\epsilon'} n)$ for some constant $0 < \epsilon' < 1$. The space of our CST is in between, with typical time $O(\log^\lambda n)$ for any constant $\lambda > \epsilon + \epsilon'$. This can be sublogarithmic when $\epsilon + \epsilon' < 1$. To achieve this, the space used in the entropy-related part will be larger than $2(1 + \log\frac{1}{H_k})nH_k$. With less than that space our CST is slower than the smaller RCST, but using more than that space our CST can achieve sublogarithmic times (except for level ancestor queries), being the only compressed suffix tree achieving it within $o(n)$ extra space.

Still, we remark that our scheme is not so attractive on large alphabets. If $\sigma = \Theta(n^\beta)$ for constant β, then our extra space includes a term $\Theta(n\log\log n)$, just as in the CST, while the latter is clearly faster.

Acknowledgments. JF wishes to thank Volker Heun and Enno Ohlebusch for interesting discussions on this subject.

References

1. Abouelhoda, M., Kurtz, S., Ohlebusch, E.: Replacing suffix trees with enhanced suffix arrays. J. Discrete Algorithms 2(1), 53–86 (2004)
2. Apostolico, A.: The myriad virtues of subword trees. In: Combinatorial Algorithms on Words. NATO ISI Series, pp. 85–96. Springer, Heidelberg (1985)
3. Berkman, O., Schieber, B., Vishkin, U.: Optimal doubly logarithmic parallel algorithms based on finding all nearest smaller values. J. Algorithms 14(3), 344–370 (1993)
4. Cole, R., Kopelowitz, T., Lewenstein, M.: Suffix trays and suffix trists: structures for faster text indexing. In: Bugliesi, M., Preneel, B., Sassone, V., Wegener, I. (eds.) ICALP 2006. LNCS, vol. 4051, pp. 358–369. Springer, Heidelberg (2006)
5. Delpratt, O., Rahman, N., Raman, R.: Engineering the louds succinct tree representation. In: Àlvarez, C., Serna, M.J. (eds.) WEA 2006. LNCS, vol. 4007, pp. 134–145. Springer, Heidelberg (2006)
6. Ferragina, P., Manzini, G., Mäkinen, V., Navarro, G.: Compressed representations of sequences and full-text indexes. ACM TALG (article 20) 3(2) (2007)
7. Fischer, J., Heun, V.: A new succinct representation of RMQ-information and improvements in the enhanced suffix array. In: Chen, B., Paterson, M., Zhang, G. (eds.) ESCAPE 2007. LNCS, vol. 4614, pp. 459–470. Springer, Heidelberg (2007)
8. Fischer, J., Heun, V.: Range median of minima queries, super cartesian trees, and text indexing (2007) (manuscript),
   www.bio.ifi.lmu.de/~fischer/fische101range.pdf
9. Geary, R., Rahman, N., Raman, R., Raman, V.: A simple optimal representation for balanced parentheses. Theoretical Computer Science 368, 231–246 (2006)
10. González, R., Navarro, G.: Compressed text indexes with fast locate. In: Ma, B., Zhang, K. (eds.) CPM 2007. LNCS, vol. 4580, pp. 216–227. Springer, Heidelberg (2007)
11. Grossi, R., Gupta, A., Vitter, J.: High-order entropy-compressed text indexes. In: Proc. 14th SODA, pp. 841–850 (2003)
12. Grossi, R., Vitter, J.: Compressed suffix arrays and suffix trees with applications to text indexing and string matching. SIAM J. on Computing 35(2), 378–407 (2006)
13. Gusfield, D.: Algorithms on Strings, Trees and Sequences: Computer Science and Computational Biology. Cambridge University Press, Cambridge (1997)
14. Jacobson, G.: Space-efficient static trees and graphs. In: Proc. 30th FOCS, pp. 549–554 (1989)
15. Kärkkäinen, J., Rao, S.: Full-text indexes in external memory. In: Meyer, U., Sanders, P., Sibeyn, J.F. (eds.) Algorithms for Memory Hierarchies. LNCS, vol. 2625, ch.7, pp. 149–170. Springer, Heidelberg (2003)
16. Ko, P., Aluru, S.: Optimal self-adjusting trees for dynamic string data in secondary storage. In: Ziviani, N., Baeza-Yates, R. (eds.) SPIRE 2007. LNCS, vol. 4726, pp. 184–194. Springer, Heidelberg (2007)
17. Kurtz, S.: Reducing the space requirements of suffix trees. Software: Practice and Experience 29(13), 1149–1171 (1999)
18. Mäkinen, V., Navarro, G.: Succinct suffix arrays based on run-length encoding. Nordic J. of Computing 12(1), 40–66 (2005)
19. Manzini, G.: An analysis of the Burrows-Wheeler transform. J. of the ACM 48(3), 407–430 (2001)
20. Munro, I.: Tables. In: Chandru, V., Vinay, V. (eds.) FSTTCS 1996. LNCS, vol. 1180, pp. 37–42. Springer, Heidelberg (1996)

21. Munro, I., Raman, V., Rao, S.: Space efficient suffix trees. J. of Algorithms 39(2), 205–222 (2001)
22. Navarro, G., Mäkinen, V.: Compressed full-text indexes. ACM Computing Surveys (article 2) 39(1) (2007)
23. Raman, R., Raman, V., Rao, S.: Succinct indexable dictionaries with applications to encoding k-ary trees and multisets. In: Proc. 13th SODA, pp. 233–242 (2002)
24. Russo, L., Navarro, G., Oliveira, A.: Fully-compressed suffix trees. In: Proc. 8th LATIN 2008. LNCS, vol. 4957, pp. 362–373. Springer, Heidelberg (2008)
25. Sadakane, K.: New text indexing functionalities of the compressed suffix arrays. J. of Algorithms 48(2), 294–313 (2003)
26. Sadakane, K.: Compressed suffix trees with full functionality. Theory of Computing Systems (to appear, 2007), doi:10.1007/s00224-006-1198-x

On Compact Representations of All-Pairs-Shortest-Path-Distance Matrices*

Igor Nitto and Rossano Venturini

Department of Computer Science, University of Pisa
{nitto,rossano}@di.unipi.it

Abstract. Let G be an unweighted and undirected graph of n nodes, and let $\mathbf{D}$ be the $n \times n$ matrix storing the All-Pairs-Shortest-Path distances in G. Since $\mathbf{D}$ contains integers in $[n] \cup +\infty$, its plain storage takes $n^2 \log(n + 1)$ bits. However, a simple counting argument shows that $(n^2 - n)/2$ bits are necessary to store $\mathbf{D}$. In this paper we investigate the question of finding a succinct representation of $\mathbf{D}$ that requires $O(n^2)$ bits of storage and still supports constant-time access to each of its entries. This is asymptotically optimal in the worst case, and far from the information-theoretic lower-bound by a multiplicative factor $\log_2 3 \simeq 1.585$. As a result $O(1)$ bits per pairs of nodes in G are enough to retain constant-time access to their shortest-path distance. We achieve this result by reducing the storage of $\mathbf{D}$ to the succinct storage of labeled trees and ternary sequences, for which we properly adapt and orchestrate the use of known compressed data structures.

1 Introduction

The study of succinct data structures has recently attracted a lot of interest in the research arena. A data structure is called *succinct* [9] when its space is *close* to the information-theoretic lower bound, and all of its operations can be supported without any slowdown with respect to the corresponding *plain* (un-succinct) data structure. The term "close to" (the information-theoretic lower bound) usually means either "equal plus some low-order terms", or "up to a constant factor from" (the information-theoretic lower bound), where the constant is pretty much close to 1. Nowadays there exist succinct versions of various data structures and data types: bitmap vectors [4,16,17], dictionaries [8], strings [14], (un)labeled trees [3,5,10], binary relations and graphs [12,1], etc.. In this paper we contribute to the design of new succinct data structures by investigating the field of compact representations of All-Pairs-Shortest-Path-Distance matrices of unweighted and undirected graphs. Formally, let G be an unweighted and undirected graph of n nodes, and let $\mathbf{D}$ be the $n \times n$ matrix that stores in its entry $\mathbf{D}[u, v]$ the length of the shortest path connecting node u to node v in G (or $+\infty$ when u and v

---

* This work has been partially supported by the Italian MIUR grants PRIN Main-Stream and Italy-Israel FIRB "Pattern Discovery Algorithms in Discrete Structures, with Applications to Bioinformatics", and by the Yahoo! Research grant on "Data compression and indexing in hierarchical memories".

P. Ferragina and G. Landau (Eds.): CPM 2008, LNCS 5029, pp. 166–177, 2008.

are not connected). $\mathbf{D}$ is called the matrix of All-Pairs-Shortest-Path distances in G (or distance matrix, for brevity) and it is typically stored in $O(n^2)$ memory words, thus taking $n^2 \log(n+1)$ bits in total.[1]

Various authors have investigated the problem of designing succinct graph encodings for supporting the retrieval of either the *adjacency list* of a node (see [12,13] and references therein), or the *approximate distance* between node pairs in various types of graphs (see [19,18] and references therein). When *exact* distances are needed, it is still open whether it is possible to deploy the intrinsic structure of matrix $\mathbf{D}$ to devise a representation which uses $o(n^2 \log n)$ bits and is as much close as possible to the information-theoretic lower bound of $n^2/2$ bits.[2] In our paper we show how to match *asymptotically* the above lower bound, by providing a succinct storage scheme for $\mathbf{D}$ which achieves a bit-space complexity that is far from the information-theoretic minimum by a multiplicative factor $\log_2 3 \simeq 1.585$, and is still able to retrieve in constant time any node-pair distance in G. We remark that the interest in space-efficient representations of shortest path distances for such a simple (undirected and unweighted) graphs is driven by applications in the field of graph layouts via Multi-Dimensional Scaling [15]. Here the distance matrix is deployed to produce a layout of the graph in the plane that closely preserves the shortest-path metric. Technically, our paper is based on an algorithmic reduction (detailed in Theorem 2) which turns the storage of $\mathbf{D}$ into the succinct storage of (ternary) labeled trees and (ternary) sequences, for which we properly adapt and orchestrate known compressed data structures. Using this algorithmic scheme we obtain two results: a simple compact representation of $\mathbf{D}$ requiring $(\log_2 3)n^2 + o(n^2)$ bits of storage and $O(1)$ access time to any of its entry (Corollary 2), and a more sophisticated one which reduces the space complexity to $(\frac{1}{2} \log_2 3) n^2 + o(n^2)$ bits (Corollary 3) without slowing down the access time.

2   Some Basic Facts

We assume the standard RAM model with memory words of $\Theta(\log n)$ bits, where n is the number of nodes in G.

Let $S[1, n]$ be a sequence drawn from the alphabet $\Sigma = \{a_1, \ldots, a_\sigma\}$. For each symbol $a_i \in \Sigma$, we let n_i be the number of occurrences of a_i in S. Let $\{P_i = n_i/n\}_{i=1}^{\sigma}$ be the empirical probability distribution for the sequence S. The zero-th order *empirical* entropy of S is defined as: $H_0(S) = -\sum_{i=1}^{\sigma} P_i \log P_i$. Recall that $|S|H_0(S)$ provides an information-theoretic lower bound to the output size of any compressor that encodes each symbol of S with a fixed codeword.

The Wavelet Tree [7] is an elegant and powerful data structure that supports rank/select primitives over sequences drawn from arbitrarily large alphabets, and achieves entropy-bounded space occupancy.

---

[1] Throughout this paper we assume that all logarithms are taken to the base 2, whenever not explicitly indicated, and we assume $0 \log 0 = 0$.

[2] This lower bound comes from the observation that there is a one-to-one correspondence between unweighted undirected graphs and their distance matrices. Thus the number of $n \times n$ distance matrices is $2^{n(n-1)/2}$.

Theorem 1. *Given a sequence $S[1, n]$ drawn from an arbitrary alphabet Σ, the Wavelet Tree built on S takes $nH_0(S) + o(n)$ bits to support the following queries in $O(\log |\Sigma|)$ time:*

- *Retrieve character $S[i]$;*
- *$Rank_c(S, i)$: compute the number of times character $c \in \Sigma$ occurs in $S[1, i]$;*
- *$Select_c(S, i)$: compute the position of the i-th occurrence of character $c \in \Sigma$ in S.*

In addition to rank/select primitives, the design of our compact representations will need to support fast *prefix sums* over integer sequences drawn from potentially large (integer) alphabets. We therefore state the following result which is an easy consequence of [11]:

Lemma 1. *Let $S[1, n]$ be a sequence drawn from the integer alphabet $\Sigma = \{-l, \ldots, 0, \ldots, l\}$. There exists an encoding of S that takes $n \lceil \log (2l + 1) \rceil + o(n \log l)$ bits and supports prefix-sum queries in $O(1)$ time.*

An essential fact in our technique will be also the availability of a storage scheme for a string S which is space succinct and is able to decode in $O(1)$ time any *short* substring of S having length logarithmic in n. To this aim, we use the following result which is an easy corollary of [6].

Corollary 1. *Given a sequence $S[1, n]$ drawn from a constant-size alphabet Σ, there is a succinct data structure that stores S in $n \log |\Sigma| + o(n)$ bits and supports the retrieval in constant time of any substring of S of length $O(\log n)$ bits.*

In the rest of this paper, we will also make use of the following two strong structural properties of the distance matrix $\mathbf{D}$:

Symmetry: $\mathbf{D}[u, v] = \mathbf{D}[v, u]$
Triangle inequality: $|\mathbf{D}[u, v] - \mathbf{D}[w, v]| \leq \mathbf{D}[u, w]$

where u, v, w are any triplets of nodes in the graph G. Note that the triangle inequality has been rewritten in a form that will help future references and intuitions. We finally notice that we can safely assume the graph G to be *connected*. Otherwise we can associate every connected-component of G with its distance matrix and then assign proper node labels in a way that takes constant-time to check whether two nodes are in the same connected component. The additional storage for these labels is $O(n \log n) = o(n^2)$ bits, thus resulting bounded above by the other terms occurring in the space bounds of our representation.

3   From Matrix D to Labeled (Spanning) Trees of G

In this section we show how to reduce the problem of succinctly representing the distance-matrix $\mathbf{D}$ into the problem of finding a succinct data structure that encodes a (ternary) labeled tree and supports in constant time a kind of *path-sum query* over its structure. To explain how this algorithmic reduction works, we introduce some useful notation and terminology.

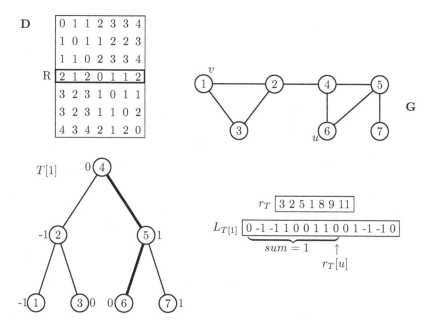

Fig. 1. (Top) A graph G and its distance-matrix $\mathbf{D}$. (Bottom) An example of labeled tree $T[1]$, relative to node $1 \in G$, and the associated arrays $L_{T[1]}$ and $r_{T[1]}$. According to Lemma 3 the sum of the labels on $\pi(6)$ is equal to the prefix-sum in $L_{T[1]}[1, r_{T[1]}[6]] = L_{T[1]}[1, 9]$ which correctly returns the value 1.

Let T be a spanning tree of the graph G and root T at anyone of its nodes, say r. Given that G is connected, T spans all n nodes of G. For each node u of T (and thus of G), we denote with:

- $\ell(u)$ an integer *label* in $\{-1, 0, 1\}$, associated to u;
- $\texttt{pre}(u)$ the rank of u in the preorder visit of T (i.e., integer in $[n]$).
- $\pi(u)$ the downward path in T which connects r to u.
- $f(u)$ the father of node u in T, and with $f^i(u)$ the ith ancestor of u in T (where $f^0(u) = u$).

Among all the possible ternary labellings ℓ of T, we consider the ones induced by the pairwise distances in G. Specifically, for any node $v \in T$ we define a labeling ℓ_v such that $\ell_v(u) = \mathbf{D}[u, v] - \mathbf{D}[f(u), v]$, where $u \in T$. This is a ternary labelling because of the triangle inequality and the adjacency of u and $f(u)$ in G. The labeled tree resulting by the ternary labelling ℓ_v applied to T is hereafter denoted by $T[v]$. An illustrative example is given in Fig. 1.

The labeled tree $T[v]$ offers an interesting property:

Lemma 2 (Path-sum Query). *For any node u, the sum of the labels on the downward path $\pi(u)$ in $T[v]$ is equal to $\mathbf{D}[u, v] - \mathbf{D}[r, v]$.*

Proof. Note that this is actually a telescopic sum:

$$\sum_{w \in \pi(u)} \ell_v(w) = \sum_{i=0,\ldots,|\pi(u)|-1} \mathbf{D}[f^i(u), v] - \mathbf{D}[f^{i+1}(u), v] = \mathbf{D}[u, v] - \mathbf{D}[r, v]. \quad \square$$

As an example, consider again Fig. 1 and sum the (ternary) labels on the downward path $\pi(6)$ in $T[1]$. The result is $0 + 1 + 0 = 1$ which is equal to $\mathbf{D}[6, 1] - \mathbf{D}[4, 1] = 3 - 2 = 1$.

Lemma 2 can be actually rephrased by saying that the computation of the distance $\mathbf{D}[u, v]$ between any pair of nodes $u, v \in G$, boils down to sum the value $\mathbf{D}[r, v]$ to the result of the path sum-query over $\pi(u)$ in $T[v]$. This is the key idea underlying the theorem below which details our reduction from the succinct storage of matrix $\mathbf{D}$ to the succinct storage of a set of path-sum query data structures built upon the labeled trees $T[v]$, for all nodes $v \in G$.

Theorem 2. *Let T be a tree of n nodes, $E(T)$ be an encoding of T's structure, and let ℓ be a labelling of T's nodes over the ternary alphabet $\{-1, 0, 1\}$. Suppose that there exists a succinct data structure $D(E(T), \ell)$ that occupies $S(n)$ bits to store ℓ and answers path-sum queries over the labeled tree $\ell(T)$ in $T(n)$ time.*

Then the distance matrix $\mathbf{D}$ of an unweighted undirected graph G of n nodes can be encoded in at most $nS(n) + |E(T)| + o(n^2)$ bits, and the distance between any pair of nodes in G can be computed in $T(n) + O(1)$ time.

Proof. Let T be the spanning tree of G rooted at node r. For each node $v \in T$, we define the labeling ℓ_v as detailed above, namely: for any node u, we set $\ell_v(u) = \mathbf{D}[u, v] - \mathbf{D}[f(u), v]$. We call $T[v]$ the tree T labeled with ℓ_v. We then represent the distance matrix $\mathbf{D}$ of graph G via the following three data structures:

- The array $R[1, n]$ which stores the shortest-path distance between r and every other node in G. Namely, R is the r-th row of matrix $\mathbf{D}$.
- The data structures $D(E(T), \ell_v)$, for any node v.
- The tree encoding $E(T)$ of T which allows the constant-time retrieval of the location of $\ell_v(u)$ inside $D(E(T), \ell_v)$, for any node-pair u, v.

The first two data structures occupy $|E(T)| + o(n^2)$ bits. The n path-sum data structures require $nS(n)$ bits, because v ranges over all n nodes in T. The claimed space bounds therefore follows.

To compute $\mathbf{D}[u, v]$ we execute a path-sum query on $D(E(T), \ell_v)$ and retrieve the sum of the labels along the path $\pi(u)$ in $T[v]$. From Lemma 2, this sum equals $\mathbf{D}[u, v] - \mathbf{D}[r, v]$, so that it suffices to add the value $R[v] = \mathbf{D}[r, v]$ to get the final result. Therefore, any distance query takes $T(n)$ time to compute the path-sum plus $O(1)$ arithmetic and table-lookup operations. $\quad \square$

4   Path-Sum Queries Boil Down to Prefix-Sum Queries

Theorem 2 allows us to shift our attention to the design of an efficient data structure that supports path-sum queries over (ternary) labeled trees. Here we go one step further and show that finding such a data structure boils down to finding an encoding of a *ternary sequence* that supports fast prefix-sum computations.

Let T be an n-node tree and let ℓ be a ternary labeling of its nodes. We visit T in preorder and build the following two arrays (see Fig. 1):

- $L_T[1, 2n]$ is the ternary sequence obtained by appending the integer label $\ell(u)$ when the pre-visit of node u starts, and the integer label $-\ell(u)$ when the pre-visit of node u ends (i.e., its subtree has been completely visited).
- $r_T[1, n]$ is the array that maps T's nodes to their positions in L_T. Hence $r_T[u]$ stores the preorder-time instant of u's visit. This way, $L_T[r_T[u]] = \ell(u)$.

The sequence L_T has the following, easy to prove, property (see Figure 1):

Lemma 3. *Let T be an n-node tree labeled with (positive and negative) integers. For any node u, the sum of the labels on path $\pi(u)$ in T can be computed as the prefix-sum of the integers in $L_T[1, r_T[u]]$.*

Theorem 2 and Lemma 3 provide us with all the algorithmic machinery we need to succinctly encode the distance matrix $\mathbf{D}$. What we really need now are succinct data structures to perform constant-time prefix-sum queries over integer sequences (namely $L_{T[v]}$, for all $v \in G$), and suitable succinct encodings of the tree T (namely $E(T)$). The following two sections will detail two possible solutions, one very simple and already asymptotically optimal, the other more sophisticated and closer to the information-theoretic lower bound.

5   Our First Solution

The labeled trees we are interested in succinctly encodings, are the ternary-labeled trees $T[v]$ introduced in the proof of Theorem 2, as a result of the ternary labeling ℓ_v. Given $T[v]$, the corresponding sequence $L_{T[v]}$ is drawn from the ternary alphabet $\{-1, 0, 1\}$. In order to compute efficiently the prefix-sum queries over $L_{T[v]}$, we use the wavelet tree data structure (see Theorem 1). This way, the prefix-sum query over $L_{T[v]}[1, r_T[u]]$ can be computed by counting (i.e., *ranking*) the number of -1 and 1 in the queried prefix of $L_{T[v]}$. By Theorem 1, this counting takes constant time and the space required to store the wavelet tree is $2(\log 3)n + o(n)$ bits (since $|\Sigma| = 3$ and $H_0(S) \leq \log |\Sigma|$).

We are therefore ready to detail our first simple solution to the succinct encoding of $\mathbf{D}$. For each node $v \in T$, we consider the labeling ℓ_v, the resulting labeled tree $T[v]$, and the corresponding ternary sequence $L_{T[v]}$. We then set the tree encoding $E(T) = r_T$ and build $D(E(T), \ell_v)$ as the wavelet tree of the ternary sequence $L_{T[v]}$. By plugging these data structures into Theorem 2, and exploiting Lemmas 2–3, we obtain:

Theorem 3. *Let G be an undirected and unweighted graph of n nodes, and let $\mathbf{D}$ be its $n \times n$ matrix storing all-pairs-shortest-path distances. There exists a succinct representation of $\mathbf{D}$ that uses at most $2n^2(\log 3) + o(n^2)$ bits, and takes constant-time to access any of its entries.*

For a running example of Theorem 3 we refer the reader to Fig. 1. Assume that we wish to compute $\mathbf{D}[6,1] = 3$. According to Lemma 2, we need to compute the path-sum over $\pi(6)$ in $T[1]$, which equals to $\mathbf{D}[6,1] - \mathbf{D}[4,1] = 1$, and then add to this value $R[1] = \mathbf{D}[4,1] = 2$ (given that T's root is node 4). By Lemma 3, the path-sum computation boils down to the prefix-sum of $L_{T[1]}[1, r_T[6]]$, which correctly gives the result 1.

In Section 1, we noted that the information-theoretic lower bound for storing the distance matrix $\mathbf{D}$ is $\frac{n^2}{2}$ bits. Therefore the solution proposed in Theorem 3 is asymptotically space- and time-optimal in the worst case, and far from such lower bound of a multiplicative factor $4 \log 3 \simeq 6.34$. This simple approach proves that a succinct encoding taking $O(1)$ bits per pairwise-distance of G and $O(1)$ time per distance computation does exist.

A non-trivial issue is now to reduce the amount of bits spent to encode every entry of $\mathbf{D}$, by exploiting some structural properties of G and T, in order to come as much close as possible to the lower bound 0.5. A first step in this direction is obtained by exploiting the symmetry of matrix $\mathbf{D}$, and thus storing just the suffix $L_{T[v]}[1, r_T[v]]$ for every ternary sequence $L_{T[v]}$. This way, when we query $\mathbf{D}[u,v]$, if $\mathtt{pre}(u) \leq \mathtt{pre}(v)$ we proceed as detailed above (because $r_T[u] \leq r_T[v]$). Otherwise, we swap the role of u and v, and proceed as before. Using this simple trick we halve the space complexity and obtain:

Corollary 2. *There exists a representation for* $\mathbf{D}$ *that uses at most* $n^2(\log 3) + o(n^2)$ *bits, and takes constant-time to access any one of its entries.*

6   Our Second Solution

In this section we show how to further halve the space complexity by deploying the structure of T. We proceed in two steps. First, we exhibit a path-sum data structure for an n-node ternary labeled tree that takes $(\log 3)n + o(n)$ bits and supports path-sum queries in $O(1)$ time (Theorem 4). The core of this technique is a well-known approach to the decomposition of arbitrary trees in suitable subtrees, called macro-micro tree partitioning (see e.g. [2]). Second, we deploy again the "symmetry in $\mathbf{D}$", and get our final result (Corollary 3).

Let T be a tree labeled over $\{-1, 0, 1\}$, and set $\mu = \lceil (\log n)/4 \rceil$. A node $v \in T$ is called a *jump* node, if it has at least μ descendants in T but every child of v has strictly less than μ descendants. A node v is called a *macro* node, if it has at least one jump node among its descendants. The root is assumed to be a macro node. Any other node of T that is neither jump nor macro is called a *micro* node. Note that all descendants of micro nodes are micro nodes too, so that we define a *micro*-tree as any maximal subtree of micro nodes in T.

Let $Q_1, \ldots, Q_t$ be the sequence of micro-trees in T ordered by preorder rank of their roots, and let T^* be the subtree of T induced by its macro and jump nodes. Of course, trees $T^*, Q_1, \ldots, Q_t$ form a partition of T (see Figure 2). Since every micro node has at most μ descendants, the size of each micro tree is upper bounded by μ. This decomposition is usually called *macro-micro* partition of T.

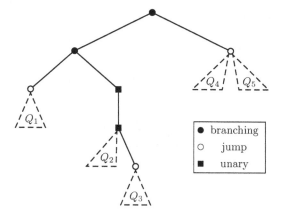

Fig. 2. Macro-micro tree partition

In this section we will show how to deploy this decomposition to further reduce the space-encoding of **D**.

Let us concentrate on the subtree T^*, formed by jump and macro nodes. Note that jump nodes form the leaves of this tree, and are $O(n/\log n)$ in number. The macro nodes are internal in T^* and can be then divided into *branching* nodes, if they have at least two children in T^*, or *unary* nodes. The number of branching nodes is upper bounded by the number of leaves in T^* (i.e., jump nodes), and thus it is $O(n/\log n)$. To deal with long chains of unary nodes in T^*, we sample them by taking one out of $\lceil \log n \rceil$ consecutive nodes in any maximal unary path of T^*. This way we sample $O(n/\log n)$ unary nodes. The set of nodes formed by jump nodes, branching nodes, and sampled unary nodes is called *breaking nodes*, and has size $O(n/\log n)$. By definition, the distance between any non-breaking node and its closest breaking ancestor in T^* is at most $\lceil \log n \rceil$.

Given the notion of breaking nodes, we define T_F as the tree T^* *contracted* to include only the breaking nodes: i.e., u has parent u' in T_F iff u, u' are breaking nodes and u' is the lowest breaking ancestor of u in T^*. Since we wish to execute path-sum queries over T^* by deploying T_F, we need to reflect the contraction process onto the tree labeling too. This is done as follows. We label every node $u \in T_F$ with the integer $\ell_F(u) = \sum_{w \in \pi(u',u)} \ell(w)$, where u' is the father of u in T_F, $\pi(u', u)$ is the path in T^* connecting u to its father u', and ℓ is the labeling of T (and thus of T^*). Given the sampling over the unary macro-nodes, and since ℓ is assumed to be a ternary labeling, the label $\ell_F(u)$ is an integer less than $\lceil \log n \rceil$ (in absolute value). At this point, we note that the path-sum leading to any breaking node u can be equally computed either in T or in T_F.

To apply Theorem 2, we need a succinct path-sum data structure that we design here based on the macro-micro decomposition of the ternary labeled tree T. Specifically, let us assume that we wish to answer a path-sum query on a node $u \in T$, we distinguish three cases depending on whether u is micro or not.

1. Node u is non-micro and breaking. As observed above, we can compute the path-sum over $\pi(u)$ by acting on the contracted tree T_F.

2. Node u is non-micro and non-breaking. Since u is not a node of T_F, we pick z as the lowest breaking ancestor of u in T^*. Hence $z \in T_F$. The path $\pi(u)$ lies in T^* and can then be decomposed into two subpaths: one connecting T's root r to the breaking node z, and the other being a unary path connecting z to u (and formed by all non-breaking nodes). The first path-sum can be executed in T_F, whereas the other path-sum needs some specific data structure over the unary paths of T^* (formed by non-breaking nodes).

3. Node u is micro. Let r_j be the root of its enclosing micro-tree Q_j. The parent of r_j, say $f(r_j)$, is a jump node (and thus $f(r_j) \in T_F$), by definition. Therefore the path $\pi(u)$ can be decomposed in two subpaths: one lies in T_F and connects its root r to $f(r_j)$, the other lies in Q_j and connects r_j to u. Consequently, the first path-sum can be executed in T_F, whereas the other path-sum can be executed in Q_j.

We are therefore left with the design of succinct data structures to support constant-time path-sum queries over the contracted tree T_F, the unary paths in T^*, and the micro-trees Q_js. We detail their implementation below.

Path-sum over the T_F. Given the labeled tree T_F, we build the integer sequence L_{T_F} and the array r_{T_F}, similarly as done in Section 4. Since there are $O(n/\log n)$ breaking nodes, $|L_{T_F}| = O(n/\log n)$ and its elements are in the range $[-\log n, +\log n]$. Now we define K as the data structure of Theorem 1 built on sequence L_{T_F} (here $l = O(\log n)$), thus taking $O(n \log \log n / \log n) = o(n)$ bits. By Lemma 3, the path-sum query involving a breaking node in T_F can then be answered in constant time using K and r_{T_F}.

Path-sum over the unary paths in T^*. We serialize the unary paths in T^* according to the pre-order visit of this tree. Let us denote by P_{T^*} the resulting sequence of ternary labels of those (serialized) nodes. Notice that P_{T^*} is similar in vein to L_{T^*}, but it avoids the double storage of the node labels. Nonetheless path-sum queries over unary paths of T^* can still be executed as prefix-sum queries over P_{T^*}; but with the additional advantage of saving a factor 2 in the space complexity. More specifically, any path-sum query over a unary path in T^* actually boils down to a *range-sum query* over the sequence P_{T^*}, because the paths are unary and node labels are written in P_{T^*} according to a pre-visit of T^*. Additionally, a range-sum query over P_{T^*} can be implemented as a difference of two prefix-sum queries over the same sequence. As a result, we build a wavelet tree on P_{T^*} (see Theorem 1) taking $(\log 3)|P_{T^*}| + o(|P_{T^*}|)$ bits of space (since $|\Sigma| = 3$ and $H_0(P_{T^*}) \leq \log |\Sigma|$). Given this wavelet tree and an array $\mathrm{pre}_{T^*}[1, n]$, which stores the rank of the macro-nodes in the preorder visit of T^*, the path-sum queries over the unary paths in T^* can be answered in constant time.

Path-sum over the micro-trees. Here we exploit the fact that micro-trees are small enough, so that we can explicitly store the answer to all possible path-sum queries over all of them in succinct space. We note that any path-sum query over a micro-tree Q can be uniquely specified by a triple $\langle Q, \ell(Q), i \rangle$, where Q denotes the micro-tree structure, $\ell(Q)$ denotes the ternary labeling of Q, and i

is the pre-order rank in Q of the queried node (hence $i \leq \mu$). We then build a table C that tabulates all possible path-sum queries over micro-trees, indexed by triplets $\langle Q, \ell(Q), i \rangle$. To access C, we need an encoding for the triplet: i.e., we encode the Q's structure via any succinct tree encoding of at most 2μ bits (see e.g. [9,12]), and encode $\ell(Q)$ via the string P_Q which consists of no more than μ ternary labels (obtained by visiting in pre-order Q, see above). Consequently, C consists of $2^{2\mu} \times 3^{\mu} \times \mu$ entries, each storing an integer smaller than μ in absolute value. Table C thus takes less than $O(n \log n \log \log n)$ bits. As a result, a path-sum query over a micro-tree Q can be answered in constant time, provided that we have constant-time access to its micro-tree encoding and labeling. To this aim, we store all structural encodings of the Q_i's in one string, thus taking $O(n)$ bits overall. Also, we create the string S_ℓ, obtained by juxtaposing the encodings of the labellings $\ell(Q_i)$ (i.e., the strings P_{Q_i}), for all micro-trees Q_i of T. Note that S_ℓ depends on the labeling ℓ of T. Finally we compress and index S_ℓ via the succinct data structure of Corollary 1. This way, we can retrieve any $\ell(Q_i)$ in constant time, taking a total of $|S_\ell| \log 3 + o(|S_\ell|)$ bits.

To complete the description of our solution we just need to store some other auxiliary arrays which take $O(n \log n) = o(n^2)$ bits overall:

- the array encoding the node type– (non)micro, breaking.
- the array of parent-pointers of T's nodes (useful to execute path-sums in micro-trees);
- the arrays storing for each micro node the root of its micro-tree and its pre-order rank inside it (useful to execute path-sums in micro-trees).
- the array storing for each unary non-breaking node the top node in its maximal unary path (useful to execute path-sums of non-micro and non-breaking nodes).

At this point, we are left with the orchestration of all data structures sketched above in order to provide a succinct data structure for performing path-sum queries over the ternary labeled tree T, and then apply Theorem 2. We indeed use the above macro-micro tree decomposition on T (and its labeling ℓ) and define:

- the succinct data structures $D(E(T), \ell)$, as the combination of data structure K built on T_F, the wavelet tree built on P_{T^*}, and the compressed indexing of S_ℓ. These data structures take $(\log 3)(|P_{T^*}| + |S_\ell|) + o(|P_{T^*}| + |S_\ell| + n) = (\log 3)n + o(n)$ bits.
- the encoding $E(T)$ as the combination of the table C, the encodings of the micro-tree structures, and all other auxiliary arrays, for a total of $o(n^2)$ bits.

We then plug this data structure to Theorem 2, and get the following result:

Theorem 4. *There exists a representation for* **D** *that uses at most* $n^2(\log 3) + o(n^2)$ *bits, and takes constant-time to access any of its entries.*

Proof. The space bound has been proved above. The time bound derives from the three-cases analysis made above and the use of $D(E(T), \ell)$ data structure which guarantees constant-time prefix-sum queries.   □

The previous solution does not deploy the symmetry-idea sketched at the end of Section 5. We then apply it to further halve the above space occupancy:

Corollary 3. *There exists a representation for $\mathbf{D}$ that uses at most $n^2(\frac{\log 3}{2}) + o(n^2)$ bits, and takes constant-time to access any of its entries.*

7 Conclusion and Open Problems

We have studied the problem of succinctly encoding the All-Pair-Shortest-Path matrix of an n-node unweighted and undirected graph. We have designed compact representations which are asymptotically time- and space-optimal, and result close to the information-theoretic lower bound by a small constant factor.

We leave two interesting open problems. The first one concerns with (dis) proving the existence of a succinct data structure that achieves $n^2/2 + o(n^2)$ bits of space occupancy and supports distance-queries in constant time. The second question deals with the design of a solution whose space complexity depends on the number m of edges in the graph G, and still guarantees constant time to compute *exactly* the shortest-path distance between any pair of its nodes. In fact, in the case of sparse graphs, the information-theoretic lower bound is $2m \log \frac{n}{m} - \Omega(m) \ll n^2$ bits. Such a solution would be of big practical relevance in applications that manage very sparse large graphs.

Acknowledgments. The authors wish to thank Paolo Ferragina for useful comments and his help in improving the exposition of this paper.

References

1. Barbay, J., He, M., Munro, J.I., Srinivasa Rao, S.: Succinct indexes for string, bynary relations and multi-labeled trees. In: Proc. 18th ACM-SIAM Symposium on Discrete Algorithms (SODA) (2007)
2. Bender, M.A., Farach-Colton, M.: The lca problem revisited. In: Gonnet, G.H., Viola, A. (eds.) LATIN 2000. LNCS, vol. 1776, pp. 88–94. Springer, Heidelberg (2000)
3. Benoit, D., Demaine, E., Munro, I., Raman, R., Raman, V., Rao, S.: Representing trees of higher degree. Algorithmica 43, 275–292 (2005)
4. Brodnik, A., Munro, I.: Membership in constant time and almost-minimum space. SIAM Journal on Computing 28(5), 1627–1640 (1999)
5. Ferragina, P., Luccio, F., Manzini, G., Muthukrishnan, S.: Structuring labeled trees for optimal succinctness, and beyond. In: Proc. 46th IEEE Symposium on Foundations of Computer Science (FOCS), pp. 184–193 (2005)
6. Ferragina, P., Venturini, R.: A simple storage scheme for strings achieving entropy bounds. Theor. Comput. Sci. 372(1), 115–121 (2007)
7. Grossi, R., Gupta, A., Vitter, J.: High-order entropy-compressed text indexes. In: Proc. 14th ACM-SIAM Symposium on Discrete Algorithms (SODA), pp. 841–850 (2003)
8. Gupta, A., Hon, W.K., Shah, R., Vitter, J.S.: Dynamic rank/select dictionaries with applications to XML indexing. Technical Report Purdue University (2006)

9. Jacobson, G.: Space-efficient static trees and graphs. In: Proc. 30th IEEE Symposium on Foundations of Computer Science (FOCS), pp. 549–554 (1989)
10. Jansson, J., Sadakane, K., Sung, W.K.: Ultra-succinct representation of ordered trees. In: Proc. 18th ACM-SIAM Symposium on Discrete Algorithms (SODA) (2007)
11. Mäkinen, V., Navarro, G.: Rank and select revisited and extended. Theor. Comput. Sci. 387(3) (2007)
12. Munro, I., Raman, V.: Succinct representation of balanced parentheses, static trees and planar graphs. In: Proc. of the 38th IEEE Symposium on Foundations of Computer Science (FOCS), pp. 118–126 (1997)
13. Munro, I., Raman, V.: Succinct representation of balanced parentheses and static trees. SIAM J. Computing 31, 762–776 (2001)
14. Navarro, G., Mäkinen, V.: Compressed full-text indexes. ACM Comput. Surv. 39(1) (2007)
15. Working Group on Algorithms for Multidimensional Scaling. Algorithms for multidimensional scaling. DIMACS Web Page, http://dimacs.rutgers.edu/Workshops/Algorithms/AlgorithmsforMultidimensionalScaling.html
16. Pagh, R.: Low redundancy in static dictionaries with constant query time. SIAM Journal on Computing 31(2), 353–363 (2001)
17. Raman, R., Raman, V., Srinivasa Rao, S.: Succinct indexable dictionaries with applications to encoding k-ary trees and multisets. In: Proc. 13th ACM-SIAM Symposium on Discrete Algorithms (SODA), pp. 233–242 (2002)
18. Thorup, M.: Compact oracles for reachability and approximate distances in planar digraphs. J. ACM 51(6), 993–1024 (2004)
19. Thorup, M., Zwick, U.: Approximate distance oracles. In: STOC, pp. 183–192 (2001)

Computing Inverse ST in Linear Complexity

Ge Nong[1,*], Sen Zhang[2], and Wai Hong Chan[3,**]

[1] Computer Science Department, Sun Yat-Sen University, P.R.C.
issng@mail.sysu.edu.cn
[2] Dept. of Math., Comp. Sci. and Stat., SUNY College at Oneonta, U.S.A.
zhangs@oneonta.edu
[3] Department of Mathematics, Hong Kong Baptist University, Hong Kong
dchan@hkbu.edu.hk

Abstract. The Sort Transform (ST) can significantly speed up the block sorting phase of the Burrows-Wheeler transform (BWT) by sorting only limited order contexts. However, the best result obtained so far for the inverse ST has a time complexity $O(N \log k)$ and a space complexity $O(N)$, where N and k are the text size and the context order of the transform, respectively. In this paper, we present a novel algorithm that can compute the inverse ST in an $O(N)$ time/space complexity, a linear result independent of k. The main idea behind the design of the linear algorithm is a set of cycle properties of k-order contexts we explored for this work. These newly discovered cycle properties allow us to quickly compute the longest common prefix (LCP) between any pair of adjacent k-order contexts that may belong to two different cycles, leading to the proposed linear inverse ST algorithm.

1 Introduction

Since the Burrows-Wheeler transform (BWT) was introduced in 1994 [1], it has been successfully used in a wide range of data compression applications. Inspired by the success of the BWT, many variants have also been proposed by the research community during the past decade. Among them, one noticeable is the Sort Transform (ST) that was introduced in 1997 by Schindler [2,3], which can speed up the block sorting phase of the BWT by sorting only a portion of the rotating matrix. The main idea of the ST is to limit sorting to the first k columns only, instead of the full matrix sorted by the BWT. Specifically, given the same rotating matrix as defined in the BWT, the k-order ST will lexicographically sort all the rows of the matrix according to their k-order contexts first; in case that there are any two identical k-order contexts, the tie will be resolved by preserving the relative order between them in the original rotating matrix, i.e., the sorting in ST is stable. With only a relatively small adjustment to the sorting size of the matrix, the ST can be expected to perform much faster than the BWT, yet

* Nong was partially supported by the National Natural Science Foundation of P.R.C. (Project No. 60573039).
** Chan was partially supported by the Faculty Research Grant (FRG/06-07/II-28), Hong Kong Baptist University and the CERG (HKBU210207), RGC, Hong Kong SAR.

P. Ferragina and G. Landau (Eds.): CPM 2008, LNCS 5029, pp. 178–190, 2008.

retaining high compression ratios. Schindler has built a fast compression software called *szip* [2] using the ST.

Problem

A major tradeoff caused by the ST's partial sorting scheme is that the inverse ST is more complex than the inverse BWT. This is because although each of the unlimited context is unique, the uniqueness of any limited order context considered by the partial sorting scheme in ST can no longer be guaranteed. To deal with the duplicated k-order contexts, Schindler proposed a hash table based approach in which the text retrieval has to rely on a hash table driven context lookup and the context lookup has to rely on the complete restoration of all the k-order contexts [4], resulting in an $O(kN)$ time/space complexity for inverting the ST of a size-N string. Noticing that neither the full restoration of contexts nor the hash-based context lookup is required by the inverse BWT, we have proposed an auxiliary vectors based framework [5,6,7], which is similar to that used for the inverse BWT [1], but different from any possible hash table based approaches suggested by Schindler [4], Yokoo [8] and Bird [9]. This framework requires only $O(N)$ space complexity, however, the time complexity of our previous solutions remains to be superlinear. The time complexity achieved in [5] is $O(kN)$, which we have recently reduced to $O(N \log k)$ in [6,7]. Nevertheless, all the time complexities of the existing solutions for the inverse ST involve the context order k, one way or another. In contrast, the inverse BWT has a linear time/space complexity of $O(N)$. Therefore, the question of our particular interest here is whether the inverse ST is linear computable.

Answer

We present here a positive answer to this question by introducing a novel algorithm that explores a set of properties about longest common prefixes (LCPs) and cycles in the k-order contexts to compute the inverse ST for any context order $k \in [1, N]$. The new algorithm has a linear time/space complexity $O(N)$, independent of k.

Section 2 introduces some basic definitions and general notations. Our linear inverse ST algorithm is developed and analyzed in section 3.

2   Preliminary

A text S of length N is denoted as $x_1 x_2 x_3 ... x_{N-1}\$$, where each character $x_i \in \Sigma$, $i \in [1, N-1]$ and Σ is the alphabet. The last character \$ of the text is a *sentinel*, which is the unique lexicographically greatest[1] character in S. (Appending a sentinel to the original text has been used in many previous publications; readers may refer to papers [10,11] for more details.) Given S, according to the cyclic rotation scheme, we call $S[i]$ the immediate preceding character of $S[i+1]$ where $i \in [1, N-1]$, and $S[N]$ the immediate preceding character of $S[1]$. Cyclic rotating S a total number of N times, we obtain the original matrix M_0 for computing the ST of S.

---

[1] Symmetrically, the sentinel can be assumed as the smallest.

$$\begin{bmatrix} i & p & p & i & \$ & m & i & s & s & i & s & s \\ i & s & s & i & s & s & i & p & p & i & \$ & m \\ i & s & s & i & p & p & i & \$ & m & i & s & s \\ i & \$ & m & i & s & s & i & s & s & i & p & p \\ m & i & s & s & i & s & s & i & p & p & i & \$ \\ p & i & \$ & m & i & s & s & i & s & s & i & p \\ p & p & i & \$ & m & i & s & s & i & s & s & i \\ s & i & s & s & i & p & p & i & \$ & m & i & s \\ s & i & p & p & i & \$ & m & i & s & s & i & s \\ s & s & i & s & s & i & p & p & i & \$ & m & i \\ s & s & i & p & p & i & \$ & m & i & s & s & i \\ \$ & m & i & s & s & i & s & s & i & p & p & i \end{bmatrix}$$

Fig. 1. The matrix M_2 for the 2-order ST, where the last column is the transformed text

Definition 1. *(The original matrix M_0). The $N \times N$ symmetric matrix originally constructed from the texts obtained by rotating the text S. Specifically, the first row of the matrix M_0 is assigned to be S, denoted by S_1; and for each of the remaining rows, a new text S_i is obtained by cyclically shifting the previous text S_{i-1} one column to the left.*

Each row in M_0 is a text, where the first k characters is called the k-order context of the last character, for $k \in [1, N]$. When $k = N$, the k-order context is also called the unlimited order context; or else a limited order context. Lexicographically sorting all the rows of M_0, we get a new matrix M_k, which last column is the ST of S. For an example of the ST, we give in Fig. 1 the matrix M_2 for $S = \begin{bmatrix} m & i & s & s & i & s & s & i & p & p & i & \$ \end{bmatrix}$. By taking the transpose of the last column of M_2 and locating the row position of the orginal text S which is 5 in this case, the transform result can be denoted as a couplet ($\begin{bmatrix} s & m & s & p & \$ & p & i & s & s & i & i & i \end{bmatrix}$, 5).

For presentation simplicity, we introduce the following notations. Let $Z[N_r, N_c]$ represent a two-dimensional array Z consisting of N_r rows and N_c columns. To specify an array's subscript range in each dimension, we use the notation of $a : b$. For example, $Z[a : b, c : d]$ represents a 2-D sub-array of $Z[N_r, N_c]$ covering the rows from a to b and the columns from c to d, where $1 \le a \le b \le N_r$ and $1 \le c \le d \le N_c$. In case $a = b$ and/or $c = d$, the simpler forms of $Z[a, c : d]$ or $Z[a : b, c]$ are used instead, respectively. From M_k, we define $F_k = M_k[1 : N, 1]^T$ and $L_k = M_k[1 : N, N]^T$, i.e. the transposes of the first and the last columns, respectively, where $k \in [1, N]$. When $k = N$, the simpler forms of M, F and L can be used for M_k, F_k and L_k instead, respectively.

3   The Linear Inverse ST Algorithm

3.1   Basis

Let's define two vectors to establish an one-to-one mapping among the characters of F_k and L_k, as below.

Definition 2. *(P_k and Q_k). P_k and Q_k are two size-N row vectors, where the former satisfies*

$$\begin{cases} F_k[P_k[i]] = L_k[i], \text{ for } i \in [1, N]; \\ P_k[i] < P_k[j], \quad \text{ for } (1 \leq i < j \leq N \text{ and } L_k[i] = L_k[j]) \text{ or } (L_k[i] < L_k[j]). \end{cases}$$

and the later satisfies

$$\begin{cases} L_k[Q_k[i]] = F_k[i], \text{ for } i \in [1, N]; \\ Q_k[i] < Q_k[j], \quad \text{ for } 1 \leq i < j \leq N \text{ and } F_k[i] = F_k[j]. \end{cases}$$

P_k maps the index of each character of L_k to its index at F_k, and Q_k maps the index of each character at F_k to its index at L_k. Furthermore, P_k and Q_k are reciprocal to each other, i.e. $Q_k[P_k[i]] = i$ and $P_k[Q_k[i]] = i$. When $k = N$, the simpler forms of P and Q can be used for P_k and Q_k instead. The solution for the inverse BWT [1] is re-stated in Fig. 2, where S is restored by *backward retrieval* using the vector P to iteratively retrieve the immediate preceding character $S[i-1]$ for each known $S[i]$, utilizing the following properties.

Property 1. Given L, F can be obtained by sorting all the characters of L.

Property 2. In both F and L, the relative orders of any two identical characters are consistent.

Property 3. Given $L[i] = S[j]$, we have $S[j-1] = L[P[i]]$ for $j \in [2, N]$, and $S[N] = L[P[i]]$ for $j = 1$.

3.2 Algorithm Framework

We previously proposed in [5,7] an auxiliary vector based framework to inverse ST using no hashing table. The most complex part in that framework is computing the k-order context switch vector D, which is defined as following.

To denote the k-order contexts in M_k, we define the k-order context vector CT_k, which is a size-kN vector with each $CT_k[i]$ denoting the k-order context of $L_k[i]$, i.e. $CT_k[i] = M_k[i, 1:k]$, where $i \in [1, N]$. From CT_k, we further define the k-order context switch vector D as below.

```
IBWT(char *L, int start, int n) {
// restore n characters starting from L[i]
j=start;
for(i=n; i>0; i--) {
  S[i]=L[j]; // restore the ith character.
  j=P[j]; //update index for backward retrieving the preceding character.
}
return S;
}
```

Fig. 2. Algorithm for the inverse BWT

Definition 3. *(k-order context switch vector D). A size-N row vector with each* $D[i]$, $i \in [1, N]$ *defined as*

$$D[i] = \begin{cases} 0, \text{ for } CT_k[i] = CT_k[i-1]; \\ 1, \text{ for } CT_k[i] \neq CT_k[i-1]. \end{cases}$$

For CT_k, we say there is a k-order context switch from row $i-1$ to row i if k-order context is new if either it is the first context (when i=1) or it is different from that at the $(i-1)$th row (for $i \in [2, N]$). Suppose that D has been known, we need the following two size-N vectors C_k and T_k to use with D together to restore S from L_k, in a fashion of backward retrieval. C_k is called the counter vector, which is a size-N vector recording the occurrence of each unique k-order context at its first position in the M_k. If $C_k[i] > 0$, the k-order context at the ith row is new and is repeated for $C_k[i]$ times starting from the ith row consecutively up to the $(i + C_k[i] - 1)$th row; otherwise, the k-order context at the ith row repeats the one at the previous row. T_k is called the index vector, which is a size-N vector pointing to the starting row of each unique k-order context. $M_k[T_k[i], 2 : k]$ is the $(k-1)$-order context of the character $L_k[i]$, where $i \in [1, N]$. Given $L_k[i]$, T_k tells that starting from the $T_k[i]$th row in M_k, there are $C_k[T_k[i]]$ consecutive rows sharing the same k-order context with $L_k[i]$ being the first character.

The complexity of calculating D constitutes the bottleneck of the whole framework. We have previously shown the best result for computing D from L requires an $O(N \log k)$ time complexity [7] and using a space $O(N)$. In the next subsection, we will present an even more efficient linear algorithm, which has a time/space complexity of $O(N)$, independent of the context order k.

3.3   Computing D in $O(N)$ Time/Space

We first show the relationship between the vector D and the lengths of the longest common prefixes (LCPs) between any two adjacent rows in $M_k[1 : N, 1 : k]$, then present a linear algorithm that can compute D in a k-independent time/space complexity of $O(N)$ by exploring the cycle and the LCP properties. Let $lcp(i, j)$ denote the longest common prefix (LCP) between $CT_k[i]$ and $CT_k[j]$, where $i, j \in [1, N]$. Further, let $Height$ be a size-N vector, where $Height[i]$ denotes the height of $L_k[i]$ that equals to the length of *the LCP between the two k-order contexts of* $CT_k[i-1]$ *and* $CT_k[i]$, i.e. $Height[i] = \|lcp(i-1, i)\|$ ($\| \cdot \|$ is the cardinality operator for a set, returning the set's size.).

Lemma 1. *Given* M_k, *the following items regarding* D *and* $Height$ *are equivalent.*

- $D[i] = 0$ *(as opposed to* $D[i] = 1$*);*
- $Height[i] = k$ *(as opposed to* $Height[i] \in [0, k)$*).*

From the above lemma, it is easy to see that once the vector $Height$ is available, D can be easily computed in $O(N)$ by traversing $Height$ once. Intuitively, this implies that we can convert the problem of solving D to finding an efficient solution for the computation of $Height$.

Cycle of Characters

Now, we introduce the definition of *cycle*, which builds the foundation for developing our algorithm to compute the vector $Height$ in linear time/space.

Definition 4. *Cycle $\alpha(i)$: the list of characters consisting of a subset of the characters in L_k, satisfying*

$$\alpha(i) = \begin{cases} \alpha(i)[1] = L_k[i] = L_k[Q_k[j]], \text{ for } \alpha(i)[\|\alpha(i)\|] = L_k[j]; \\ \alpha(i)[x+1] = L_k[Q_k[j]], \quad \text{ for } \alpha(i)[x] = L_k[j] \text{ and } x \in [1, \|\alpha(i)\| - 1]; \end{cases}$$

Given $\alpha(i)$, calling the function $IBWT(L_k, i, \|\alpha(i)\|)$ will backward retrieve all the characters in $\alpha(i)$ one by one with a period length of $\|\alpha(i)\|$. In this sense, we term $\alpha(i)$ a cycle. From the definition of cycle, we immediately see this property.

Property 4. Any two cycles are disjoint.

Finding All Cycles

Because of the existence of cycles, we want to discover all the cycles from L_k first and then we use them to compute the heights for all characters in L_k in a linear complexity of $O(N)$. To achieve this goal, we introduce three one-dimension size-N arrays X_0, X_1 and Y to store all the cycles. An algorithm for finding all the cycles from L_k in $O(N)$ time/space is given below.

1. Initially, mark all items of L_k as unvisited.
2. Traverse L_k once from left to right. For each unvisited item $L_k[i]$, retrieve the cycle $\alpha(i)$ using Q_k in $O(\alpha(i))$ time, and mark all the characters in this cycle as visited. All the characters of the found cycle $\alpha(i)$ are consecutively stored into X_0. To help separate two neighbor cycles stored in X_0, we maintain the relative head and tail positions of each cycle by X_1. The array X_1 contains two different kinds of values: non-negative values and negative values. If $X_1[i] \geq 0$, $X_0[i + X_1[i]]$ is the end character of the cycle; otherwise, $X_0[i + X_1[i]]$ is the head character of the cycle. To show where the characters in X_0 comes from L_k, the array Y is used to map each character in L_k to its position in X_0, i.e. $L_k[Q_k[i]] = X_0[Y[i]]$. Doing in this way, extracting all the cycles from L_k and compute the arrays X_0, X_1 and Y can be done in a total time/space complexity of $O(N)$.

Let the immediate predecessor of S_i in M_0 be S_{i-1} for $i \in [2, N]$, and S_N for $i = 1$, respectively. To generalize these definitions, in M_0, we call the jth row the xth generation successor of the ith row and the ith row the xth generation predecessor of the jth row, if the jth row is x rows cyclically below the ith row. Let Q_k^x be the power notation of Q_k, which maps each row in M_k to its xth generation successor (recalling that Q_k is the vector mapping each row in M_k to its immediate successor), for instance, $Q_k^3[i] = Q_k[Q_k[Q_k[i]]]$. Similarly, we can define $P_k^x[i]$ to map row i in M_k to its xth generation predecessor. From the definitions of CT_k, L_k, Q_k, P_k and cycle, we observed these properties.

Property 5. For any $k \in [1, N]$, $i \in [1, N]$, we have (1) $Q_k^{\|\alpha(i)\|}[i] = P_k^{\|\alpha(i)\|}[i] = i$; and (2) $CT_k[i] = [L_k[Q_k[i]], L_k[Q_k[Q_k[i]]], ..., L_k[Q_k^k[i]]]$;

Property 6. The k-order context of $L_k[i]$ is the first k characters of the string made up of the unlimited repetitions of cycle $\alpha(Q_k[i])$.

Property 6 describes a relationship between the context of a character in L_k and the cycle that the character belongs to. According to the definition of cycle, we have $\alpha(Q_k[i]) = [L_k[Q_k[i]], L_k[Q_k[Q_k[i]]], ..., L_k[Q_k^{\|\alpha(Q_k[i])\|}[i]]]$; from Property 5, we have $CT_k[i] = [L_k[Q_k[i]], L_k[Q_k[Q_k[i]]], ..., L_k[Q_k^k[i]]]$. Comparing $CT_k[i]$ and $\alpha(i)$, this property is immediately true because the cycle $\alpha(Q_k[i])$ will repeat itself in CT_k at each position $j \in [1, k]$ satisfying $(j - 1)\%\|\alpha(Q_k[i])\| = 0$, where $\%$ is the integer modulo operator. Once we have all the cycles extracted from L_k and have them saved in the three arrays X_0, X_1 and Y, retrieving the character that is l character(s) cyclic far away from the character $X_0[i]$ is a simple algebra problem, which can be trivially done in time $O(1)$. In other words, for any given character $L_k[i]$, using X_0, X_1 and Y, we can retrieve the character that is l character(s) cyclic away from it in the original text S in time $O(1)$, where $l \in [0, k - 1]$.

Computing Heights for a Cycle

We establish a theorem for inductive computing the heights of all the characters in a cycle, as below.

Theorem 1. $Height[Q_k[i]] \geq Height[i] - 1$ *for any* $k \in [1, N]$, $i \in [1, N]$ *and* $Height[i] \geq 1$.

Proof. Given that $Height[i] = \|lcp(i - 1, i)\| \geq 1$, according to the definition of Q_k, we have $Q_k[i-1] < Q_k[i]$ and it is trivial to see that $Height[j] \geq Height[i]-1$ for any $j \in [Q_k[i - 1], Q_k[i]]$.

Given the notations in this paper and Theorem 1, we can derive from [12] the linear algorithm GetHeight [2] in Fig. 3 for computing the LCPs for the characters in *a single cycle*, where Pos is a size-N vector with each $Pos[i]$ giving the position index of $L_k[i]$ in the original text S, i.e. $S[Pos[i]] = L_k[i]$. However, neither S nor Pos is known, for how to retrieve S is the ultimate goal of the problem in our hands. To solve this problem, we can utilize X_0, X_1 and Y to do the same job instead. Notice that in GetHeight, there is $h < k$, which implies that both $S[Pos[j] + h]$ and $S[Pos[j - 1] + h]$ in line 6, as well as $Height[Pos[j]]$ in line 10, always can be retrieved/maintained in $O(1)$ time using X_0, X_1 and Y. In order words, we can derive another logically equivalent alternative for GetHeight that can do the same job without knowing S and Pos. (Details are omitted here due to space limit.)

The complexity of GetHeight is dominated by the execution time of line 8 in the inner loop. If we comment out lines 11-12, it is obvious that GetHeight

---

[2] Please notice that GetHeight can not be used to compute the LCPs for any two characters belonging to two different cycles! How to compute the LCPs for any two characters in two different cycles is the critical part of our solution.

```
GetHeight(int start, int n) {
1   j=start; h=0;
2   for(i=1; i<=n; i++)
3   {
4     while(h<k) // k-order LCP is at most k.
5     {
6       if(j==1 || S[Pos[j]+h]!=S[Pos[j-1]+h])
7         break;
8       h++;
9     }
10    Height[Pos[j]]=h; // save the LCP for the char S[Pos[j]].
11    if(h>0)
12      h--; // decrease for computing the LCP of the succeeding char.
13    j=Qk[j]; // update the index of the succeeding char.
14  }
15  return Height;
}
```

Fig. 3. Algorithm for computing the heights for all the characters in a single cycle $\alpha(start)$

has a time complexity of $O(k)$, for h can be increased one at a time for at most k times by line 8. By adding lines 11-12 back to GetHeight, it can trigger at most n more times running of line 8. This is because line 12 can be executed at most n time, and each running of line 12 can enable one more running of line 8, therefore at most n times running of line 8 can be introduced to the running time complexity. As the combined consequence of line 8 and lines 11-12, the total time complexity of GetHeight is $O(k + n)$, which is k-dependent.

Theorem 2. *For any cycle $\alpha(i)$, $i \in [1, N]$, the heights of all characters in $\alpha(i)$ can be computed in a time complexity of $O(k + \|\alpha(i)\|)$.*

Proof. The correctness of this theorem comes directly from the above analysis for the time complexity of GetHeight, for $n = \|\alpha(i)\|$ in this case.

From the above result, we know that for each cycle β stored in X_0, we can compute the heights of all characters in β in a time complexity of $O(k + \|\beta\|)$. Hence, the total time complexity of computing the heights for all the characters in all cycles is $\sum_{\beta_i}(k + n_i)$, where $n_i = \|\beta_i\|$, which can be decomposed into two parts as below: one for the cycles not longer than ck and another for those longer than ck, where $c > 0$,

$$O(\sum_{n_i \leq ck}(k + n_i)) + O(\sum_{n_j > ck}(k + n_j)).$$

For a cycle longer than ck, k in the complexity can be safely ignored from $O(\sum_{n_j > ck}(k + n_j))$, resulting in $O(\sum_{n_j > ck}(k + n_j)) = O(\sum_{n_j > ck} n_j) = O(N_{ck})$, where N_{ck} is the total number of characters in all cycles longer than ck (cf. Property 4). However, for a cycle not longer than ck, k in the complexity can not be ignored, because the complexity is $O(\sum_{n_i \le ck}(k + n_i)) = O(\sum_{n_i \le ck} k)$, which is a function of k. To exclude k from the complexity, we need to explore more properties of cycles.

Lemma 2. *For a cycle β longer than ck, where $c > 0$ is a constant, the heights of all its characters can be computed in a time complexity of $O(\|\beta\|)$.*

Proof. According to Theorem 2, we can compute the heights for all characters in the cycle β with a time complexity of $O(k + \|\beta\|)$. Given that $\|\beta\| > ck$, we have $O(k + \|\beta\|) = O(\|\beta\|)$.

Computing Heights for All Cycles

Let CH_0 denote the subset of the characters in all the cycles longer than ck, where $c > 0$ is a constant. In addition, let CH_1 denote the subset of the characters in L_k, satisfying that for each $L_k[i] \in CH_1$, $L_k[i-1]$ is in CH_0, where $i \in [2, N]$. Following the definitions of CH_0 and CH_1, it is trivial to derive from Lemma 2 the below lemma about the complexity of computing the heights of all characters in $CH_0 \cup CH_1$.

Lemma 3. *The heights of all characters in $CH_0 \cup CH_1$ can be computed in a time complexity of $O(\|CH_0\| + \|CH_1\|)$.*

Proof. Referring to the algorithm GetHeight in Fig.3, at line 6, the character $S[Pos[j] + h]$ is compared with $S[Pos[j-1] + h]$ to compute the height of $L_k[i]$. Similarly, if we revise Line 6 to be "if(j==N || S[Pos[j]+h]!= S[Pos[j+1]+h])", i.e. to compare $S[Pos[j] + h]$ with $S[Pos[j + 1] + h]$ instead of $S[Pos[j-1] + h]$, the algorithm GetHeight can compute the heights of all the characters in CH_1. Hence, according to Lemma 2, the heights of all characters in set $CH_0 \cup CH_1$ can be computed in a time complexity of $O(\|CH_0\| + \|CH_1\|)$.

Having solved the issue of computing the heights of all the characters in the set $CH_0 \cup CH_1$ in a linear time complexity, the only pending problem is how to compute the heights of all the other remaining characters in the set $\{L_k[i] | i \in [1, N]\} - \{CH_0 \cup CH_1\}$ in a linear time complexity independent of k. For this purpose, we establish the below theorems and lemmas.

Theorem 3. *For any $i \in [2, N]$, $CT_k[i] = CT_k[i - 1]$, $\|\alpha(i)\| \le \lfloor k/2 \rfloor$ and $\|\alpha(i - 1)\| \le \lfloor k/2 \rfloor$, we have $\alpha(i) = \alpha(i - 1)$.*

Proof. Without loss of generality, let's suppose $\|\alpha(i-1)\| \ge \|\alpha(i)\|$. We continue the proof as follows:

1. Given the condition for the theorem in question, we have that $M_k[i, 1 : \|\alpha(i)\|] = M_k[i - 1, 1 : \|\alpha(i)\|]$ and $M_k[i, \|\alpha(i - 1)\| + 1 : \|\alpha(i - 1)\| + \|\alpha(i)\|] = M_k[i - 1, \|\alpha(i - 1)\| + 1 : \|\alpha(i - 1)\| + \|\alpha(i)\|]$. Because that $M_k[i - 1, 1 : \|\alpha(i)\|] = M_k[i - 1, \|\alpha(i - 1)\| + 1 : \|\alpha(i - 1)\| + \|\alpha(i)\|]$, we have $M_k[i, 1 : \|\alpha(i)\|] = M_k[i, \|\alpha(i - 1)\| + 1 : \|\alpha(i - 1)\| + \|\alpha(i)\|]$.

Hence, the preceding character of $M_k[i,1]$ is $M_k[i, \|\alpha(i-1)\|]$, i.e. $L_k[i] = M_k[i, \|\alpha(i-1)\|] = M_k[i-1, \|\alpha(i-1)\|] = L_k[i-1]$. Given $L_k[i] = L_k[i-1]$ and $CT_k[i] = CT_k[i-1]$, we further have $CT_k[P_k[i]] = CT_k[P_k[i-1]]$, i.e. the two k-order contexts at rows $P_k[i]$ and $P_k[i-1]$ in M_k are equivalent.

2. From the above analysis (in this proof) and Property 5, we see $L_k[P_k^{\|\alpha(i)\|}[i]] = L_k[i]$ as well as $L_k[P_k^{\|\alpha(i)\|}[i-1]] = L_k[i-1]$. Furthermore, for any character $L_k[j] \in \alpha(i)$, we have $L_k[j-1] \in \alpha(i-1)$, $L_k[j] = L_k[j-1]$ and $CT_k[j] = CT_k[j-1]$. Hence, $\alpha(i)$ and $\alpha(i-1)$ are two equivalent cycles.

The above theorem says that for any two characters i and $i-1$ in L_k, if the k-order contexts of the two characters are equal and both cycles $\alpha(i)$ and $\alpha(i-1)$ are not longer than $\lfloor k/2 \rfloor$, then the two cycles must be equivalent.

Lemma 4. *For any $i \in [2, N]$, if $\|\alpha(L[i-1])\| \le \lfloor k/2 \rfloor$ and $\|\alpha(L[i])\| \le \lfloor k/2 \rfloor$, then $CT_k[i-1] = CT_k[i]$ only if the lengths of the two cycles $\alpha(i-1)$ and $\alpha(i)$ are equal.*

Proof. Given the condition, in case that $CT_k[i-1] = CT_k[i]$, from Theorem 3, we have $\alpha(i-1) = \alpha(i)$, which implies that $\|\alpha(i-1)\| = \|\alpha(i)\|$.

Lemma 4 suggests that if two cycles $\alpha(i)$ and $\alpha(i-1)$ are not longer than $\lfloor k/2 \rfloor$ and their lengths are different, we can immediately determine that the two k-order contexts of $L_k[i]$ and $L_k[i-1]$ are different in a complexity of $O(1)$, as stated below.

Corollary 1. *For any two cycles $\alpha(i-1)$ and $\alpha(i)$ not longer than $\lfloor k/2 \rfloor$, $i \in [2, N]$, we have $CT_k[i-1] \ne CT_k[i]$ if the lengths of two cycles are different.*

Proof. According to Lemma 4, provided that the two cycles are not longer than $\lfloor k/2 \rfloor$, the two k-order contexts of $L_k[i]$ and $L_k[i-1]$ can be identical only if the lengths of the two cycles are equal. Hence, if the two cycles $\alpha(i-1)$ and $\alpha(i)$ have different lengths, we must have $CT_k[i-1] \ne CT_k[i]$.

The above corollary says that if the lengths of two cycles $\alpha(i)$ and $\alpha(i-1)$ are not longer than $\lfloor k/2 \rfloor$ and different, then the k-order contexts of $L_k[i-1]$ and $L_k[i]$ are different too.

Corollary 2. *For any $i \in [2, N]$, $CT_k[i] = CT_k[i-1]$, $\|\alpha(i)\| \le \lfloor k/2 \rfloor$ and $\|\alpha(i-1)\| \le \lfloor k/2 \rfloor$, we have $CT_k[j] = CT_k[j-1]$ and $\alpha(j) = \alpha(j-1)$ for any character $L_k[j] \in \alpha(i)$.*

Proof. Given $CT_k[i] = CT_k[i-1]$, $\|\alpha(i)\| \le \lfloor k/2 \rfloor$ and $\|\alpha(i-1)\| \le \lfloor k/2 \rfloor$, from the proof of Theorem 3, we know that $L_k[i] = L_k[i-1]$ and $\alpha(i) = \alpha(i-1)$. Further, according to Property 6, we have $CT_k[P_k[i]] = CT_k[P_k[i-1]]$. Because that when $k \in [1, N]$ and $L_k[i] = L_k[i-1]$, there must be $P_k[i-1] = P_k[i] - 1$. Hence, from Theorem 3 again, we have that $CT_k[P_k[i]] = CT_k[P_k[i]-1]$ and both cycles $\alpha(P_k[i])$ and $\alpha(P_k[i] - 1)$ are equal and not longer than $\lfloor k/2 \rfloor$. Repeating the induction in the same way, we have that for any character $L_k[j] \in \alpha(i)$, we have $CT_k[j] = CT_k[j-1]$ and $\alpha(j) = \alpha(j-1)$.

The above corollary says that for any $L_k[i]$ belonging to a cycle not longer than $\lfloor k/2 \rfloor$, if the cycle $\alpha(i-1)$ is also not longer than $\lfloor k/2 \rfloor$, and the two k-order contexts of $L_k[i]$ and $L_k[i-1]$ are equivalent, then for any character $L_k[j]$ belonging to the cycle $\alpha(i)$, the context of $L_k[j]$ must equal to that of $L_k[j-1]$.

Corollary 3. *For $i \in [2, N]$, $CT_k[i] \neq CT_k[i-1]$, $\|\alpha(i)\| \leq \lfloor k/2 \rfloor$ and $\|\alpha(i-1)\| \leq \lfloor k/2 \rfloor$, we have $CT_k[j] \neq CT_k[j-1]$ for any character $L_k[j] \in \alpha(i)$ seeing $\|\alpha(j-1)\| \leq \lfloor k/2 \rfloor$.*

Proof. We prove it by contradiction. Suppose that there exists a character $L_k[j] \in \alpha(i)$ seeing $CT_k[j] = CT_k[j-1]$ and $\|\alpha(j-1)\| \leq \lfloor k/2 \rfloor$. Because $\|\alpha(j)\| = \|\alpha(i)\| \leq \lfloor k/2 \rfloor$, from Corollary 2, we have that for any $L_k[x] \in \alpha(j)$, there must be $CT_k[x] = CT_k[x-1]$. This contradicts to the assumption $CT_k[i] \neq CT_k[i-1]$ of this corollary, where $L_k[i] \in \alpha(j)$.

The above corollary says that for any $L_k[i]$ in a cycle not longer than $\lfloor k/2 \rfloor$, if the cycle $\alpha(i-1)$ is also not longer than $\lfloor k/2 \rfloor$ and the k-order contexts of $L_k[i]$ and $L_k[i-1]$ are different, then for any character $L_k[j] \in \alpha(i)$ and $\alpha(j-1)$ is not longer than $\lfloor k/2 \rfloor$, the two k-order contexts of $L_k[j]$ and $L_k[j-1]$ must be different too.

Theorem 4. *For any cycle β not longer than $\lfloor k/2 \rfloor$, the values in D for all characters in set $\eta = \{L[j] \in \beta | j \in [2, N]$ and $\|\alpha(j-1)\| \leq \lfloor k/2 \rfloor\}$ can be computed in a complexity of $O(\|\beta\|)$.*

Proof. For each character $L_k[j]$ in β, there are two cases with respect to the length of cycle $\alpha(j-1)$: longer than $\lfloor k/2 \rfloor$ or not. The former case is for the characters in CH_1 (cf. Lemma 3), thus has been considered there. As for the later case, according to Theorem 3 and Corollary 1, we need at most $O(\|\beta\|)$, instead of $O(k + \|\beta\|)$, steps to compare all the characters in the two contexts of $L_k[j]$ and $L_k[j-1]$ to determine if they are equal or not when $\|\alpha(j)\| = \|\alpha(j-1)\|$. Based on the comparison result, from Corollary 2 and 3, we can extend the result to all the other characters in η with at most $\|\beta\| - 1$ steps, where each step has a complexity of $O(1)$. Hence, we complete the proof.

Making the Solution

Now, we apply the analysis results established in the previous subsections to build a solution for computing D in linear time/space. The key idea is to decompose the problem of computing $D[i]$ for each $L_k[i]$, $i \in [1, N]$, into two sub-problems, depending on whether $\alpha(i)$ is longer than $\lfloor ck \rfloor$. In particular, we choose the constant $c = 1/2$ in our algorithm design. The algorithm consists of the following 3 steps:

1. Compute P_k and Q_k from L_k, which can be done in $O(N)$ time.
2. Find all cycles from L_k using P_k and Q_k, and record by X_0, X_1 and Y, which can be done in $O(N)$ time.
3. Now, let's initialize all the items of L_k as unvisited and then traverse L_k once. For each unvisited $L_k[i]$, we mark all the characters in $\alpha(i)$ as visited

in $O(\alpha(i))$ time. Further, according to whether $\alpha(i)$ is longer than ck (which can be determined in time $O(1)$ using X_0, X_1 and Y) or not, we do according to the following two cases:

- The cycle $\alpha(i)$ is longer than ck. We calculate $D[i]$ for any character $L_k[i]$ in the cycle together with $D[i + 1]$ for $L_k[i + 1]$. The total time for this step is well bounded by the number of characters in all the cycles longer than ck (cf. Lemma 3), thus bounded by $O(N)$.
- For any $L_k[j]$ with $\alpha(j)$ not longer than ck, depending on whether the cycle $\alpha(j-1)$ is longer than ck or not, we further consider two subcases. (1) True. This has been considered in the previous case for the characters in cycles longer than ck. (2) False. We will look into the lengths of both cycles $\alpha(j)$ and $\alpha(j-1)$.
  - If the two cycles' lengths are different from each other, we can determine that the two contexts $CT[j]$ and $CT[j-1]$ must not equal to each other (cf. Corollary 1), in a time complexity $O(1)$. To propagate the same result to all the other characters in the cycle whose corresponding values in vector D have not been computed (cf. Corollary 3), the time complexity is $O(\|\alpha(j)\|)$.
  - If both cycles $\alpha(j)$ and $\alpha(j-1)$ have the same length, to compute $D[j]$, we only need to consider a length up to the size of the cycle $\alpha(j)$ (cf. Property 6), instead of the context order k. Once $D[j]$ has been computed, the same value of $D[j]$ can be populated to the other characters in cycle $\alpha(j)$ whose corresponding values in vector D have not been computed (cf. Corollary 2 and 3), in a time complexity $O(\|\alpha(j)\|)$.

Because all cycles are disjoint, the aggregated number of all characters in all such kinds of cycles is bounded by $O(N)$, the time complexity for this case is thus bounded by $O(N)$.

Hence, the total time complexity counted for this case is $O(N)$.

The algorithm described above can always compute D within $O(N)$ time complexity, no matter what kind of combination of cycles we have. The space complexity is obvious $O(N)$, for only a constant number of size-N arrays are required. As a result of the above analysis, we have the following theorem to state the linearity of our algorithm for computing D as well as the inverse ST for any given L_k.

Theorem 5. *Given L_k, we can restore the original text of S in $O(N)$ time/space for any $k \in [1, N]$.*

The presented algorithm has been coded in C and validated, which is available upon request.

Acknowledgment

The authors wish to thank the anonymous reviewers for this paper and its under-review journal version, for their constructive suggestions and insightful comments that have helped improve the presentation of this paper.

References

1. Burrows, M., Wheeler, D.J.: A block-sorting lossless data compression algorithm. Technical Report SRC Research Report 124, Digital Systems Research Center, California, USA (May 1994)
2. Schindler, M.: The sort transformation, http://www.compressconsult.com
3. Schindler, M.: A fast block-sorting algorithm for lossless data compression. In: Proceedings of DCC 1997, p. 469 (1997)
4. Schindler, M.: Method and apparatus for sorting data blocks. Patent in United States (6199064) (March 2001)
5. Nong, G., Zhang, S.: Unifying the Burrows-Wheeler and the Schindler transforms. In: Proceedings of DCC 2006, March 2006, p. 464 (2006)
6. Nong, G., Zhang, S.: An efficient algorithm for the inverse ST problem. In: Proceedings of DCC 2007, p. 397 (2007)
7. Nong, G., Zhang, S.: Efficient algorithms for the inverse sort transform. IEEE Transactions on Computers 56(11), 1564–1574 (2007)
8. Yokoo, H.: Notes on block-sorting data compression. Electronics and Communications in Japan (Part III: Fundamental Electronic Science) 82(6), 18–25 (1999)
9. Bird, R.S., Mu, S.C.: Inverting the burrows-wheeler transform. Journal of Functional Programming 14(6), 603–612 (2004)
10. Manzini, G.: The Burrows-Wheeler transform: theory and practice. In: Kutyłowski, M., Wierzbicki, T., Pacholski, L. (eds.) MFCS 1999. LNCS, vol. 1672, pp. 34–47. Springer, Heidelberg (1999)
11. Balkenhol, B., Kurtz, S.: Universal data compression based on the Burrows-Wheeler transformation: theory and practice. IEEE Transactions on Computers 49(10), 1043–1053 (2000)
12. Kasai, T., Lee, G., Arimura, H., et al.: Linear-time longest-common-prefix computation in suffix arrays and its applications. In: Amir, A., Landau, G.M. (eds.) CPM 2001. LNCS, vol. 2089, pp. 181–192. Springer, Heidelberg (2001)

Dynamic Fully-Compressed Suffix Trees

Luís M.S. Russo[1,3,*], Gonzalo Navarro[2,**], and Arlindo L. Oliveira[1]

[1] INESC-ID / IST, R. Alves Redol 9, 1000 Lisboa, Portugal
aml@algos.inesc-id.pt
[2] Dept. of Computer Science, University of Chile
gnavarro@dcc.uchile.cl
[3] Dept. of Computer Science, University of Lisbon, Portugal
lsr@di.fc.ul.pt

Abstract. Suffix trees are by far the most important data structure in stringology, with myriads of applications in fields like bioinformatics, data compression and information retrieval. Classical representations of suffix trees require $O(n \log n)$ bits of space, for a string of size n. This is considerably more than the $n \log_2 \sigma$ bits needed for the string itself, where σ is the alphabet size. The size of suffix trees has been a barrier to their wider adoption in practice. A recent so-called fully-compressed suffix tree (FCST) requires asymptotically only the space of the text entropy. FCSTs, however, have the disadvantage of being static, not supporting updates to the text. In this paper we show how to support dynamic FCSTs within the same optimal space of the static version and executing all the operations in polylogarithmic time. In particular, we are able to build the suffix tree within optimal space.

1 Introduction and Related Work

Suffix trees are extremely important for a large number of string processing problems. Their many virtues have been described by Apostolico [1] and Gusfield [2]. The combinatorial properties of suffix trees have a profound impact in the *bioinformatics* field, which needs to analyze large strings of DNA and proteins with no predefined boundaries. This partnership has produced several important results, but it has also exposed the main shortcoming of suffix trees. Their large space requirements, together with their need to operate in main memory to be useful in practice, renders them inapplicable in the cases where they would be most useful, that is, on large texts.

The space problem is so important that it originated a plethora of research results, ranging from space-engineered implementations [3] to novel data structures that simulate suffix trees, most notably suffix arrays [4]. Some of those space-reduced variants give away some functionality in exchange. For example suffix arrays miss the important suffix link navigational operation. Yet, all

* Supported by the Portuguese Science and Technology Foundation by grant SFRH/BPD/34373/2006 and project ARN, PTDC/EIA/67722/2006.
** Partially funded by Millennium Institute for Cell Dynamics and Biotechnology, Grant ICM P05-001-F, Mideplan, Chile.

these classical approaches require $O(n \log n)$ bits, while the indexed string requires only $n \log \sigma$ bits (we write log for $\log_2$), n being the size of the string and σ the size of the alphabet. For example the human genome requires 700 Megabytes, while even a space-efficient suffix tree on it requires at least 40 Gigabytes [5], and the reduced-functionality suffix array requires more than 10 Gigabytes. This is particularly evident in DNA because $\log \sigma = 2$ is much smaller than $\log n$.

These representations are also much larger than the size of the *compressed* string. Recent approaches [6] combining data compression and succinct data structures have achieved spectacular results for the pattern search problem. For example Ferragina *et al.* [7] presented an index that requires $nH_k + o(n \log \sigma)$ bits and counts the occurrences of a pattern of length m in time $O(m(1 + (\log_\sigma \log n)^{-1}))$. Here nH_k denotes the k-th order empirical entropy of the string [8], a lower bound on the space achieved by any compressor using k-th order modeling. As that index is also able of reproducing any text substring, its space is asymptotically optimal in the sense that no k-th order compressor can achieve asymptotically less space to represent the text.

It turns out that it is possible to use this kind of data structures, that we will call *compressed suffix arrays* (CSAs)[1], and, by adding a few extra structures, support all the operations provided by suffix trees. Sadakane presented the first *compressed suffix tree* (CST) [5], adding $6n$ bits on top of the CSA. Recently Russo *et al.* [9] achieved a *fully-compressed suffix tree* (FCST), which works over the smallest existing CSA [7], adding only $o(n \log \sigma)$ bits to it. Hence the FCST breaks the $\Theta(n)$ extra-bits space barrier and retains asymptotic space optimality.

Albeit very interesting as a first step, the FCST has the limitation of being static, and moreover of being built from the uncompressed suffix tree. CSAs have recently overcome this limitation, starting with the structure by Chan *et al.* [10]. In its journal version this work included the first dynamic CST, which builds on Sadakane's (static) CST [5] and retains its $\Theta(n)$ extra space penalty. On the other hand, the smallest existing CSA [7] was made dynamic within the same space by Mäkinen *et al.* [11], which was recently improved by González *et al.* [12] so as to achieve logarithmic time slowdown. In this paper we make the FCST dynamic by building on this latter dynamic CSA. We retain the optimal space complexity and polylogarithmic time for all the operations.

A comparison between Chan *et al.*'s CST and our FCST is shown in Table 1. Our FCST is not significantly slower, yet it requires much less space (e.g. one can realistically predict 25% of Chan *et al.*'s CST space on DNA). For the table we chose the smallest existing dynamic CSA, so that we show the time complexities that can be obtained within the smallest possible space for both CSTs.

All these dynamic structures, as well as ours, indeed handle a *collection* of texts, where whole texts are added/deleted to/from the collection. Construction in compressed space is achieved by inserting a text into an empty collection.

---

[1] These are also called compact suffix arrays, FM-indexes, etc., see [6].

Table 1. Comparing compressed suffix tree representations. The operations are defined along Section 2. Time complexities, but not space, are big-O expressions. We give the generalized performance (assuming $\Psi, t, \Phi \geq \log n$) and an instantiation using $\delta = (\log_\sigma \log n) \log n$. For the instantiation we also assume $\sigma = O(\text{polylog}(n))$, and use the dynamic FM-Index variant of González *et al.* [12] as the compressed suffix array (CSA), for which the space holds for any $k \leq \alpha \log_\sigma(n) - 1$ and any constant $0 < \alpha < 1$.

	Chan *et al.* [10]	Ours
Space in bits	$\|CSA\| + \mathbf{O(n)} + o(n)$ $= nH_k + \mathbf{O(n)} + o(n \log \sigma)$	$\|CSA\| + O((n/\delta) \log n)$ $= nH_k + o(n \log \sigma)$
SDep	$\Phi \quad = (\log_\sigma \log n) \log^2 n$	$\Psi\delta \quad = (\log_\sigma \log n) \log^2 n$
Count/ Ancestor	$\log n \quad = \log n$	$1 \quad\quad\quad\quad\quad\quad = 1$
Parent	$\log n \quad = \log n$	$(\Psi + t)\delta \quad = (\log_\sigma \log n) \log^2 n$
SLink	$\Psi \quad\quad = \log n$	$(\Psi + t)\delta \quad = (\log_\sigma \log n) \log^2 n$
SLink^i	$\Phi \quad = (\log_\sigma \log n) \log^2 n$	$\Phi + (\Psi + t)\delta \quad = (\log_\sigma \log n) \log^2 n$
Letter / Locate	$\Phi \quad = (\log_\sigma \log n) \log^2 n$	$\Phi \quad\quad\quad = (\log_\sigma \log n) \log^2 n$
LCA	$\log n \quad = \log n$	$(\Psi + t)\delta \quad = (\log_\sigma \log n) \log^2 n$
FChild/ NSib	$\log n \quad = \log n$	$(\Psi + t)\delta + \Phi \log \delta + (\log n) \log(n/\delta)$ $= ((\log_\sigma \log n) \log^2 n) \log \log n$
Child	$\Phi \log \sigma \quad = (\log \log n) \log^2 n$	$(\Psi + t)\delta + \Phi \log \delta + (\log n) \log(n/\delta)$ $= ((\log_\sigma \log n) \log^2 n) \log \log n$
WeinerLink	$t \quad\quad\quad = \log n$	$t \quad\quad\quad\quad\quad = \log n$
Insert(T) / Delete(T)	$\|T\|(\Psi + t)\delta$ $= \|T\|(\log_\sigma \log n) \log^2 n$	$\|T\|(\Psi + t)\delta \quad = \|T\|(\log_\sigma \log n) \log^2 n$

2   Basic Concepts

Fig. 1 illustrates the concepts in this section. We denote by T a **string**; by Σ the **alphabet** of size σ; by $T[i]$ the symbol at position $(i \bmod n)$ (so the first symbol is $T[0]$); by $T.T'$ concatenation; by $T = T[..i-1].T[i..j].T[j+1..]$ respectively a **prefix**, a **susbtring** and a **suffix**; by $\text{PARENT}(v)$ the parent node of node v; by $\text{TDEP}(v)$ its tree-depth; by $\text{ANCESTOR}(v, v')$ whether v is an ancestor of v'; by $\text{LCA}(v, v')$ the **lowest common ancestor**.

The **path-label** of a node v in a labeled tree is the concatenation of the edge-labels from the root down to v. We refer indifferently to nodes and to their path-labels, also denoted by v. The i-th letter of the path-label is denoted as $\text{LETTER}(v, i) = v[i]$. The **string-depth** of a node v, denoted by $\text{SDEP}(v)$, is the length of its path-label. $\text{CHILD}(v, X)$ is the node that results of descending from v by the edge whose label starts with symbol X, if it exists. The **suffix tree** of T is the deterministic compact labeled tree for which the path-labels of the leaves are the suffixes of T. We assume that T ends in a terminator symbol \$ that does not belong to Σ. The **generalized suffix tree** of a collection $\mathcal{C}$ of texts is the tree that results from merging the respective suffix trees. Moreover each text is assumed to have a distinct terminator. For a detailed explanation see Gusfield's book [2]. The **suffix-link** of a node $v \neq \text{ROOT}$ of a suffix tree, denoted $\text{SLINK}(v)$, is a pointer to node $v[1..]$. Note that $\text{SDEP}(v)$ of a leaf v identifies the suffix of T

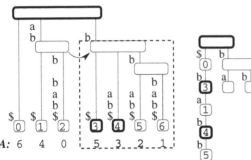

Fig. 3. Parentheses representations of trees. The parentheses on top represent the suffix tree, those in the middle the sampled tree, and those on the bottom the sampled tree when b is also sampled along with the B bitmap. The numbers are not part of the representation; they are shown for clarity. The rows labeled i: give the index of the parentheses.

Fig. 1. Suffix tree $\mathcal{T}$ of string *abbbab*, with the leaves numbered. The arrow shows the SLINK between node ab and b. Below it we show the suffix array. The portion of the tree corresponding to node b and respective leaves interval is within a dashed box. The sampled nodes have bold outlines.

Fig. 2. Reverse tree $\mathcal{T}^R$

starting at position $n-\text{SDEP}(v) = \text{LOCATE}(v)$. For example $T[\text{LOCATE}(ab\$)..] = T[7-3..] = T[4..] = ab\$$. The **suffix array** $A[0, n-1]$ stores the LOCATE values of the leaves in lexicographical order. Note that in a generalized suffix tree LOCATE must also identify the text to which the suffix corresponds. When we use arithmetic expressions involving A and A^{-1} they are computed within a given text, *i.e.* they do not jump to another text. Moreover for simplicity we use only one text in our example and hence omit the text identifier. The *suffix tree nodes can be identified with suffix array intervals*: each node corresponds to the *range* of leaves that descend from v. The node b corresponds to the interval $[3, 6]$. Hence the node v will be represented by the interval $[v_l, v_r]$. Leaves are also represented by their left-to-right index (starting at 0). For example by $v_l - 1$ we refer to the leaf immediately before v_l, *i.e.* $[v_l-1, v_l-1]$. With this representation we can COUNT in constant time the number of leaves that descend from v. The number of leaves below b is $4 = 6 - 3 + 1$. This is precisely the number of times that the string b occurs in the indexed string T. We can also compute ANCESTOR in $O(1)$ time: $\text{ANCESTOR}(v, v') \Leftrightarrow v_l \le v'_l \le v'_r \le v_r$.

3   Static Fully-Compressed Suffix Trees and Our Plan

In this section we briefly explain the static FCST we build on [9]. The FCST consists of a compressed suffix array, a δ-sampled tree S, and mappings between these structures. We also give the road map of our plan to dynamize the FCST.

Compressed Suffix Arrays (CSAs) are compact and functional representations of suffix arrays [6]. Apart from the basic functionality of retrieving

$A[i] = \text{LOCATE}(i)$ (within a time complexity that we will call $\Phi = \Omega(\log n)$), state-of-the-art CSAs support operation $\text{SLINK}(v)$ *for leaves* v. This is called $\psi(v)$ in the literature: $A[\psi(v)] = A[v] + 1$, and thus $\text{SLINK}(v) = \psi(v)$, let its time complexity be $\Psi = \Omega(\log n)$. The iterated version of ψ, denoted ψ^i, can usually be computed faster than $O(i\Psi)$ with CSAs. This is achieved as $\psi^i(v) = A^{-1}[A[v] + i]$, let us assume that the CSA can also compute A^{-1} within $O(\Phi)$ time. CSAs might also support the $\text{WEINERLINK}(v, a)$ operation [13]: for a node v the $\text{WEINERLINK}(v, X)$ gives the suffix tree node with path-label $X.v[0..]$. This is called the LF mapping in CSAs, and is a kind of inverse of ψ, let its time complexity be $t = \Omega(\log n)$. Consider the interval $[3, 6]$ that represents the leaves whose path-labels start by b. In this case we have that $\text{LF}(a, [3, 6]) = [1, 2]$, *i.e.* by using the LF mapping with a we obtain the interval of leaves whose path-labels start by ab. We extend of LF to strings, $\text{LF}(X.Y, v) = \text{LF}(X, \text{LF}(Y, v))$.

CSAs also implement $\text{LETTER}(v, i)$ for leaves v. The easiest case is the first letter of a given suffix, $\text{LETTER}(v, 0) = T[A[v]]$. This corresponds to $v[0]$, the first letter of the path-label of leaf v. Dynamic CSAs implement $v[0]$ in time $O(\log n)$. In general, $\text{LETTER}(v, i) = \text{LETTER}(\text{SLINK}^i(v), 0)$ is implemented in $O(\Phi)$ time. CSAs are usually self-indexes, meaning that they replace the text: they can extract any substring, of size ℓ, of the indexed text in $O(\Phi + \ell\Psi)$ time.

In this paper we will use a dynamic CSA for this part [12], which implements these operations with logarithmic slowdown to its static version [7]. The dynamic CSA actually handles a collection of texts, where insertions and deletions of whole texts T are carried out in time $O(|T|(\Psi + t))$.

The δ-Sampled Tree exploits the property that suffix trees are self-similar, $\text{SLINK}(\text{LCA}(v, v')) = \text{LCA}(\text{SLINK}(v), \text{SLINK}(v'))$ whenever the expressions are well defined. This means, roughly, that the tree structure below $\text{SLINK}(v)$ contains the tree structure below v. Because of this regularity it is possible to store only a few sampled nodes instead of the whole suffix tree. A *δ-sampled tree S*, from a suffix tree $\mathcal{T}$ of $\Theta(n)$ nodes, chooses $O(n/\delta)$ nodes such that, for each node v, node $\text{SLINK}^i(v)$ is sampled for some $i < \delta$. Such a sampling can be obtained by choosing nodes with $\text{SDEP}(v) \equiv_{\delta/2} 0$ such that there is another node v' for which $v = \text{SLINK}^{\delta/2}(v')$. For such a sampling Lemma 1 holds, where $\text{LCSA}(v, v')$ is the *lowest common sampled ancestor* of v and v':

Lemma 1. *Let v, v' be nodes such that $\text{SLINK}^r(\text{LCA}(v, v')) = \text{ROOT}$, and let $d = \min(\delta, r + 1)$. Then*
$\text{SDEP}(\text{LCA}(v, v')) = \max_{0 \leq i < d}\{i + \text{SDEP}(\text{LCSA}(\text{SLINK}^i(v), \text{SLINK}^i(v')))\}.$

By itself however this property leads to an entangled loop of operations, because LCA depends on SLINK and $\text{SLINK}(v) = \text{LCA}(\psi(v_l), \psi(v_r))$ depends on LCA. Using CSAs and observing that $\text{LCA}(v, v') = \text{LCA}(\min\{v_l, v'_l\}, \max\{v_r, v'_r\})$ we can simplify this equation to $\text{SDEP}(\text{LCA}(v, v')) = \max_{0 \leq i < d}\{i + \text{SDEP}(\text{LCSA}($ $\psi^i(\min\{v_l, v'_l\}), \psi^i(\max\{v_r, v'_r\})))\}$.

Therefore the kernel operations can be computed as:

$\text{SDEP}(v) = \text{SDEP}(\text{LCA}(v, v)) = \max_{0 \leq i < d}\{i + \text{SDEP}(\text{LCSA}(\psi^i(v_l), \psi^i(v_r)))\}$
$\text{LCA}(v, v') = \text{LF}(v[0..i-1], \text{LCSA}(\psi^i(\min\{v_l, v'_l\}), \psi^i(\max\{v_r, v'_r\})))$

from which SLINK is obtained as well. The i in the last equation is the one that maximizes SDEP(LCA(v, v')). Operation PARENT(v) is easily computed on top of LCA. These operations take time $O((\Psi + t)\delta)$, except that SDEP takes $O(\Psi\delta)$.

Note that we have to solve LCSA. This requires to solve LCA$_S$, that is, LCA queries on the sampled tree S, and also to map nodes to sampled nodes using operation LSA (see later). The sampled tree also needs to solve PARENT$_S$ and store SDEP$_T$. The rest is handled by the CSA.

For the dynamic version, we first show how the suffix tree $\mathcal{T}$ changes upon insertion and deletion of texts T to the collection. Then we show how to maintain the sampling properties of S under those updates of the (virtual) $\mathcal{T}$. This will require some more data to be stored in the sampled nodes. Finally, we will make use of a dynamic parentheses representation for the sampled tree, which will already give us LCA$_S$ and PARENT$_S$, as well as a way to associate data to nodes and insert/delete nodes. Note that we just have to show how to provide this basic tree functionality, as the remaining operations are obtained as in the static version.

To support TDEP, however, they add other $O(n/\delta)$ nodes to the sampling, such that for any node v the node PARENT$^j(v)$ is sampled, for some $0 \leq j < \delta$. We have not found a way to efficiently maintain this second sampling in a dynamic scenario. As a consequence, our dynamic FCST does not support operation TDEP nor those that require it [9]: LAQ$_T$ and LAQ$_S$. The basic navigation operations FCHILD and NSIB also require TDEP, but we will present a different idea that solves them together with CHILD and a generalization of it, using just the CSA.

Mapping Between the CSA and the Sampled Tree. For every node v of the sampled tree we need to obtain the corresponding interval $[v_l, v_r]$. On the other hand, given a CSA interval $[v_l, v_r]$ representing node v of $\mathcal{T}$, the *lowest sampled ancestor* LSA(v) gives the lowest sampled tree node containing v. With LSA we can compute LCSA(v, v') = LCA$_S$(LSA(v), LSA(v')).

In this paper we introduce a new method to implement these mappings that is efficient and simpler than the one presented in the static version [9].

4   Updating the Suffix Tree and Its Sampling

In this section we explain how to modify a suffix tree to reflect changes caused by inserting and removing a text T to/from the suffix tree.

The CSA of Mäkinen *et al.* [11], on which we build, inserts T in right-to-left order. It first determines the position of the new terminator[2] and then uses LF

---

[2] This insertion point is arbitrary in that CSA, thus there is no order among the texts. Moreover, all the terminators are the same in the CSA, yet it can be easily modified to handle different terminators.

to find the consecutive positions of longer and longer suffixes, until the whole T is inserted. This right-to-left method perfectly matches with Weiner's algorithm [13] to build the suffix tree of T: it first inserts suffix $T[i+1..]$ and then suffix $T[i..]$, finding the points in the tree where the node associated to the new suffix is to be created if it does not already exist. The node is found by using PARENT until the WEINERLINK operation returns a non-empty interval. This requires one PARENT and one WEINERLINK amortized operation per symbol of T. This algorithm has the important invariant that the intermediate data structure is a suffix tree. Hence, by carrying it out in synchronization with the CSA insertion algorithm, we can use the current CSA to implement ≤ PARENT and WEINERLINK.

To maintain the property that the intermediate structure is a suffix tree, deletion of a text T must proceed by first locating the node of T that corresponds to T, and then using SLINKs to remove all the nodes corresponding to its suffixes in T. We must simultaneously remove the leaves in the CSA (Mäkinen *et al.*'s CSA deletes a text right-to-left as well, but it is easy to adapt to use Ψ instead of LF to do it left-to-right).

We now explain how to update the sampled tree S whenever nodes are inserted or deleted from the (virtual) suffix tree T. The sampled tree must maintain, at all times, the property that for any node v there is an $i < \delta$ such that $\mathrm{SLINK}^i(v)$ is sampled. The following concept from Russo *et al.* [14] is fundamental to explain how to obtain this result.

Definition 1. *The **reverse tree** T^R of a suffix tree T is the minimal labeled tree that, for every node v of T, contains a node v^R denoting the reverse string of the path-label of v.*

We note we are *not* maintaining nor sampling T^R, we just use it as a conceptual device. Fig. 2 shows a reverse tree. Observe that since there is a node with path-label ab in T there is a node with path-label ba in T^R. We can therefore define a mapping R that maps every node v to v^R. Observe that for any node v of T, except for the ROOT, we have that $\mathrm{SLINK}(v) = R^{-1}(\mathrm{PARENT}(R(v)))$. This mapping is partially shown in Figs. 1 and 2 by the numbers. Hence the reverse tree stores the information of the suffix links. By $\mathrm{HEIGHT}(v^R)$ we refer to the distance between v and its farthest descendant leaf. For a regular sampling we choose the nodes for which $\mathrm{TDEP}(v^R) \equiv_{\delta/2} 0$ and $\mathrm{HEIGHT}(v^R) \geq \delta/2$. This is equivalent to our sampling rules on T (Section 3): Since the reverse suffixes form a prefix-closed set, T^R is a non-compact trie, *i.e.* each edge is labeled by a single letter. Thus, $\mathrm{SDEP}(v) = \mathrm{TDEP}(v^R)$. The rule for $\mathrm{HEIGHT}(v^R)$ is obviously related to that on $\mathrm{SLINK}(v)$ by R. See Fig. 2 for an example of this sampling.

Likewise, stating that there is an $i < \delta$ for which $\mathrm{SLINK}^i(v)$ is sampled is the same as stating that there is an $i < \delta$ for which $\mathrm{TDEP}(\mathrm{PARENT}^i(v^R)) \equiv_{\delta/2} 0$ and $\mathrm{HEIGHT}(\mathrm{PARENT}^i(v^R)) \geq \delta/2$. Since $\mathrm{TDEP}(\mathrm{PARENT}^i(v^R)) = \mathrm{TDEP}(v^R) - i$, the first condition holds for exactly two i's in $[0, \delta[$. Since HEIGHT is strictly increasing the second condition holds for sure for the largest i. Notice that since every sampled node has at least $\delta/2$ descendants that are not sampled, this means that we sample at most $\lfloor 4n/\delta \rfloor$ nodes from a suffix tree with $\leq 2n$ nodes.

Notice that whenever a node is inserted or removed from a suffix tree it never changes the SDep of the other nodes in the tree, hence it does not change any TDep in $\mathcal{T}^R$. This means that whenever the suffix tree is modified the only nodes that can be inserted or deleted from the reverse tree are the leaves. In $\mathcal{T}$ this means that when a node is inserted it does not break a chain of suffix links; it is always added at the beginning of such a chain. Weiner's algorithm works precisely by appending a new leaf to a node of $\mathcal{T}^R$.

Assume that we are using Weiner's algorithm and decide that the node $X.v$ should be added and we know the representation of node v. All we need to do to update the structure of the sampled tree is to verify that if by adding $(X.v)^R$ as a child of v^R in $\mathcal{T}^R$ we increase the HEIGHT of some of ancestor, in $\mathcal{T}^R$, that will now become sampled. Hence we must scan upwards in $\mathcal{T}^R$ to verify if this is the case. Notice that we already carry out this scanning as a side effect of computing SLINK(v), which also gives us the required SDep information. Also, we do not need to maintain HEIGHT values. Instead, if the distance from $(X.v)^R$ to the closest sampled node $(v')^R$ is exactly $\delta/2$ and TDep$((v')^R) \equiv_{\delta/2} 0$, then we know that v' meets the sampling condition and we sample it.

Deleting a node (*i.e.* a leaf in $\mathcal{T}^R$) is slightly more complex and involves some reference counting. This time assume we are deleting node $X.v$, again we need to scan upwards, this time to decide whether to make a node non-sampled. However SDep(v) − SDep(v') $< \delta/2$ is not enough, as it may be that HEIGHT(v'^R) $\geq \delta/2$ because of some other descendant. Therefore every sampled node v' counts how many descendants it has at distance $\delta/2$. A node becomes non-sampled only when this counter reaches zero. Insertions and deletions of nodes in $\mathcal{T}$ must update these counters, by increasing/decreasing them whenever inserting/deleting a leaf at distance exactly $\delta/2$ from nodes.

Hence to INSERT or DELETE a node requires $O((\Psi + t)\delta)$ time, plus the time to manipulate the structure that holds the topology of S: we need to carry out insertions/deletions of nodes, while maintaining information associated to them (SDep, reference counts). Section 5.2 shows that those operations do not dominate the time $O((\Psi + t)\delta)$ needed to maintain the sampling conditions.

5 Dynamic Fully-Compressed Suffix Trees

In this section we present the compact data structures we use and create to handle our dynamic structures: the CSA, the sampled tree, and the mappings.

5.1 Dynamic Compressed Suffix Arrays

To maintain a dynamic CSA we use the following result by González *et al.* [12], which is an improvement upon those of Mäkinen *et al.* [11]:

Theorem 1. *A dynamic CSA over a collection $\mathcal{C}$ of texts can be stored within $nH_k(\mathcal{C}) + o(n \log \sigma)$ bits, for any $k \leq \alpha \log_\sigma(n) - 1$ and any constant $0 < \alpha < 1$, supporting all the operations with times $t = \Psi = O(((\log_\sigma \log n)^{-1} +$*

1) $\log n$), $\Phi = O((\log_\sigma \log n) \log^2 n)$, and inserting/deleting texts T in time $O(|T|(t + \Psi))$.

Note that for a collection with p texts it is necessary to store the positions of the texts in A. This requires $O(p \log n)$ bits but it is not an issue unless the texts are very short [11].

Therefore, the problem of maintaining a dynamic CSA is already solved, except that we promised to support operation CHILD_T (and some derivatives) directly on the CSA. Indeed, $\text{CHILD}_T(v, X)$ can be easily computed in $O(\Phi \log n)$ time by binary searching for the interval of $v = [v_l, v_r]$ formed by those v' where $\text{LETTER}(v', \text{SDEP}(v) + 1) = X$. Similarly, $\text{FCHILD}(v)$ can be determined by computing $X = \text{LETTER}(v_l, \text{SDEP}(v) + 1)$ and then $\text{CHILD}_T(v, X)$. To compute $\text{NSIB}(v)$ the process is similar: If $\text{PARENT}(v) = [v'_l, v'_r]$ and $v'_r > v_r$, then we compute $X = \text{LETTER}(v'_r + 1, \text{SDEP}(v) + 1)$ and do $\text{CHILD}_T(v, X)$. All the time complexities are thus dominated by that of CHILD_T.

Now we show how CHILD_T can be computed in a more general and efficient way. The *generalized branching* for nodes v_1 and v_2 consists in determining the node with path-label $v_1.v_2$ if it exists. A simple solution is to binary search the interval of v_1 for the sub-interval of the v''s such that $\psi^m(v') \in v_2$, where $m = \text{SDEP}(v_1)$. This approach requires $O(\Phi \log n)$ time and it was first considered using CSA's by Huynh *et al.* [15]. Thus we are able to generalize $\text{CHILD}_T(v, X)$, which uses v_2 as the sub-interval of A of the suffixes starting with X.

This general solution can be improved by noticing that we are using SLINK^i at arbitrary positions of the CSA for the binary search. Recall that SLINK^i is solved via A and A^{-1}. Thus, we could sample A and A^{-1} regularly so as to store their values explicitly. That is, we explicitly store the values $A[j\delta]$ and $A^{-1}[j\delta]$ for all j. To solve a generalized branching, we start by building a table of ranges $D[0] = v_2$ and $D[i] = \text{LF}(v_1[m - i..m - 1], v_2)$, for $1 \le i < \delta$. If $m < \delta$ the answer is $D[m]$. Otherwise, we binary search the interval of v_1, accessing only the sampled elements of A. To determine the branching we should compute $\psi^m(j\delta) = A^{-1}[A[j\delta] + m]$ for some $j\delta$ values in v_1. To use the cheaper sampled A^{-1} as well, we need that $A[j\delta] + m$ be divisible by δ, thus we instead compute $\psi^{m'}$ for $m' = \lfloor (A[j\delta] + m)/\delta \rfloor \delta - A[j\delta]$. Hence instead of verifying that $\psi^m(j\delta) \in v_2$, we verify that $\psi^{m'} \in D[m - m']$. After this process we still have to binary search an interval of size $O(\delta)$, which is carried out naively.

The overall process requires time $O(\Phi + (\Psi + t)\delta)$ to access the last letters of v_1 and build D, plus $O((\log n) \log(n/\delta))$ for binary searching the samples; plus $O(\Phi \log \delta)$ for the final binary searches. We have assumed $O(\log n)$ time to access the sampled A and A^{-1} values in a dynamic scenario, whereas in a static scenario[3] it would be $O(1)$.

In fact in a dynamic scenario we do not store exactly the $A[j\delta]$ values; instead we guarantee that for any k there is a k' such that $k - \delta < k' \le k$ and $A[k']$ is sampled, and the same for A^{-1}. Still the sampled elements of A and the m' to use can be easily obtained in $O(\log n)$ time. Those sampled sequences are not hard to maintain. For example, Mäkinen *et al.* [11, Sec. 7.1 of journal version] describes

---

[3] This speedup immediatly improves the results of Huynh *et al.* [15].

how to maintain A^{-1} (called S_C in there), and essentially how to maintain A (called S_A in there; the only missing point is to maintain approximately spaced samples in A, which can be done exactly as for A^{-1}).

5.2   Dynamic Sampled Trees

The sampled tree contains only $O(n/\delta)$ nodes. As such it could be stored with pointers using only $O((n/\delta)\log n)$ bits. Instead we use a dynamic parentheses data structure given by Chan *et al.* [10], which already supports LCA.

Theorem 2. *A list of $O(n/\delta)$ balanced parentheses can be maintained in $O(n/\delta)$ bits supporting the following operations in $O(\log n)$ time:*

- FINDMATCH(u), *finds the matching parenthesis of u;*
- ENCLOSE(u), *finds the nearest pair of matching parentheses that encloses u;*
- DOUBLEENCLOSE(u, u'), *finds the nearest pair of parentheses that encloses both u and u';*
- INSERT(u, u'), DELETE(u, u'), *inserts or deletes the matching parentheses located at u, u'.*

The ENCLOSE primitive computes PARENT$_S$ in the sampled tree. Likewise the DOUBLEENCLOSE primitive computes the LCA$_S$ operation. In Section 5.3 we explain how to update the parentheses sequence when a node becomes sampled or non-sampled (*i.e.* , how to maintain the mapping with the CSA). Operations RANK and SELECT on the sequence of parentheses S can also be used to store information on the nodes, by mapping between the parentheses sequence and their preorder values and vice versa: RANK$'_{('}(S, i)$ gives the preorder number of the node identified by the opening parenthesis at $S[i]$, while SELECT$'_{('}(S, j)$ identifies the j-th node (in preorder) in S. RANK and SELECT over the parentheses bitmap can be handled using the following theorem.

Theorem 3 ([11]). *A bitmap of n bits supporting RANK, SELECT, INSERT and DELETE in $O(\log n)$ time can be maintained in $nH_0 + O(n/\sqrt{\log n})$ bits.*

Each node of S must also store its SDEP. This is not complicated because the SDEP of the nodes of T does not change, at least using Weiner's algorithm. Thus we maintain a balanced tree where the SDEP values can be read, inserted, and deleted, at the positions given by RANK$'_{('}(S, i)$. When a node becomes sampled/non-sampled we insert/delete in this sequence. A similar mechanism is used to store the reference counts used for the sampling; in this case the stored values can be modified as well. Thus $O(\log n)$ time suffices for simulating the tree operations on S.

5.3   Mapping from CSA to the Sampled Tree and Back

The *lowest sampled ancestor* LSA is the way to map from the CSA to S. LSA is computed by using an operation REDUCE(v), that receives the numeric representation of leaf v and returns the position, in the parentheses representation of the sampled tree, where that leaf should be. Consider for example the leaf numbered 5 in Fig. 3. This leaf is not sampled, but in the original tree it appears somewhere between leaf 4 and the end of the tree, more specifically between parenthesis ')' of 4 and parenthesis ')' of the ROOT. We assume REDUCE returns the first parenthesis, *i.e.* REDUCE(5) = 4. In this case since the parenthesis we obtain is a ')' we know that LSA should be the parent of that node. Hence we compute LSA as follows:

$$\text{LSA}(v) = \begin{cases} \text{REDUCE}(v) & \text{, if } S[\text{REDUCE}(v)] = '(' \\ \text{PARENT}(\text{REDUCE}(v)) & \text{, otherwise} \end{cases}$$

We present a new way to compute REDUCE in $O(\log n)$ time and $o(n)$ bits (cf. [9]). We use a bitmap B initiated with n bits all equal to 0. Now for every node $v = [v_l, v_r]$ we insert a 1 at $\text{SELECT}_0(B, v_l)$ and after $\text{SELECT}_0(B, v_r)$, which yields a bitmap with $n + O(n/\delta)$ bits. In our example it is 1000101101001, see Fig. 3. Hence we have the following relation $\text{REDUCE}(v) = \text{RANK}_1(B, \text{SELECT}_0(B, v+1)) - 1$. We do not store B uncompressed, but rather using Theorem 3, which requires only $O((n/\delta) \log n)$ bits as there are few 1's in B. When a node $[v_l, v_r]$ becomes sampled we insert matching parentheses at $S[\text{REDUCE}(v_l)]$ and after $S[\text{REDUCE}(v_r)]$. Also, it is necessary to insert the new 1's in B as before. Fig. 3 illustrates the effect of sampling $b = [3, 6]$.

Updating S when a sampled node v becomes non-sampled is easy, as we can obtain the parentheses u, u' to delete. We must also delete the corresponding 1's in B; note that the relative position of a 1 in a run of 1's is irrelevant. Therefore REDUCE can be computed in $O(\log n)$ time. According to our previous explanation, so can LSA and LCSA, for leaves.

To map in the other direction, each node in the sampled tree must know its corresponding interval $[v_l, v_r]$. This is also easy to obtain from B. Let u be the position in S of the opening parenthesis that identifies sampled node v. The corresponding closing parenthesis is $u' = \text{FINDMATCH}(u)$. Now $v_l = \text{RANK}_0(B, \text{SELECT}_1(B, u+1))$ and $v_r = \text{RANK}_0(B, \text{SELECT}_1(B, u'+1)) - 1$.

6   Putting All Together

The following theorem summarizes our result.

Theorem 4. *It is possible to represent the suffix tree of a dynamic text collection within the space and time bounds given in Table 1. The space and the variables Ψ, Φ, t, can be instantiated to the values of Theorem 1 for $\delta = \omega(\log_\sigma n)$, or to another dynamic CSA supporting ψ, A, A^{-1}, LF, and $T[A[v]]$, in times $O(\Psi)$, $O(\Phi)$, $O(\Phi)$, $O(t)$, and $O(\log n)$, respectively, provided texts are inserted in right-to-left order and deleted in left-to-right order within the given time bounds.*

We note that Theorem 4 assumes that $\lceil \log n \rceil$ is fixed, and so is δ. This assumption is not uncommon in dynamic data structures, even if it affects assertions like that of pointers taking $O(\log n)$ bits. The CSA used in Theorem 1 can handle varying $\lceil \log n \rceil$ within the same worst-case space and complexities, and the same happens with Theorem 3, which is used for the mapping bitmap B. The only remaining part is the sampled tree. We discuss now how to cope with it while retaining the same space and worst-case time complexities.

We use $\delta = \lceil \log n \rceil \cdot \lceil \log_\sigma \lceil \log n \rceil \rceil$, which will change whenever $\lceil \log n \rceil$ changes (sometimes will change by more than 1). Let us write $\delta = \Delta(\ell) = \ell \lceil \log_\sigma \ell \rceil$. We maintain $\ell = \lceil \log n \rceil$. As S is small enough, we can afford to maintain three copies of it: S sampled with δ, S^- with $\delta^- = \Delta(\ell - 1)$, and S^+ sampled with $\delta^+ = \Delta(\ell + 1)$. When $\lceil \log n \rceil$ increases (*i.e.* n doubles), S^- is discarded, the current S becomes S^-, the current S^+ becomes S, we build a new S^+ sampled with $\Delta(\ell + 2)$, and ℓ is increased. A symmetric operation is done when $\lceil \log n \rceil$ decreases (*i.e.* n halves due to deletions), so let us focus on increases from now on. Note this can occur in the middle of the insertion of a text, which must be suspended, and then resumed over the new set of sampled trees.

The construction of the new S^+ can be done by retraversing all the suffix tree $\mathcal{T}$ deciding which nodes to sample according to the new δ^+. An initially empty parentheses sequence and a bitmap B initialized with zeros would give the correct insertion points from the chosen intervals, as both structures are populated. To ensure that we consider each node of $\mathcal{T}$ once, we process the leaves in order (*i.e.* $v = [0,0]$ to $v = [n-1, n-1]$), and for each leaf v we also consider all its ancestors $[v_l, v_r]$ (using $\text{PARENT}_\mathcal{T}$) as long as $v_r = v$. For each node $[v_l, v_r]$ we consider, we apply SLINK at most δ^+ times until either we find the first node $v' = \text{SLINK}^i([v_l, v_r])$ which either is sampled in S^+, or $\text{SDEP}(v') \equiv_{\delta^+/2} 0$ and $i \geq \delta^+/2$. If v' was not sampled we insert it into S^+, and in both cases we increase its reference count if $i = \delta^+/2$ (recall Section 4).

All the δ^+ suffix links in $\mathcal{T}$ are computed in $O(\delta^+(\Psi + t))$ time, as they form a single chain. Therefore the solution maintains the current complexities, yet only in an amortized sense.

Deamortization can be achieved by the classical method of interleaving the normal operations of the data structure with the construction of the new S^+. By performing a constant number of operations on the new S^+ for each insertion/deletion operation over $\mathcal{C}$, we can ensure that the new S^+ will be ready in time. The challenge is to maintain the consistency of the traversal of $\mathcal{T}$ while texts are inserted/deleted.

As we insert a text, the operations that update $\mathcal{T}$ consist of insertion of leaves, and possibly creation of a new parent for them. Assume we are currently at node $[v_l, v_r]$ in our traversal of $\mathcal{T}$ to update S^+. If a new node $[v'_l, v'_r]$ we are inserted is behind the current node in our traversal order (that is, $v'_r < v_r$, or $v'_r = v_r$ and $v'_l > v_l$), then we consider $[v'_l, v'_r]$ immediately; otherwise we leave this for the moment when we will reach $[v'_l, v'_r]$ in our traversal. Recall from Section 4 that those new insertions do not affect the existing SDEPs nor suffix link paths, and hence can be considered independently of the current traversal process. Similarly, deleted nodes that fall behind the current node are processed immediately, and the others left for the traversal to handle it.

If ℓ decreases while we are still building S^+, we can discard it even before having completed its construction. Note that in general discarding a tree when ℓ changes involves freeing several data structures. This can also be done progressively, interleaved with the other operations.

7   Conclusions

We presented the first dynamic fully-compressed representation of suffix trees (FCSTs). Static FCSTs broke the $\Theta(n)$ bits barrier of previous representations at a reasonable (and in some cases no) time complexity penalty, while retaining a surprisingly powerful set of operations. Dynamic FCSTs permit not only managing dynamic collections, but also building static FCSTs within optimal space. Hence the way is open to practical implementations of this structure, which can run in main memory for very large texts.

We also gave some relevant results for the static case, as we improved or simplified the operations REDUCE and CHILD. A challenge for future work is to obtain operations TDEP, LAQT, and LAQS, which we were not able to maintain in a dynamic scenario.

Acknowledgments. We are grateful to Veli Mäkinen and Johannes Fisher for pointing out the generalized branching problem to us.

References

1. Apostolico, A.: Combinatorial Algorithms on Words. In: The myriad virtues of subword trees. NATO ISI Series, pp. 85–96. Springer, Heidelberg (1985)
2. Gusfield, D.: Algorithms on Strings, Trees and Sequences. Cambridge University Press, Cambridge, UK (1997)
3. Giegerich, R., Kurtz, S., Stoye, J.: Efficient implementation of lazy suffix trees. Softw. Pract. Exper. 33(11), 1035–1049 (2003)
4. Manber, U., Myers, E.W.: Suffix arrays: A new method for on-line string searches. SIAM J. Comput. 22(5), 935–948 (1993)
5. Sadakane, K.: Compressed suffix trees with full functionality. Theory Comput. Syst. 41, 589–607 (2007), http://dx.doi.org/10.1007/s00224-006-1198-x
6. Navarro, G., Mäkinen, V.: Compressed full-text indexes. ACM Comp. Surv. 39(1), 2 (2007)
7. Ferragina, P., Manzini, G., Mäkinen, V., Navarro, G.: Compressed representations of sequences and full-text indexes. ACM Trans. Algor. 3(2), 20 (2007)
8. Manzini, G.: An analysis of the Burrows-Wheeler transform. J. ACM 48(3), 407–430 (2001)
9. Russo, L., Navarro, G., Oliveira, A.: Fully-Compressed Suffix Trees. In: LATIN. LNCS, vol. 4957, pp. 362–373. Springer, Heidelberg (2008)
10. Chan, H.-L., Hon, W.-K., Lam, T.-W., Sadakane, K.: Compressed indexes for dynamic text collections. ACM Trans. Algorithms 3(2) (2007)
11. Mäkinen, V., Navarro, G.: Dynamic entropy-compressed sequences and full-text indexes. In: Lewenstein, M., Valiente, G. (eds.) CPM 2006. LNCS, vol. 4009, pp. 307–318. Springer, Heidelberg (to appear in ACM TALG, 2006)
12. González, R., Navarro, G.: Improved dynamic rank-select entropy-bound structures. In: LATIN. LNCS, vol. 4957, pp. 374–386. Springer, Heidelberg (2008)
13. Weiner, P.: Linear pattern matching algorithms. In: IEEE Symp. on Switching and Automata Theory, pp. 1–11 (1973)
14. Russo, L., Oliveira, A.: A compressed self-index using a Ziv-Lempel dictionary. In: Crestani, F., Ferragina, P., Sanderson, M. (eds.) SPIRE 2006. LNCS, vol. 4209, pp. 163–180. Springer, Heidelberg (2006)
15. Huynh, T.N.D., Hon, W.-K., Lam, T.W., Sung, W.-K.: Approximate string matching using compressed suffix arrays. Theor. Comput. Sci. 352(1-3), 240–249 (2006)

A Linear Delay Algorithm for Building Concept Lattices

Martin Farach-Colton and Yang Huang*

Department of Computer Science, Rutgers University, Piscataway, NJ 08854
farach@cs.rutgers.edu, yahuang@cs.rutgers.edu

Abstract. Concept lattices (also called Galois lattices) have been applied in numerous areas, and several algorithms have been proposed to construct them. Generally, the input for lattice construction algorithms is a binary matrix with size $|G||M|$ representing binary relation $I \subseteq G \times M$. In this paper, we consider polynomial delay algorithms for building concept lattices. Although the concept lattice may be of exponential size, there exist polynomial delay algorithms for building them. The current best delay-time complexity is $O(|G||M|^2)$. In this paper, we introduce the notion of *irregular concepts*, the combinatorial structure of which allows us to develop a linear delay lattice construction algorithm, that is, we give an algorithm with delay time of $O(|G||M|)$. Our algorithm avoids the union operation for the attribute set and does not require checking if new concepts are already generated. In addition, we propose a compact representation for concept lattices and a corresponding construction algorithm. Although we are not guaranteed to achieve optimal compression, the compact representation can save significant storage space compared to the full representation normally used for concept lattices.

1 Introduction

Concept lattices have proved useful in many areas, such as knowledge representation [13], information retrieval [3], web document management [4,8], software engineering [16] and bioinformatics [6,11]. Of particular importance in these applications is the structure of the lattice, i.e. the *Hasse diagram* of the concept lattices. For example, the immediate predecessors and successors of a concept are used in browsing web documents [8] or to infer the class hierarchy of a program [16]. The edge information of Hasse diagrams can be used to compare two concept lattices in which gene expression information has been coded [6].

A concept lattice can be briefly defined as follows. Given a binary relation between an object set and an attribute set, a *concept* is a pair of object set A and attribute set B, denoted as (A, B), where A contains all objects sharing every attribute in B and B contains all attributes shared by every object in A. A concept lattice is the partially ordered set of all concepts, in which the order

---

* Yang Huang is currently at NCBI/NLM/NIH, 8600 Rockville Pike, Room 8N811I, Bethesda, MD 20894. This work was done when he was at Rutgers University.

P. Ferragina and G. Landau (Eds.): CPM 2008, LNCS 5029, pp. 204–216, 2008.

is defined using subset order on object sets (which is equivalent to containment order on attribute sets).

In real-world applications, we often find that it is necessary to construct a concept lattice from large amount of input data. It is known that the number of concepts in a concept lattice can be exponential in the size of the input binary matrix. The problem of deciding the number of concepts has been shown to be #P-complete [19].

Constructing a concept lattice is a type of *enumeration problem*. Algorithms for enumeration problems are typically measured both by *total time complexity* and *delay-time complexity* [12]. The running time of algorithms with polynomial total time is a polynomial in the size of the input and output. Polynomial delay time means that there is a polynomial in the input size that bounds the time to the first entity outputted as well as the delay between any two consecutive output entities. An algorithm with polynomial delay-time complexity has total polynomial running time, which is the polynomial delay time multiplied by the output size. Algorithms with polynomial delay time are often preferred because they allow us to predict the time to get the next entity. They allow the procedure of entity processing to follow immediately. They also allow to generate a subset of entities without generating others.

Related Work. The problem of generating the set of concepts is closely related to two other important problems: generating all maximal bipartite cliques in a given bipartite graph $\mathcal{G}_b = (V_1, V_2, E)$ and generating all frequent closed itemsets in a transaction database [21]. Note that generating all maximal bipartite cliques or all frequent closed itemsets is not enough for generating a concept lattice since a concept lattice requires the partial order among all concepts be recovered. A maximal bipartite clique corresponds to a concept if we consider $\mathcal{G}_b$ as a representation for the matching relation between two parts of $\mathcal{G}_b$'s vertices. A frequent closed itemset corresponds to a concept whose object set size is larger than a certain threshold. Eppstein [7] showed that the number of all maximal bipartite cliques is $O(|V(\mathcal{G})|)$ in a graph with bounded arboricity and gave a linear total time algorithm, where $V(\mathcal{G})$ is the vertex set of $\mathcal{G}$. An algorithm for generating all maximal bipartite cliques in any bipartite graph was designed by Makino and Uno [14]. It takes $O(\Delta^2)$ polynomial delay, where Δ is the maximum degree of $\mathcal{G}_b$. Given $|V_1| = |G|$ and $|V_2| = |M|$, then $\Delta = \max(|G|, |M|)$. CLOSET+ [18], CHARM [20] and LCM2 [17] are among state-of-the-art algorithms for generating the set of frequent closed itemsets.

However, though many algorithms are available for generating the set of concepts and some of them are quite fast, few algorithms compute the edge structure of the lattice. Bordat's algorithm [2] uses a trie to store and retrieve concepts with delay-time complexity $O(|G||M|^2)$, where G is the input object set and M is the input attribute set. Without loss of generality, we will assume $|G| \geq |M|$. Depending on the value of Δ, the delay-time complexity of Makino and Uno's algorithm [14] may be better than $O(|G||M|)$. However, it can not be bound by $O(|G||M|)$ in the worst case. The best polynomial delay-time complexity of algorithms for constructing a concept lattice in terms of $|G|$ and $|M|$ is $O(|G||M|^2)$.

Godin *et al.* [10] proposed an incremental algorithm that dynamically updates the structure of the concept lattice as new rows or columns are added to the input matrix. The algorithm by Nourine and Raynaud [15] has the best known total time complexity $O(|G||M||\mathcal{B}|)$, where $\mathcal{B}$ is the set of all concepts. But it is not a polynomial delay algorithm. Recently, Choi [5] proposed an efficient concept lattice construction algorithm with complexity $O(\sum_{a \in ext(\mathcal{C})} |cnbr(a)|)$, where $ext(\mathcal{C})$ is the object set of the concept $\mathcal{C}$ and $cnbr(a)$ is a reduced attribute set of a. However, it seems to us that the condition used in the algorithm, which is to check if a newly generated pair of object set and attribute set is a concept, is not sufficient. Berry *et al.* [1] suggested constructing concept lattices by searching non-dominating maxmods in a co-bipartite graph. The complexity is $O(|G||M|)$ per concept plus $O(|G||M|^2)$ per traversed maximal chain of the lattice.

Our Results. In this paper, we propose a concept lattice construction algorithm with delay $O(|G||M|)$, which is linear in the size of the input matrix. Though the total time complexity of our algorithm, $O(|G||M||\mathcal{B}|)$, ties with that of Nourine and Raynaud, our algorithm is a polynomial delay algorithm, which their algorithm is not. By introducing the set of irregular concepts, we ensure that when we compute the union of several attribute sets they are disjoint. Our algorithm also avoids the operation to check if a newly generated pair of object set and attribute set is a concept or if it is going to be subsumed, as most previous algorithms do.

The usual way to represent a concept is by a pair of its object set and attribute set, which contain a lot of redundant information. We call this the *full representation*. The space required for storing the full representation of all concepts is $O(|G||\mathcal{B}|)$. We propose a *compact representation* for concept lattices, in which we represent a concept in terms of the set difference between its object/attribute set and the one in one of its predecessors. In the optimal case, a compact representation only requires $O(|\mathcal{B}|)$ space for all concepts, which reaches the lower bound. Given the compact representation of a concept lattice, we can easily recover the full representation in linear time. We modify our algorithm for the full representation to construct a compact representation.

From now on, we will refer to concept lattices as lattices from time to time when the context is clear. The remaining of the paper is organized as follows: We introduce the basics of lattices in section 2. In section 3 we present some characterization for lattices. We introduce our algorithm for the full representation in section 4 and the modified version for the compact representation in section 5. Finally, we conclude in section 6 and discuss some future research direction.

2   Preliminary

In this section we will give a brief overview for lattices. For a complete introduction, please refer to the book [9]. Many of our notations follow the ones used in the book. Given a context (G, M, I) where G is the *object* set and M is the *attribute* set, a binary matrix R is used to represent the relation $I \subseteq G \times M$, i.e. $R_{i,j} = 1$ if $(g_i, m_j) \in I$ where $g_i \in G$ and $m_j \in M$ and $R_{i,j} = 0$ otherwise. For $g_i \in G$, we define $g_i' = \{m_j | R_{i,j} = 1\}$. Furthermore, for an object set $A \subseteq G$, we denote $A' = \cap_{g_i \in A} g_i'$. Dually, we define $m_j' = \{g_i | R_{i,j} = 1\}$ for $m_j \in M$ and $B' = \cap_{m_j \in B} m_j'$ for $B \subseteq M$. With the above notation we are ready to define the concept.

Definition 2.1. *The* concept *is a pair* (A, B) *where* $A \subseteq G$, $B \subseteq M$, $A = B'$ *and* $B = A'$. A *is called* extent *and* B *is called* intent *of the concept.*

For a concept $C = (A, B)$, we denote A as $ext(C)$ and B as $int(C)$. We call a set A *closed* if $A = A''$. The extent and intent of a concept are closed sets. It can be seen that a closed object set A or a closed attribute set B uniquely determines a concept (A, A') or (B', B).

A partial order $\preceq$ is defined on $\mathcal{B}$, the set of all concepts:

Definition 2.2. *If* $ext(C) \subseteq ext(D)$ $(int(D) \subseteq int(C))$, *then* $C \preceq D$. C *is called* successor *of* D *and* D *is called* predecessor *of* C.

Please note that definition of predecessors and successors in a concept lattice may be somewhat counter-intuitive. However, this way successors will be placed in a lower lever, below its predecessors, in the diagram representing a concept lattice. The diagram will be shown next. According to the definition, a concept is a predecessor and a successor of itself. In particular, if C is a successor of D other than D itself and $\forall \mathcal{E}$ such that $C \preceq \mathcal{E} \preceq D$ implies $\mathcal{E} = C$ or $\mathcal{E} = D$, then C is an *immediate successor* of D and D is an *immediate predecessor* of C.

Definition 2.3. *The partially ordered set* $\mathcal{L}(G, M, I) = \langle \mathcal{B}, \preceq \rangle$ *is called* concept lattice *or* Galois lattice.

The diagram representing a partially ordered set is called *Hasse diagram*, where a vertex represents a concept, and two concepts are connected by an edge if one concept is an immediate successor of the other. We show a lattice example in the Figure 1.

Later we will need the definition of the infimum.

Definition 2.4. *The infimum of a subset* S *of a partially ordered set* $(P, \preceq)$, *denoted as* $\wedge S$, *is an element* l *of* P *such that*

1. $\forall x \in S, l \preceq x$, *and*
2. *for any* $p \in P$ *such that* $\forall x \in S, p \preceq x$, *it holds that* $p \preceq l$.

3   Some Characterization of Lattices

In this section, we will present some characterization of lattices, which will help us design the lattice construction algorithm. Due to the limit of space, all the proof is omitted.

Fig. 1. (a) A binary matrix representing I, where the entry corresponding to g_i and m_j is x iff $(g_i, m_j) \in I$. $G = \{a, b, c, d\}$ and $M = \{1, 2, 3, 4\}$. (b) The Hasse diagram of the lattice constructed from (G, M, I). The lattice is represented in full representation.

We need the following known result:

Proposition 3.1. *[9] For a concept $C \in \mathcal{B}$ in the lattice $\mathcal{L}$, C and all of its successors forms a concept lattice, denoted by $\mathcal{L}^C$.*

In the following, we will first define regular and irregular concepts. Then we will present a lemma on the concept C and its irregular successors.

Suppose $\mathcal{D}$ is a concept in the lattice $\mathcal{L}^C$, and denote the set of its immediate predecessors that are successors of C by $IP_{\mathcal{D}}^C$. In addition, suppose $IP_{\mathcal{D}}^C = \{\mathcal{D}_i | i \in [1..n]\}$, we denote the set $\bigcup_{i=1}^n int(\mathcal{D}_i)$, the union of the intent of concepts in $IP_{\mathcal{D}}^C$, by $int(IP_{\mathcal{D}}^C)$.

Definition 3.1. *If $int(\mathcal{D}) = int(IP_{\mathcal{D}}^C)$, $\mathcal{D}$ is called* regular *concept of C. If $int(\mathcal{D}) \supset int(IP_{\mathcal{D}}^C)$, $\mathcal{D}$ is called* irregular *concept of C.*

Note that an irregular concept $\mathcal{D}$ of C is not necessarily *meet-irreducible* in $\mathcal{L}^C$, where $\mathcal{D}$ is meet-irreducible if $\mathcal{D} = \wedge\{\mathcal{E}, \mathcal{F}\} \Rightarrow \mathcal{D} = \mathcal{E}$ or $\mathcal{D} = \mathcal{F}$, because $\mathcal{D}$ can have more than one immediate predecessor in $\mathcal{L}^C$.

Any immediate successor of C is an irregular concept of C. We denote the set of all C's irregular concepts by IR^C. The following proposition will help us identify immediate successors of C from IR^C.

Proposition 3.2. *Given $C_i \in IR^C$, it is an immediate successor of C if and only if there is no $C_j \in IR^C, j \neq i$ such that $C_i \preceq C_j$.*

The following lemma shows one of important properties of IR^C:

Lemma 3.1. *Suppose $IR^C = \{C_i | i \in T\}$, where T is an index set. Given $C_i \in IR^C$, let $V_i = int(IP_{C_i}^C)$ and let $B_i = int(C_i) \setminus V_i$. If $i, j \in T$ and $i \neq j$, then $B_i \cap B_j = \emptyset$, and $\bigcup_{i \in T} B_i = \bigcup_{g \in ext(C)} g' \setminus int(C)$.*

The above lemma indicates that $\{B_i | i \in T\}$ constitutes a partition of $\bigcup_{g \in ext(C)} g' \setminus int(C)$, which is the set of attributes belonged to some $g \in ext(C)$ but not appearing in $int(C)$. Since $B_i \cap B_j = \emptyset$, $B_i \neq \emptyset, B_j \neq \emptyset$ and $B_i \subseteq M, B_j \subseteq M$, we have a direct corollary from the lemma:

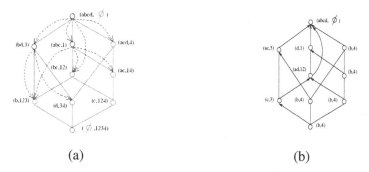

(a)   (b)

Fig. 2. (a) Irregular concepts regarding $(abcd, \emptyset)$, $(bd, 3)$ and $(abc, 1)$. The concepts pointed by an arrow and connected to $(abcd, \emptyset)$ $((bd, 3)/(abc, 1))$ by dashed lines are its irregular concepts. Note that the lattice is in the full representation. (b) The same lattice in a compact representation computed by our algorithm. For each concept, a directed edge points to its base.

Corollary 3.1. *For any concept $\mathcal{C}$, $|IR^{\mathcal{C}}| \leq |M|$.*

Since B_i is still "partial" compared to $int(\mathcal{C}_i)$, let us introduce the set $\{(ext(\mathcal{C}_i), B_i)|\mathcal{C}_i \in IR^{\mathcal{C}}\}$ as $PIR^{\mathcal{C}}$, where B_i is defined as in Lemma 3.1. Please note that the only difference between $PIR^{\mathcal{C}}$ and $IR^{\mathcal{C}}$ is that the attribute set in $PIR^{\mathcal{C}}$ is not complete yet. To see some examples of PIR, let $\mathcal{C} = (abcd, \emptyset)$, $\mathcal{C}_1 = (bd, 3)$ and $\mathcal{C}_2 = (abc, 1)$ in Figure 2 (a). Then $PIR^{\mathcal{C}} = \{(bd, 3), (abc, 1), (acd, 4), (bc, 2)\}$. $PIR^{\mathcal{C}_1} = \{(b, 1), (d, 4)\}$. And $PIR^{\mathcal{C}_2} = \{(b, 3), (bc, 2), (ac, 4)\}$.

For a concept in $\mathcal{L}^{\mathcal{C}}$, the following corollary makes it easy to identify its predecessors in $PIR^{\mathcal{C}}$.

Corollary 3.2. *Given $\mathcal{C}$, $\forall(A_i, B_i) \in PIR^{\mathcal{C}}$, $\forall \mathcal{D} \preceq \mathcal{C}$, if $int(\mathcal{D}) \cap B_i \neq \emptyset$, then $\mathcal{D} \preceq \mathcal{E}$, where $\mathcal{E} = (A_i, A_i')$.*

Since $IR^{\mathcal{C}}$ contains all $\mathcal{C}$'s immediate successors, we will be able to generate the sublattice $\mathcal{L}^{\mathcal{C}}$ if we can generate $IR^{\mathcal{C}}$. Actually we can obtain $IR^{\mathcal{C}}$ by augmenting the attribute set B_i in $PIR^{\mathcal{C}}$ in a certain way. The theorem that will be shown next provides the basis for the processing.

Given $\mathcal{C}_j \in IR^{\mathcal{C}} = \{\mathcal{C}_i | i \in T\}$, where T is an index set, we define an equivalence relation $\sim_j$ on the set $T \setminus \{j\}$. $i \sim_j k$ if $ext(\mathcal{C}_i) \cap ext(\mathcal{C}_j) = ext(\mathcal{C}_k) \cap ext(\mathcal{C}_j) \neq \emptyset$ for $i, k \in T \setminus \{j\}$. Let the resulting equivalence classes on T be $[j_1], [j_2], \ldots, [j_r]$. For each equivalence class $[j_h], h \in [1..r]$, let us denote $A_{j_h} = ext(\mathcal{C}_i) \cap ext(\mathcal{C}_j), i \in [j_h]$ and $B_{j_h} = \bigcup_{i \in [j_h]} B_i$.

When we proceed to our main theorem in this section, the following proposition will become useful:

Proposition 3.3. $\{B_{j_h} | h \in [1..r]\}$ *constitutes a partition of* $\bigcup_{g \in ext(\mathcal{C}_j)} g' \setminus (int(\mathcal{C}) \cup B_j)$, *where* $[j_1], [j_2], \ldots, [j_r]$ *are equivalent classes defined above.*

Theorem 3.1. *If $\mathcal{C}_j \in IR^{\mathcal{C}}$ is an immediate successor of $\mathcal{C}$, then $PIR^{\mathcal{C}_j} = \{(A_{j_h}, B_{j_h}) | h \in [1..r]\}$.*

It is easy to extend the theorem to the case that $\mathcal{C}_j \in IR^{\mathcal{C}}$ is not an immediate successor of $\mathcal{C}$.

Corollary 3.3. *If $\mathcal{C}_j \in IR^{\mathcal{C}}$ is not an immediate successor of $\mathcal{C}$, then $PIR^{\mathcal{C}_j} \cup \{(ext(\mathcal{C}_j), int(\mathcal{C}_j) \setminus (int(\mathcal{C}) \cup B_j))\} = \{(A_{j_h}, B_{j_h}) | h \in [1..r]\}$.*

4   Algorithm for the Full Representation

We will present a lattice construction algorithm which generates a lattice in the full representation. The input for the algorithm is a binary matrix R with $|G|$ rows and $|M|$ columns representing the relation I between the object set G and the attribute set M, where $R_{i,j} = 1$ if and only if $(g_i, m_j) \in I$.

4.1   Overview

The algorithm builds the lattice while traversing it in depth first search (DFS). Each node will represent a concept. Suppose the node $\mathcal{C}$, is already visited. And suppose $PIR^{\mathcal{C}}$ is already generated, each element of which is put in a child node of $\mathcal{C}$. These child nodes are sorted in the ascending order by the size of the object set in each child node. Though the intent of those concepts represented by the child nodes is not complete yet, we will still represent the nodes by $\mathcal{C}_j$. Note that the following conditions are met: Each shadow child node, which will be introduced below, of $\mathcal{C}$ contains (A_j, s_j) where $s_j = |B_j|, (A_j, B_j) \in PIR^{\mathcal{C}}$; All unvisited child nodes are of form $(A_j, B_j) \in PIR^{\mathcal{C}}$. When the algorithm visits a child node $\mathcal{C}_j = (A_j, B_j)$ for the first time, it begins to traverse the sublattice $\mathcal{L}^{\mathcal{C}_j}$. It will first generate $PIR^{\mathcal{C}_j}$ as $\mathcal{C}_j$'s child nodes, which contains all immediate successors of $\mathcal{C}_j$. To do so, the algorithm uses GeneratePIR($\mathcal{C}_j$, $\mathcal{C}$) and SearchEquiClass($\mathcal{C}_j$, $\mathcal{C}$) to generate $PIR^{\mathcal{C}_j}$ as Theorem 3.1 indicates. In GeneratePIR, we generate two kinds of child nodes using intersection on extents. One is marked as unvisited and the other is marked as *shadow*. There is no need to call GeneratePIR for a shadow node since its corresponding concept, say $\mathcal{F}$, and all $\mathcal{F}$'s successors have already been generated at that time because of DFS traversal. As we will see, a shadow node will never enter the stack. The shadow nodes are used to prevent generating a concept more than once without losing track of its immediate successors. Each shadow node has a pointer pointing to the corresponding concept in the lattice.

After the child nodes of $\mathcal{C}_j$ are generated, the algorithm uses SearchEquiClass to find equivalence classes $[j_r]$, where each class corresponds to $\mathcal{C}_j$ or a member in $IR^{\mathcal{C}_j}$. If a class contains a shadow node, it means that the concept corresponding to the class is already visited. If not, it means that the corresponding concept is unvisited yet. In both cases, we remove the child nodes under $\mathcal{C}_j$ for memory reuse. By Proposition 3.2, we check if there exists an equivalence class with $|A_{j_h}| = |A_j|$ to determine if $\mathcal{C}_j$ is an immediate successor of $\mathcal{C}$ or not. By Proposition 3.3, when we generate B_{j_h} and $(A_j)'$ no actual union operation is required since B_is are disjoint. After constructing $\mathcal{L}^{\mathcal{C}_j}$ by repeatedly calling GeneratePIR and SearchEquiClass, the algorithm will then visit $\mathcal{C}$'s next unvisited child node which is right after $\mathcal{C}_j$ in the child node list of $\mathcal{C}$.

4.2   Implementation

The pseudo-code of GeneratePIR($\mathcal{C}_j$, $\mathcal{C}$) and SearchEquiClass($\mathcal{C}_j$, $\mathcal{C}$) is shown by Algorithms 1 and 2, respectively.

In GeneratePIR, we generate a shadow node $(A_j \cap A_i, s_i)$ under $\mathcal{C}_j$ when $\mathcal{C}_i$ is a shadow node. To obtain a pointer to the concept for $(A_j \cap A_i, s_i)$, we can either build a trie or a hash table for the object sets of concepts generated so far to facilitate the search. Note that if we are only going to generate all concepts,

---

Algorithm 1. GeneratePIR($\mathcal{C}_j$, $\mathcal{C}$)

  for each child node $\mathcal{C}_i \neq \mathcal{C}_j$ of $\mathcal{C}$ **do**
    if $\mathcal{C}_i = (A_i, B_i)$ is unvisited **then**
      put an unvisited node $(A_j \cap A_i, B_i)$ under $\mathcal{C}_j$;
      put a shadow node $(A_j \cap A_i, |B_j|)$ under $\mathcal{C}_i$; {This is a shadow node for $(A_j \cap A_i, (A_j \cap A_i)')$.}
    else if $\mathcal{C}_i = (A_i, s_i)$ is a shadow node and A_i has not intersected with A_j **then**
      put a shadow node $(A_j \cap A_i, s_i)$ under $\mathcal{C}_j$;
    end if
  end for
  mark $\mathcal{C}_j$ as visited;

Algorithm 2. SearchEquiClass($\mathcal{C}_j$, $\mathcal{C}$)

  group $\mathcal{C}_j = (A_j, B_j)$'s child nodes according to their object set to find equivalence classes $[j_h], h \in [1..r]$;
  if $|A_{j_h}| == |A_j|$ for the largest $|A_{j_h}|$ **then**
    $A'_j = B_{j_h} \cup (int(\mathcal{C}) \cup B_j)$;
  else
    $A'_j = int(\mathcal{C}) \cup B_j$;
    mark $\mathcal{C}_j$ as an immediate successor of $\mathcal{C}$;
  end if
  for each equivalence class $[j_h]$ where $|A_{j_h}| \neq |A_j|$ **do**
    if all child nodes (A_{j_h}, B_i) of $\mathcal{C}_j$ are unvisited **then**
      remove all (A_{j_h}, B_i) from $\mathcal{C}_j$'s child node list;
      put an unvisited node (A_{j_h}, B_{j_h}) under $\mathcal{C}_j$; {This node represents a new concept.}
    else
      remove all (A_{j_h}, s_i) and (A_{j_h}, B_i) from $\mathcal{C}_j$'s child node list;
      $s = \sum_{(A_{j_h}, s_i)} s_i + \sum_{(A_{j_h}, B_i)} |B_i|$;
      put a shadow node (A_{j_h}, s) under $\mathcal{C}_j$; {Suppose the node is the shadow of $\mathcal{E} = (A_{j_h}, (A_{j_h})')$.}
      if $s + |A'_j| = |(A_{j_h})'|$ **then**
        mark $\mathcal{E}$ as an immediate successor of $\mathcal{C}_j$;
      end if
    end if
  end for
  sort $\mathcal{C}_j$'s child nodes by their object set size in the ascending order;

Algorithm 3. Construct a lattice in the full representation

generate $PIR^{\mathcal{U}}$ as $\mathcal{U}$'s child nodes;
sort $\mathcal{U}$'s child nodes by their object set size in the ascending order;
initialize an empty stack S;
push$((\mathcal{E}, \mathcal{U}), S)$ where $\mathcal{E}$ is the first node in $\mathcal{U}$'s sorted child node list;
while S is not empty **do**
$\quad (\mathcal{C}_j, \mathcal{C}) = \text{pop}(S)$;
$\quad$GeneratePIR$(\mathcal{C}_j, \mathcal{C})$;
$\quad$SearchEquiClass$(\mathcal{C}_j, \mathcal{C})$;
$\quad$**if** there is an unvisited node $\mathcal{C}_k$ after $\mathcal{C}_j$ among $\mathcal{C}$'s child nodes **then**
$\quad\quad$push$((\mathcal{C}_k, \mathcal{C}), S)$;
$\quad$**end if**
$\quad$push$((\mathcal{E}, \mathcal{C}_j), S)$ where $\mathcal{E}$ is the first unvisited node among $\mathcal{C}_j$'s child nodes;
end while

---

we do not need pointers in shadow nodes or a trie or a hash table for the object sets. To find the equivalence classes in SearchEquiClass yet, for $\mathcal{C}_j$, we build a trie for object sets of its child nodes and put node ids in the leaves of the trie. Then we are able to find the equivalence classes of child nodes by just checking leaves of the trie.

With GeneratePIR and SearchEquiClass ready, we present our lattice construction algorithm in Algorithm 3, where the supremum of the lattice is $\mathcal{U}$.

At the beginning of Algorithm 3, we scan the input matrix once to obtain $\mathcal{U} = (G, G')$ and (A_j, m_j) for each $j \in [1..|M|]$, where m_j is the attribute shared by each member of the object set A_j. For the object sets $A_j \subset G, j \in [1..|M|]$, we build a trie for them and find equivalence classes just as we did in SearchEquiClass. This way we obtain $PIR^{\mathcal{U}}$ as the child nodes of U. Then we construct the remaining of the lattice by processing each node once with GeneratePIR and SearchEquiClass, when we traverse the lattice in DFS with the stack S. In DFS, only unvisited nodes will be pushed into S.

4.3   Algorithm Analysis

We will show that our algorithm correctly constructs the lattice and the delay-time complexity is $O(|G||M|)$.

Lemma 4.1. *After the completion of the algorithm, For each concept, its extent and intent are correctly computed and its immediate successors are correctly marked in its child nodes.*

The lemma can be proved by applying Theorem 3.1 and Corollary 3.3 in GeneratePIR and SearchEquiClass to check the correctness of $ext(\mathcal{C}_j)$, $int(\mathcal{C}_j)$, and $PIR^{\mathcal{C}_j}$.

The complexity analysis of the algorithm is shown by the following result:

Theorem 4.1. *The algorithm correctly builds the lattice with $O(|G||M|)$ delay time.*

The sketch of the proof can be outlined as follows: The initialization, one run of GeneratePIR and one run of SearchEquiClass all take time $O(|G||M|)$. At the beginning of each iteration of while loop, a new concept is obtained by popping it from the stack. During each iteration of while loop, GeneratePIR and SearchEquiClass are executed once. So it takes $O(|G||M|)$ time to complete one iteration.

As for the space required to store shadow nodes, we can analyze it as follows: Each shadow node needs space $O(|G|)$. The size of the maximal chain in the lattice is at most $|G|$. At each level of the chain, the algorithm will incur at most $O(|M|^2)$ shadow nodes. When the algorithm reaches the bottom of the chain, the size of space is maximized, which is $O(|G|^2|M|^2)$. Other previously used shadow nodes are already recycled in SearchEquiClass. So we only need $O(|G|^2|M|^2)$ space for storing shadow nodes.

5 Algorithm for the Compact Representation

We will define the compact representation for lattices and modify the above algorithm to construct the lattice in a compact representation.

5.1 Compact Representation

Usually a lattice is represented in the full representation as in Figure 2 (a). It is easy to see there is much redundant information in this representation. Suppose (A_1, B_1) is a successor of (A_2, B_2) and $A_1 \subset A_2$. Given (A_2, B_2), we can represent the concept (A_1, B_1) as $(A_2 \setminus A_1, B_1 \setminus B_2)$. Following this idea, we define the compact representation of the lattice $\mathcal{L}$ as follows:

Definition 5.1. *For each concept* $\mathcal{C} = (A_1, B_1)$ *in* $\mathcal{L}$, *the* compact representation *of* $\mathcal{C}$ *regarding* $\mathcal{D} = (A_2, B_2)$ *is* $(A_2 \setminus A_1, B_1 \setminus B_2)$, *where* $\mathcal{C} \preceq \mathcal{D}$. $\mathcal{D}$ *is called* $\mathcal{C}$'s base.

We denote such a compact representation of $\mathcal{C}$ as $(CR_{\mathcal{D}}(ext(\mathcal{C})), CR_{\mathcal{D}}(int(\mathcal{C})))$. In Figure 2 (b), we show a concept lattice in a compact representation.

Note that the compact representation is not unique and depends on how we choose the base for each concept. As long as we keep the identity of the base for each concept, we are able to recover the full representation from the compact one, in which we only need to perform two set operations for each concept in a top-down manner.

5.2 Implementation

First, let us consider the compact representation for extents. In [22] a technique using set difference, called diffset, was applied to speed up computation of closed itemsets. The diffset can be used to compute the compact presentation for extents as well.

Suppose nodes $\mathcal{C}_i$ and $\mathcal{C}_j$ are two successors of $\mathcal{C}$. We restate an observation by Zaki *et al.* as follows:

Proposition 5.1. *[22] Suppose $\mathcal{F} = \wedge\{\mathcal{C}_i, \mathcal{C}_j\}$. For $\mathcal{C}_i$ and $\mathcal{C}_j$, if neither of them is an immediate successor of the other, then*

$CR_{\mathcal{C}_j}(ext(\mathcal{F})) = CR_{\mathcal{C}}(ext(\mathcal{C}_i)) \setminus CR_{\mathcal{C}}(ext(\mathcal{C}_j))$.

If $\mathcal{C}_i$ is an immediate successor of $\mathcal{C}_j$, i. e. $\mathcal{F} = \mathcal{C}_i$, then

$CR_{\mathcal{C}_j}(ext(\mathcal{F})) = CR_{\mathcal{C}}(ext(\mathcal{C}_j)) \setminus C\mathring{R}_{\mathcal{C}}(ext(\mathcal{C}_i))$.

To generate a compact representation, we will modify the algorithms for the full representation. In the beginning of Algorithm 3, we will represent the concepts in $IR^{\mathcal{U}}$ in a compact representation with their base to be $\mathcal{U}$. Note that there is no base for $\mathcal{U}$. In GeneratePIR($\mathcal{C}_j$, $\mathcal{C}$), suppose $\mathcal{C}_j = (CR_{\mathcal{C}}(ext(\mathcal{C}_j)), B_j)$ and $\mathcal{C}_i = (CR_{\mathcal{C}}(ext(\mathcal{C}_i)), B_i)$ and they are two child nodes under $\mathcal{C}$. Moreover, suppose $\mathcal{F} = \wedge\{\mathcal{C}_i, \mathcal{C}_j\}$. We only need to do the following: We will put $CR_{\mathcal{C}_j}(ext(\mathcal{F}))$ into the node under $\mathcal{C}_j$, and put $CR_{\mathcal{C}_i}(ext(\mathcal{F}))$ into the node under $\mathcal{C}_i$. The computation of $CR_{\mathcal{C}_j}(ext(\mathcal{F}))$ and $CR_{\mathcal{C}_i}(ext(\mathcal{F}))$ will use $CR_{\mathcal{C}}(ext(\mathcal{C}_j))$ and $CR_{\mathcal{C}}(ext(\mathcal{C}_i))$. Details about the computation will be provided later. In SearchEquiClass($\mathcal{C}_j$, $\mathcal{C}$), at the beginning we compute $int(\mathcal{C}_j)$ and thus generate the concept $\mathcal{C}_j$. To compute $\mathcal{C}_j$'s compact representation, $\mathcal{C}$ is set as its base. $CR_{\mathcal{C}}(ext(\mathcal{C}_j))$ is already obtained in previous run of GeneratePIR. And $CR_{\mathcal{C}}(int(\mathcal{C}_j)) = B_j$. In addition, the operation related to the size of object sets needs to be modified. For example, child nodes should be sorted in descending order by the size of the compact representation of their extents.

Actually, the diffset technique can be improved to make it more efficient to compute extents in a compact representation with the help of shadow nodes. Suppose there are 3 child nodes $\mathcal{C}_i, i \in [1..3]$ under a node. $|ext(\mathcal{C}_3)|$ is the smallest among the three extents. Furthermore, suppose $\mathcal{D} = \wedge\{\mathcal{C}_1, \mathcal{C}_3\}$, $\mathcal{D}^* = \wedge\{\mathcal{C}_2, \mathcal{C}_3\}$ and $\mathcal{E} = \wedge\{\mathcal{C}_1, \mathcal{C}_2\}$. When we visit the node $\mathcal{C}_3$, we will put a shadow node corresponding $\mathcal{D}$ containing $CR_{\mathcal{C}_1}(ext(\mathcal{D}))$ under $\mathcal{C}_1$ and a shadow node corresponding $\mathcal{D}^*$ containing $CR_{\mathcal{C}_2}(ext(\mathcal{D}^*))$ under $\mathcal{C}_2$. We may continue to apply Proposition 5.1 to compute $CR_{\mathcal{C}_j}(ext(\mathcal{E}))$ (j is 1 or 2) by using $CR_{\mathcal{C}}(ext(\mathcal{C}_1))$ and $CR_{\mathcal{C}}(ext(\mathcal{C}_2))$. However, since $|CR_{\mathcal{C}_1}(ext(\mathcal{D}))| \leq |CR_{\mathcal{C}}(ext(\mathcal{C}_1))|$ and $|CR_{\mathcal{C}_1}(ext(\mathcal{D}^*))| \leq |CR_{\mathcal{C}}(ext(\mathcal{C}_2))|$, we are interested in how to compute $CR_{\mathcal{C}_j}(ext(\mathcal{E}))$ (j is 1 or 2) more efficiently with the help of two shadow nodes. As the following lemma shows, when $\mathcal{D} = \mathcal{D}^*$, which often occurs during lattice construction, we can completely avoid using $CR_{\mathcal{C}}(ext(\mathcal{C}_1))$ and $CR_{\mathcal{C}}(ext(\mathcal{C}_2))$.

Lemma 5.1. *Suppose $\mathcal{D} = \wedge\{\mathcal{C}_1, \mathcal{C}_3\} = \wedge\{\mathcal{C}_2, \mathcal{C}_3\}$, where $\mathcal{C}_i, i \in [1..3]$ are concepts, $|ext(\mathcal{C}_3)| \leq |ext(\mathcal{C}_1)|$ and $|ext(\mathcal{C}_3)| \leq |ext(\mathcal{C}_2)|$. If $\mathcal{E} = \wedge\{\mathcal{C}_1, \mathcal{C}_2\}$, then*

$CR_{\mathcal{C}_1}(ext(\mathcal{E})) = CR_{\mathcal{C}_1}(ext(\mathcal{D})) \setminus CR_{\mathcal{C}_2}(ext(\mathcal{D}))$,

$CR_{\mathcal{C}_2}(ext(\mathcal{E})) = CR_{\mathcal{C}_2}(ext(\mathcal{D})) \setminus CR_{\mathcal{C}_1}(ext(\mathcal{D}))$.

Clearly, the complexity for constructing the compact representation for a lattice is the same as the one for constructing the full representation.

6   Conclusion and Future Direction

Because of many applications of lattices in various areas, it has become an important question to construct lattices efficiently. Previously, the best delay-time complexity is $O(|G||M|^2)$ for lattice construction algorithms. In this paper, we propose a linear delay algorithm for constructing a lattice with the input matrix of size $|G||M|$. Other advantages of the algorithm include that it does not need the union operation for computing intents of concepts. And it does not check against all generated concepts to see if a new pair of object set and attribute set is a new concept or will be subsumed. In addition, we propose to represent concept lattices in a compact representation, which eliminates redundant information in the full representation. The algorithm for the full representation is modified with improved diffset technique to build a compact representation for lattices.

The lower bound of delay-time complexity for lattice construction algorithms is still unknown. Our future work will focus on finding this lower bound and designing new algorithms to match the bound. With the efficient lattice construction algorithm we also like to apply lattices in more areas.

References

1. Berry, A., Bordat, J.-P., Aigayret, A.: Concepts can't afford to stammer. In: INRIA Proceedings of the International Conference, Journées de l'Informatique Messine (JIM 2003) (2003)
2. Bordat, J.-P.: Calcul pratique du treillis de galois dune correspondance. Mathmatique, Informatique et Sciences Humaines 24, 31–47 (1986)
3. Carpineto, C., Romano, G.: A lattice conceptual clustering system and its application to browsing retrieval. Machine Learning 24, 95–122 (1996)
4. Carpineto, C., Romano, G.: Concept Data Analysis: Theory and Applications. Wiley, Chichester (2004)
5. Choi, V.: Faster algorithms for constructing a concept (galois) lattice (2006), http://arxiv.org/pdf/cs.DM/0602069
6. Choi, V., Huang, Y., Lam, V., Potter, D., Laubenbacher, R., Duca, K.: Using formal concept analysis for microarray data comparison. In: Proceedings of the 5th Asia Pacific Bioinformatics Conference, pp. 57–66 (2006)
7. Eppstein, D.: Arboricity and bipartite subgraph listing algorithm. Information Processing Letters 54, 207–211 (1994)
8. Everts, T.J., Park, S.S., Kang, B.H.: Using formal concept analysis with an incremental knowledge acquisition system for web document management. In: Proceedings of the 29th Australasian Computer Science Conference, pp. 247–256 (2006)
9. Ganter, B., Wille, R.: Formal concept analysis: Mathematical Foundations. Springer, Heidelberg (1999)
10. Godin, R., Missaoui, R., Alaoui, H.: Incremental concept formation algorithms based on galois (concept) lattices. Computational Intelligence 11, 246–267 (1995)
11. Huang, Y., Farach-Colton, M.: Lattice based clustering of temporal gene-expression matrices. In: Proceedings of the 7th SIAM International Conference on Data Mining (2007)
12. Johnson, D.S., Yannakakis, M., Papadimitriou, C.H.: On generating all maximal independent sets. Information Processing Letters 27, 119–123 (1988)
13. Kalfoglou, Y., Dasmahapatra, S., Chen-Burger, Y.: Fca in knowledge technologies: Experiences and opportunities. In: Proceedings of 2nd International Conference on Formal Concept Analysis, pp. 252–260. Springer, Heidelberg (2004)

14. Makino, K., Uno, T.: New algorithms for enumerating all maximal cliques. In: Proceedings of 9th Scand. Workshop on Algorithm Theory, pp. 260–272 (2004)

15. Nourine, L., Raynaud, O.: A fast algorithm for building lattices. Information Processing Letters 71, 199–204 (1999)

16. Snelting, G., Tip, F.: Reengineering class hierarchies using concept analysis. In: Proceedings of the 6th ACM SIGSOFT International Symposium on Foundations of Software Engineering, pp. 99–110 (1998)

17. Uno, T., Asai, T., Uchida, Y., Arimura, H.: Lcm ver. 2: Efficient mining algorithms for frequent/closed/maximal itemsets. In: IEEE ICDM Workshop on Frequent Itemset Mining Implementation (2004)

18. Wang, J., Han, J., Pei, J.: Searching for the best strategies for mining frequent closed itemsets. In: Proceedings of the 10th ACM SIGKDD International Conference on Knowledge Discovery and Data Mining, pp. 344–353 (2004)

19. Yang, G.: The complexity of mining maximal frequent itemsets and maximal frequent patterns. In: Proceedings of the 10th ACM SIGKDD International Conference on Knowledge Discovery and Data Mining, pp. 344–353 (2004)

20. Zaki, M.J., Hsiao, C.-J.: Efficient algorithms for mining closed itemsets and their lattice structure. IEEE Transaction on Knowledge and Data Engineering 17, 462–478 (2005)

21. Zaki, M.J., Ogihara, M.: Theoretical foundations of association rules. In: Proceedings of 3rd ACM SIGMOD Workshop on Research Issues in Data Mining and Knowledge Discovery, pp. 1–7 (1998)

22. Zaki, M.J., Gouda, K.: Fast vertical mining using diffsets. In: Proceedings of the 9th ACM SIGKDD International Conference on Knowledge Discovery and Data Mining, pp. 326–335 (2003)

Matching Integer Intervals by Minimal Sets of Binary Words with *don't cares*[*]

Wojciech Fraczak[1,3], Wojciech Rytter[2,4], and Mohammadreza Yazdani[3]

[1] Dépt d'informatique, Université du Québec en Outaouais, Gatineau PQ, Canada
[2] Inst. of Informatics, Warsaw University, Warsaw, Poland
[3] Dept. of Systems and Computer Eng., Carleton University, Ottawa ON, Canada
[4] Department of Mathematics and Informatics, Copernicus University, Torun, Poland

Abstract. An interval $[p, q]$, where $0 \le p \le q < 2^n$, can be considered as the set X of n-bit binary strings corresponding to encodings of all integers in $[p, q]$. A word w with *don't care* symbols is *matching* the set $L(w)$ of all words of the length $|w|$ which can differ only on positions containing a *don't care*. A set Y of words with *don't cares* is *matching* X iff $X = \bigcup_{w \in Y} L(w)$. For a set X of codes of integers in $[p, q]$ we ask for a minimal size set Y of words with *don't cares* matching X. Such a problem appears in the context of network processing engines using *Ternary Content Addressable Memory* (TCAM) as a lookup table for IP packet header fields. The set Y is called a *template* in this paper, and it corresponds to a TCAM representation of an interval. It has been traditionally calculated by a heuristic called "prefix match", which can produce a result of the size approximately twice larger than the minimal one. In this paper we present two fast (linear time in the size of the input and the output) algorithms for finding minimal solutions for two natural encodings of integers: the usual binary representation (lexicographic encoding) and the reflected Gray code.

1 Introduction and Motivation

The *Ternary Content Addressable Memory* (TCAM), [1,2], is a type of associative memory with a highly parallel architecture which is used for performing very fast (constant time) table look up operations. The problem of interval representation by TCAM appears in network processing engines where the header fields of each IP packet (e.g., source address, destination address, port number, etc.) should be matched under strict time constraints against the entries of an Access Control List (ACL) [3,4,5]. Often, an ACL is represented by a TCAM. Each entry of the ACL defines either a single value or an interval of values for the fields of the packet header. If an ACL entry defines only single values for all header fields, then it can be directly and very efficiently represented using a single TCAM rule. However, if the ACL entry defines some non-trivial intervals

---

[*] The research of the second author was supported by the grant of the Polish Ministry of Science and Higher Education N 206 004 32/0806.

P. Ferragina and G. Landau (Eds.): CPM 2008, LNCS 5029, pp. 217–229, 2008.

for some header fields of the packet, then it may need more than one TCAM rule [6,7].

The problem of finding a minimal TCAM-like representation, here called *template*, of an interval corresponds to the problem of finding a minimal set of words over three letter alphabet $\{0, 1, *\}$ (where $*$ plays a role of *don't care*) covering the interval.

In this paper we show that in the case of lexicographic and reflected Gray encodings the problem can be solved very efficiently. We present two fast (linear time in the size of the input and the output) algorithms for finding minimal solutions.

2   Definition of the Problem

A *template* (a TCAM array) is defined as a two dimensional array of cells, where each cell carries one of the three values 0, 1, or $*$. Each row of this array is called a *rule*. Examples of templates are shown in Figure 1.

$$
a) \begin{array}{l|l} 1 & 0001 \\ 2 & 001* \\ 3 & 01** \\ 4 & 10** \\ 5 & 110* \\ 6 & 1110 \end{array}
\qquad
b) \begin{array}{l|l} 1 & 0**1 \\ 2 & 10** \\ 3 & *10* \\ 4 & **10 \end{array}
\qquad
c) \begin{array}{l|l} 1 & 01** \\ 2 & 1*0* \\ 3 & *0*1 \\ 4 & **10 \end{array}
\qquad
d) \begin{array}{l|l} 1 & 0*1* \\ 2 & 10** \\ 3 & *1*0 \\ 4 & **01 \end{array}
\qquad
e) \begin{array}{l|l} 1 & 01** \\ 2 & 1**0 \\ 3 & *01* \\ 4 & **01 \end{array}
$$

Fig. 1. Templates for interval $[1, 14]$, where the integers are represented in the standard unsigned binary encoding over 4 bits. The first table corresponds to the template generated by the "prefix match" heuristics. The next four tables show all minimal canonical templates.

Observation. The template corresponds to a boolean formula in disjunctive normal form, in which variable x_i determines whether the i-th bit is one or zero (negation of x_i). For example the last TCAM array in Figure 1 corresponds to the boolean formula:

$$(\neg x_1 \wedge x_2) \vee (x_1 \wedge \neg x_4) \vee (\neg x_2 \wedge x_3) \vee (\neg x_3 \wedge x_4).$$

A *rule* of width n is a sequence $r = e_1 e_2 \ldots e_n$, where $e_i \in \{0, 1, *\}$ for $i \in \{1, \ldots, n\}$. It defines the following non-empty language $L(r)$:

$$L(r) \overset{\text{def}}{=} L(e_1) L(e_2) \ldots L(e_n),$$

where $L(0) \overset{\text{def}}{=} \{0\}$, $L(1) \overset{\text{def}}{=} \{1\}$, and $L(*) \overset{\text{def}}{=} \{0, 1\}$. For example, $L(0*1) = \{0\} \cdot \{0, 1\} \cdot \{1\} = \{001, 011\}$. A template $\mathcal{R}$ consisting of k rules $r_1, r_2, \ldots, r_k$ will be written as $\mathcal{R} = (r_1, r_2, \ldots, r_k)$.. The language of $\mathcal{R} = (r_1, r_2, \ldots, r_k)$ is the union of the languages defined by its rules, i.e., $L(\mathcal{R}) \overset{\text{def}}{=} L(r_1) \cup L(r_2) \cup \ldots \cup L(r_k)$.

Let $E : \{0,1\}^n \hookrightarrow \{0, 1, \ldots, 2^n - 1\}$ be an encoding of integer values by n-bit strings. Any subset $X \subseteq \{0, 1, \ldots, 2^n - 1\}$ can be represented by a template $\mathcal{R}$ of width n. More precisely, $\mathcal{R}$ represents X iff $E^{-1}(X) = L(\mathcal{R})$. For two integers x, y, we denote by $[x, y]$ the set $\{x, x + 1, \ldots, y\}$ and call it an *interval*. A set of binary strings X of length n is an *interval-set* of E if $E(X)$ is an interval.

The problem of finding a minimum size template $\mathcal{R}$ (i.e., a template with the minimum number of rules) for a given set X is known to be NP-hard (as it corresponds to the problem of finding a minimal disjunctive normal form for a Boolean expression). However, in this paper we are interested in sets of binary strings which correspond to the encodings of the numbers in a given interval. A given interval can be represented by several templates. For example, Figure 1 shows 5 templates for the interval $[1, 14]$ in the case of the lexicographic 4-bit encoding.

3 Notation and Preliminary Results

Let $\mathcal{R} = (r_1, r_2, \ldots, r_k)$ and $\mathcal{R}' = (p_1, p_2, \ldots, p_m)$ be two templates of the same width d, i.e., $r_i, p_j \in \{0, 1, *\}^d$ for $i \in \{1, 2, \ldots, k\}$ and $j \in \{1, 2, \ldots, m\}$, and $w \in \{0, 1, *\}^n$ be a rule of length n. We define:

$$\mathcal{R} + \mathcal{R}' \stackrel{\text{def}}{=} (r_1, r_2, \ldots, r_k, p_1, p_2, \ldots, p_m)$$
$$w \cdot \mathcal{R} \stackrel{\text{def}}{=} (w r_1, w r_2, \ldots, w r_k) \ .$$

Intuitively, the template $\mathcal{R} + \mathcal{R}'$ is the union of $\mathcal{R}$ and $\mathcal{R}'$, thus its width remains d and the number of its rules is $k + m$. The template $w \cdot \mathcal{R}$ is of width $d + n$ and each of its rules is the concatenation of w with a rule of $\mathcal{R}$.

For $a \in \{0, 1\}$, by $\bar{a}$ we will denote the *complement* of a, i.e., $\bar{0} \stackrel{\text{def}}{=} 1$ and $\bar{1} \stackrel{\text{def}}{=} 0$. We define a *full* tree of height n as a perfect binary tree of height n such that each pair of sibling edges are labeled 0 and 1. The assignment of labels to the edges (two alternatives per each internal node) can be chosen arbitrarily. Let T_n be a full tree of height n. The label $w \in \{0,1\}^n$ of the path from the root to i-th leaf defines the n-bit encoding of number i, with 0 corresponding to the furthest left leaf of the tree. In this way, T_n defines a bijection $T_n : \{0,1\}^n \hookrightarrow \{0, 1, \ldots, 2^n - 1\}$, which we call *dense-tree encoding*. The lexicographic encoding (i.e., standard unsigned binary encoding) and the binary reflected Gray encoding [8,9] are two important examples of dense-tree encodings. They are presented in Figure 2 in the forms of full trees. In the context of a dense-tree encoding T_n, a set $X \subseteq \{0,1\}^n$ defines both the set $T_n(X)$ of integers and a subset of leaves of T_n. X can be represented by a *skeleton tree* (see Figure 3). The skeleton tree of X is obtained from T_n by removing all edges which are not leading to the leaves of X and turning all full sub-trees into leaves.

By $\emptyset$ and $\bullet$ we will represent the empty tree and the single node tree, respectively. Let S be a tree and $w \in \{0,1\}^*$. By $S.w$ we will denote the corresponding sub-tree of S; the root of $S.w$ is the last vertex of the unique path starting from the root of S and labeled by w.

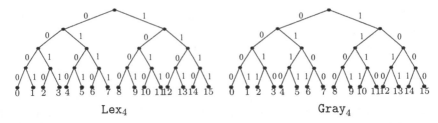

Fig. 2. Two 4-bit dense-tree encodings: the lexicographic encoding ($\texttt{Lex}_4$) and the reflected Gray encoding ($\texttt{Gray}_4$). Notice that in the case of $\texttt{Gray}$ every pair of sibling sub-trees are labeled symmetrically.

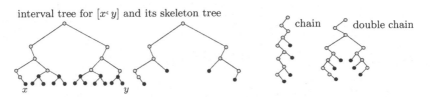

Fig. 3. The skeleton tree for an interval $[x, y]$, a chain, and a double-chain

Let $X \subseteq \{0,1\}^n$. A rule r is called an X-*limited* rule if $L(r) \subseteq X$. An X-limited rule r is said to be X-*essential* if there is no other X-limited rule r' such that $L(r) \subseteq L(r')$. In the context of two-level logic minimization, an X-essential rule is called a "prime implicant" of X (see [10]). Any coverage of X by a template with k rules can be turned into a coverage by k X-essential rules. Therefore, it is quite natural to consider only X-essential rules in the process of finding a minimal template for X; representation of a set by essential rules will be called a *canonical representation*.

We say that a skeleton tree S is a *chain*, if every vertex of S has at most one non-leaf child. A *double-chain* is a skeleton tree with at most one vertex v having two non-leaf children and such that all ancestors of v have only one child. Examples of a chain and a double-chain are illustrated in Figure 3. The skeleton tree of any interval in a dense-tree encoding is either a chain or a double-chain. A chain is called *left chain* (resp., *right chain*) if every right (resp., left) child is a unique child or it is a leaf. Intuitively, a left-chain C_L defines interval $[x, 2^n - 1]$, and a right-chain C_R interval $[0, y]$, where n is the width of the dense-tree encoding, and $x, y \in \{0, 1, \ldots, 2^n - 1\}$. We say that C_L and C_R are *complementary chains* if the intervals they define cover all values, i.e., if and only if $x \leq y + 1$.

Lemma 1. [11] *Let the skeleton tree of $X \subseteq \{0,1\}^n$ in a dense-tree encoding T_n be a chain C with k leaves. There exists a unique minimal canonical template for X, denoted by $ChainTempl_n(C)$, with k rules, which can be computed in time $O(kn)$.*

The minimal canonical template representation of a chain C can be calculated as follows:

- $ChainTempl_n(\emptyset) = ()$, i.e., the empty template;
- $ChainTempl_n(\bullet) = (*^n)$, i.e., the single rule consisting of n star symbols.

Otherwise, i.e., when $C \notin \{\emptyset, \bullet\}$:

$$ChainTempl_n(C) = \begin{cases} 1 \cdot ChainTempl_{n-1}(C.1) & \text{if } C.0 = \emptyset \\ 0 \cdot ChainTempl_{n-1}(C.0) & \text{if } C.1 = \emptyset \\ * \cdot ChainTempl_{n-1}(C.1) + 0(*^{n-1}) & \text{if } C.0 = \bullet \\ * \cdot ChainTempl_{n-1}(C.0) + 1(*^{n-1}) & \text{if } C.1 = \bullet \end{cases}$$

If the skeleton tree of a given interval is a chain C, we can directly use Lemma 1 to calculate its unique minimal canonical template independently of the dense-tree encoding T_n.

4   Intervals in the Lexicographic Encoding

Let C_L, C_R be left and right chains, respectively. Denote by $Merge(C_L, C_R)$ the interval skeleton tree which results by creating a new root and connecting C_L to the left child and C_R to the right child of the root (see S' in Figure 6).

The recursive algorithm $LexTempl(S, n)$, presented above, returns a minimal canonical template for an interval represented by its skeleton tree S in the Lex_n encoding. It employs a *top-down-reduction* approach by reducing the problem of finding a minimal template for a skeleton tree of height k to a problem of find a minimal template for a skeleton tree of height $k - 1$.

The algorithm consists of 5 parts which correspond to the treatment of 5 different types of interval skeleton trees described below (see also Figure 4):

ξ_0 – Chains;
ξ_1 – Double-chains whose roots have only one child;
ξ_2 – Double-chains whose roots have two children and two grandchildren;
ξ_3 – Double-chains whose roots have three grandchildren;
ξ_4 – Double-chains whose roots have four grandchildren.

Let us consider an interval $I = [19, 61]$ within Lex_6 encoding.

$$Lex_6(010011) = 19 \quad \text{and} \quad Lex_6(111101) = 61$$

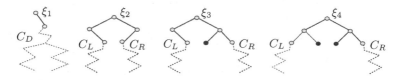

Fig. 4. Double-chains of types ξ_1, ξ_2, ξ_3, and ξ_4. The double-chain C_D, left chain C_L, and right chain C_R are non-empty. Notice that in some cases the chains cannot be trivial (i.e., a single node tree). For example, in ξ_4 both chains C_L and C_R are non-trivial.

The history of the execution of $LexTempl(S,6)$, where S is the interval skeleton tree for I, is presented in Figure 5. The final result of the computation is shown in the rightmost column of the table and consists of 6 rules.

Correctness of the algorithm is checked by analysing all five steps of the algorithm.

ALGORITHM $LexTempl(S,n)$

1. if S is a chain then **return** $ChainTempl_n(S)$;

2. if $v = root(S)$ has one child z and the edge $v \to z$ has label $a \in \{0,1\}$
   then **return** $a \cdot LexTempl(S.a, n-1)$, where $S.a$ is the tree rooted at z;

3. k := number of grandchildren of the root;
   let L, R be the leftmost and rightmost grandchildren of v, and C_L, C_R the left and the right chains rooted in L and R, respectively;

   if $k = 2$ then **return** $01 \cdot ChainTempl_{n-2}(C_L) + 10 \cdot ChainTempl_{n-2}(C_R)$;

4. S' := $Merge(C_L, C_R)$; $\mathcal{R}'$:= $LexTempl(S', n-1)$;
   Split rules of $\mathcal{R}'$ w.r.t. the first symbol, i.e., $\mathcal{R}' = 0 \cdot \mathcal{R}'_0 + 1 \cdot \mathcal{R}'_1 + * \cdot \mathcal{R}'_*$;

   if $k = 3$ and L is a right child then
   $\qquad$ **return** $01 \cdot \mathcal{R}'_0 + 1* \cdot \mathcal{R}'_1 + *1 \cdot \mathcal{R}'_* + (10*^{n-2})$;
   if $k = 3$ and L is a left child then
   $\qquad$ **return** $0* \cdot \mathcal{R}'_0 + 10 \cdot \mathcal{R}'_1 + *0 \cdot \mathcal{R}'_* + (01*^{n-2})$;

5. $(k = 4)$ $\mathcal{R}$:= $0* \cdot \mathcal{R}'_0 + *1 \cdot \mathcal{R}'_1 + ** \cdot \mathcal{R}'_* + (10*^{n-2})$;

   if (C_L, C_R) are *complementary* then **return** $\mathcal{R}$ else **return** $\mathcal{R} + (01*^{n-2})$;

Lemma 2. *Let I and I' be two intervals corresponding to the skeleton trees S and S' of Figure 6, respectively. A minimal template for I has at least k rules more than a minimal template for I', where $k \geq 1$ is the height difference between S and S'.*

Proof. Let $\mathcal{R}$ be a minimal canonical template for I and $\mathcal{R}_C$ be the set of those rules in $\mathcal{R}$ which enter into the chains C_L and C_R (i.e., rules that intersect with $0^k1** \cdots *$ or with $1^{k+1}** \cdots *$). The rules of $\mathcal{R}_C$ can be turned into a template for I' by replacing $k + 1$ initial symbols of each rule with one of 0, 1, or $*$.

If $k = 1$, then the I-essential rule $10** \cdots *$ cannot be covered by other I-essential rules. Therefore a minimal template for I has exactly one rule more than a minimal template for I'. For $k > 1$, since C_R is not a single node tree (i.e., at least $111 \cdots 1 \notin C_R$), no inner node of S at level k, except the parents of C_L and C_R, can be covered by rules from $\mathcal{R}_C$. More precisely, no string which has both a 0 and a 1 in its k first bits, 0 at position $k+1$, and then only 1 in all positions bigger than $k+1$, can be covered by rules from $\mathcal{R}_C$.

	S'''' ᴄ 1	S''' ᴄ 3	S'' ᴄ 4	S' ᴄ 5	S ᴄ 6
	ξ_0: i.e., chain	ξ_2: $\mathcal{R} =$ $01\mathcal{R}' + 10(*)$	ξ_4 (non-compl.): $\mathcal{R} = 0*\mathcal{R}'_0 + *1\mathcal{R}'_1 +$ $**\mathcal{R}'_* + (01*^2) + (10*^2)$	ξ_4 (complementary): $\mathcal{R} = 0*\mathcal{R}'_0 + *1\mathcal{R}'_1 +$ $**\mathcal{R}'_* + (10*^3)$	ξ_3 (L is right child): $\mathcal{R} = 01\cdot\mathcal{R}'_0 + 1*\cdot\mathcal{R}'_1 +$ $*1\cdot\mathcal{R}'_* + (10*^4)$
$\mathcal{R}$	1 $\boxed{1}$	1 $\boxed{011}$ 2 $\boxed{10*}$	1 $\boxed{0*11}$ 2 $\boxed{*10*}$ 3 $\boxed{01**}$ 4 $\boxed{10**}$	1 $\boxed{0**11}$ 2 $\boxed{0*1**}$ 3 $\boxed{*10**}$ 4 $\boxed{**10*}$ 5 $\boxed{10***}$	1 $\boxed{01**11}$ 2 $\boxed{01*1**}$ 3 $\boxed{1*0***}$ 4 $\boxed{*110**}$ 5 $\boxed{*1*10*}$ 6 $\boxed{10****}$
$\mathcal{R}_0$	$\emptyset$	1 $\boxed{11}$	1 $\boxed{*11}$ 2 $\boxed{1**}$	1 $\boxed{**11}$ 2 $\boxed{*1**}$	
$\mathcal{R}_1$	1 $\boxed{\varepsilon}$	1 $\boxed{0*}$	1 $\boxed{0**}$	1 $\boxed{0***}$	
$\mathcal{R}_*$	$\emptyset$	$\emptyset$	1 $\boxed{10*}$	1 $\boxed{10**}$ 2 $\boxed{*10*}$	

Fig. 5. The history of the algorithm *LexTempl* for interval $I = [19, 61]$ with Lex$_6$ encoding, presented in terms of skeleton trees together with intermediate templates. The skeleton trees from the first row of the table correspond to the arguments of recursive calls (from right to left). The second row describes the type of the skeleton tree argument as well as the formula which is used to calculate the template in which $\mathcal{R}' = 0 \cdot \mathcal{R}'_0 + 1 \cdot \mathcal{R}'_1 + * \cdot \mathcal{R}'_*$ corresponds to the result of the sub-recursive call. For convenience, all intermediate results $\mathcal{R}$ are presented explicitly (third row) and in form of $\mathcal{R}_0$, $\mathcal{R}_1$, and $\mathcal{R}_*$ (remaining rows). For example, $\mathcal{R}'_0$ in the formula of column 3 (for recursive call S'', 4), refers to the template written in column 2 (call S''', 3) row 4 ($\mathcal{R}_0$).

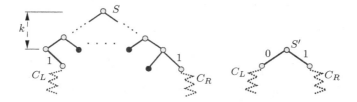

Fig. 6. Two skeleton trees S and S' from Lemma 2

In [11], it was shown that a minimal template representation of $I_k = [1, 2^k - 2]$ in Lex$_k$ needs exactly k rules. Thus, we need at least k rules outside of $\mathcal{R}_C$ to cover all those k-level inner nodes. $\square$

Step 1. For skeleton trees of type ξ_0, i.e., chains, we use the solution provided by Lemma 1.

Step 2. The case of a skeleton tree of type ξ_1 is simple since there is the one-to-one correspondence between all templates for (the interval of) S and all templates for (the interval of) $S.a$, where a is the label of the unique edge of the root of S.

Proposition 1 (Skeleton trees of type ξ_1). *Let S be an interval skeleton tree in Lex_n such that for some $a \in \{0,1\}$, $S.a = S' \neq \emptyset$ and $S.\bar{a} = \emptyset$. $\mathcal{R}$ is a minimal canonical template for S' encoded on $n - 1$ bits if and only if $a\mathcal{R}$ is a minimal canonical template for S on n bits.*

Step 3. For the skeleton trees of type ξ_2 we have:

Proposition 2 (Skeleton trees of type ξ_2). *Let S be an interval skeleton tree in Lex_n such that $S.00 = S.11 = \emptyset$, $S.01 = C_L \neq \emptyset$, and $S.10 = C_R \neq \emptyset$. If $\mathcal{R}_L$ is a minimal canonical template for C_L on $n - 2$ bits and $\mathcal{R}_R$ is a minimal canonical template for C_R on $n - 2$ bits, then $\mathcal{R} = 01\mathcal{R}_L + 10\mathcal{R}_R$ is a minimal canonical template for S encoded on n bits.*

Proof. Every rule in a template for S starts by 01 or 10. Therefore, any template covering S can be written as $01\mathcal{R}_L + 10\mathcal{R}_R$, where $\mathcal{R}_L$ and $\mathcal{R}_R$ cover C_L and C_R, respectively. $\qquad\square$

Step 4. If a skeleton tree S is of type ξ_3 then either $S.00 = \emptyset$ or $S.11 = \emptyset$. Since these two cases are very similar, in the following proposition we consider only one case, $S.00 = \emptyset$, which corresponds to the graphical representation of ξ_3 in Figure 4. Due to Lemma 2, for $k = 1$, we have the following fact.

Proposition 3 (Skeleton trees of type ξ_3). *Let S be an interval skeleton tree in Lex_n such that $S.00 = \emptyset$, $S.01 = C_L \neq \emptyset$, $S.10 = \bullet$, and $S.11 = C_R \neq \emptyset$. Also let $S' = \text{Merge}(C_L, C_R)$, i.e., S' is an interval skeleton tree in Lex_{n-1}, such that $S'.0 = C_L$ and $S'.1 = C_R$. If $\mathcal{R}' = 0\mathcal{R}'_0 + 1\mathcal{R}'_1 + *\mathcal{R}'_*$ is a minimal canonical template for S', then*

$$\mathcal{R} = 01\mathcal{R}'_0 + 1*\mathcal{R}'_1 + *1\mathcal{R}'_* + 10(*^{n-2})$$

is a minimal canonical template for S.

Step 5. In case of skeleton trees of type ξ_4, i.e., when all four grandchildren of the root are non-empty, we distinguish two cases; the chains rooted in the leftmost grandchild and the rightmost grandchild (i.e., chains C_L and C_R in ξ_4 of Figure 4) are *complementary* or are not.

Proposition 4 (Skeleton trees of type ξ_4). *Let S be an interval skeleton tree in Lex_n such that $S.00 = C_L \neq \emptyset$, $S.01 = \bullet$, $S.10 = \bullet$, and $S.11 = C_R \neq \emptyset$. Also suppose that $S' = \text{Merge}(C_L, C_R)$ is an interval skeleton tree in Lex_{n-1} with a minimal canonical template $\mathcal{R}' = 0\mathcal{R}'_0 + 1\mathcal{R}'_1 + *\mathcal{R}'_*$. If C_L and C_R are complementary chains then a minimal canonical template for S is $\mathcal{R} = 0*\mathcal{R}'_0 + *1\mathcal{R}'_1 + **\mathcal{R}'_* + 10(*^{n-2})$. Otherwise a minimal canonical template for S is: $\mathcal{R} = 0*\mathcal{R}'_0 + *1\mathcal{R}'_1 + **\mathcal{R}'_* + 10(*^{n-2}) + 01(*^{n-2})$.*

Proof. If C_L and C_R are complementary, then the construction in Figure 7 shows that template $\mathcal{R} = 0*\mathcal{R}'_0 + *1\mathcal{R}'_1 + **\mathcal{R}'_* + 10(*^{n-2})$ covers S, whenever $0\mathcal{R}'_0 +$

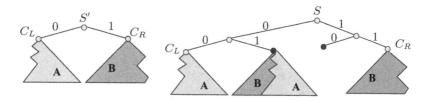

Fig. 7. Skeleton trees $S' = Merge(C_L, C_R)$ and S of Proposition 4 in case when C_L and C_R are complementary. Intervals corresponding to left chain C_L and right chain C_R are represented by trees A and B, respectively.

$1\mathcal{R}'_1 + *\mathcal{R}'_*$ covers S' and the number of rules in $\mathcal{R}$ is one more than that of $\mathcal{R}'$. Besides Lemma 2 guarantees that no smaller template can cover S. Notice that another symmetric construction for $\mathcal{R}$ is also possible: $\mathcal{R} = *0\mathcal{R}'_0 + 1*\mathcal{R}'_1 + **\mathcal{R}'_* + 01(*^{n-2})$.

If C_L and C_R are non-complementary, then $01*\cdots*$ (resp., $10*\cdots*$) is an S-essential rule which cannot be covered by other S-essential rules. Therefore, $10*\cdots*$ and $01*\cdots*$ have to be in any canonical solution for S. Notice that in case when C_L and C_R are non-complementary, there are four variants for construction of a minimal canonical template for S from a minimal canonical template for S':
$$\mathcal{R} = \alpha\mathcal{R}'_0 + \beta\mathcal{R}'_1 + **\mathcal{R}'_* + 10(*^{n-2}) + 01(*^{n-2}),$$
for $\alpha \in \{*0, 0*\}$ and $\beta \in \{*1, 1*\}$

Theorem 1. *The algorithm LexTempl computes in time $O(n + K)$ a minimal canonical template for an interval in the n-bit lexicographic encoding, where K is the total size (in bits) of the generated template.*

Proof. The algorithm obviously runs in $O(n + K)$. Correctness follows from Lemma 1 (for step 1), and Propositions 1, 2, 3, 4 (for all other steps). □

5 Intervals in the Reflected Gray Encoding

Unlike for the lexicographic encoding, not every sub-tree of the reflected Gray dense-tree encoding $\mathtt{Gray}_n$ is a reflected Gray dense-tree encoding. However, every pair of sibling sub-trees of the dense tree representing $\mathtt{Gray}_n$ are the mirror copies of each other, $\mathtt{Gray}_k$ and its mirror copy $\overline{\mathtt{Gray}}_k$, $1 \le k < n$.

Let n be the number of bits. We denote by $x' = 2^n - 1 - x$ the mirror image of x. Suppose that $I = [x, y]$ is an interval and $y < 2^n$. We say that I is *reciprocal* if $x = y'$. If I is reciprocal, then there is a $w \in \{0, 1\}^{n-1}$ such that $x = \mathtt{Gray}_n(0w)$ and $y = \mathtt{Gray}_n(1w)$. (The same holds for $\overline{\mathtt{Gray}}_n$ where 0 and 1 are interchanged.)

Lemma 3. *Let $I = [x, y]$ be a reciprocal interval with $x = \mathtt{Gray}_n(0w)$ and $y = \mathtt{Gray}_n(1w)$. Let $\mathcal{R}$ be a minimal canonical template for the interval $I' = [\mathtt{Gray}_{n-1}(w), 2^{n-1} - 1]$. The template $*\mathcal{R}$ is a minimal canonical template for the interval I. (See Figure 8.)*

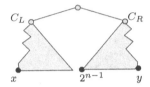

Fig. 8. Illustration of Lemma 3; y is the mirror image of x with respect to the root of the tree ($x' = y$ & $y' = x$). If $\mathcal{R}$ is a minimal template for left-chain C_L then it is also a minimal template for C_R, and $*\mathcal{R}$ is a minimal template for the whole tree.

Our algorithm for calculating a minimal canonical template for intervals in Gray encoding relies on Lemma 3; it divides a given non-reciprocal interval into two overlapping sub-intervals where one of the intervals is reciprocal and the other one is a chain. We consider one case, the others are analogous.

ALGORITHM $GrayTempl(S, n)$ (the case of S from Figure 9)

1. u := right child of the root;

   v := $LCA(x, y')$; α := $PATH_L(v)$; w := left child of v;

   let C_1, C_2 be the chains rooted at u, w, respectively;

   let a be the label of the edge $v \to w$; and d_1, d_2 be as shown in Figure 9;

2. $\mathcal{R}_1$:= $ChainTempl_{n-1}(C_1)$; $\mathcal{R}_2$:= $ChainTempl_{n-|\alpha|-1}(C_2)$;

3. $\mathcal{R}$:= $*\mathcal{R}_1 + \alpha * \mathcal{R}_2$;

4. **if** $d_1 \le d_2$ **then return** $\mathcal{R}$ **else return** $\mathcal{R} + \alpha \bar{a} * \ldots *$;

The algorithm $GrayTempl$ generates a minimal canonical template for an interval $[x, y]$ which has the skeleton tree S as shown in Figure 9. In this skeleton tree y' is the mirror image of y in the left sub-tree and $y' > x$. $LCA(x, y')$ denotes the lowest common ancestor of x and y', which is calculated by taking the greatest common prefix of x and y' in Gray_n. $PATH_L(v)$ is the label of the path from the root of S to node v, where every symbol of a left-going edge is replaced by $*$.

Observation. Let I_v be the interval covered by the subtree (chain) rooted at the right child of v. If $d_1 \le d_2$ (equivalently if C_2, C_3 are complementary) then $\mathcal{R}$ covers the whole interval, otherwise we need to add one extra rule $0\alpha\bar{a} * \ldots *$ which covers a non empty gap between $*\mathcal{R}_1$ and $0\alpha * \mathcal{R}_2$.

Correctness is based on the fact that the templates for C_1 and C_2 cover disjoint intervals. Consequently we need at least $|\mathcal{R}_1| + |\mathcal{R}_2|$ rules. In case $d_1 > d_2$ one extra rule is needed. The rigorous proof is omitted in this version.

Theorem 2. *The algorithm $GrayTempl$ computes in time $O(n + K)$ a minimal canonical template for an interval in the n-bit reflected Gray encoding, where K is the total size (in bits) of the generated template.*

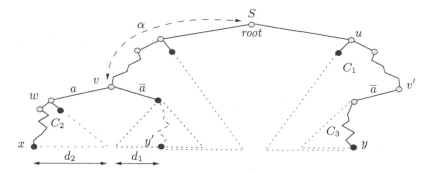

Fig. 9. The structure of the skeleton tree S for a non-reciprocal interval in **Gray**. C_1 is the subtree (a chain) rooted at u, C_2 is the chain rooted at w and C_3 is the chain rooted at the left child of v'. We have: $d_1 \leq d_2$ iff C_2, C_3 are complementary chains.

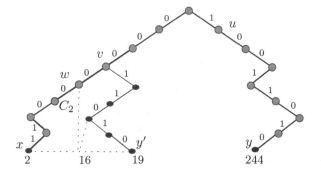

Fig. 10. An example of the structure related to the algorithm for the interval $[2, 244]$. The mirror branches are from the root to y and its mirror image y', they have the same path-labels.

Example: We show how the algorithm works for the interval $[2, 244]$, see Figure 10. We have:

- $d_2 = 15 - 2 + 1 = 14$ and $d_1 = 19 - 16 + 1 = 4$, i.e., $d_1 < d_2$;
- $C_2 = S(w)$ and $C_1 = S(u)$;
- $\alpha = 000$;
- $ChainTempl(C_1) = \mathcal{R}_1 = \{01*****, 001****, 00111**, 0011010\}$;
- $ChainTempl(C_2) = \mathcal{R}_2 = \{1***, 01**, 001*\}$.

Consequently, $GrayTempl(S, 8) = *\mathcal{R}_1 + 000 * \mathcal{R}_2$.

References

1. Kempke, R., McAuley, A.: Ternary CAM memory architecture and methodology. US Patent 5, 841–874 (August 1996)
2. Kohonen, T.: Content-Addressable Memories. Springer, New York (1980)

3. Davis, G., Jeffries, C., Lunteren, J.: Method and system for performing range rule testing in a ternary content addressable memory. US Patent 6, 886,073 (April 2005)
4. Jeong, H.J., Song, I.S., Kwon, T.G., Lee, Y.K.: A multi-dimension rule update in a tcam-based high-performance network security system. In: AINA 2006: Proceedings of the 20th International Conference on Advanced Information Networking and Applications. AINA 2006, Washington, DC, USA, vol. 2, pp. 62–66. IEEE Computer Society, Los Alamitos (2006)
5. Srinivasan, V., Varghese, G., Suri, S., Waldvogel, M.: Fast and scalable layer four switching. In: SIGCOMM 1998: Proceedings of the ACM SIGCOMM 1998 conference on Applications, technologies, architectures, and protocols for computer communication, pp. 191–202. ACM, New York (1998)
6. Lakshminarayanan, K., Rangarajan, A., Venkatachary, S.: Algorithms for advanced packet classification with Ternary CAMs. In: SIGCOMM, pp. 193–204 (2005)
7. Liu, H.: Efficient mapping of range classifier into ternary-cam. In: HOTI 2002: Proceedings of the 10th Symposium on High Performance Interconnects HOT Interconnects (HotI 2002), Washington, DC, USA, p. 95. IEEE Computer Society, Los Alamitos (2002)
8. Gilbert, E.: Gray codes and paths on the n-cube. Bell Systems Technical Journal 37, 815–826 (1958)
9. Gray, F.: Pulse code communications. US Patent 2,632,058 (March 1953)
10. Brayton, R., Hachtel, G., McMullen, C., Sangiovanni-Vincentelli, A.: Logic Minimization Algorithms for VLSI Synthesis. Kluwer Academic Publishers, Dordrecht (1984)
11. Fraczak, W., Rytter, W., Yazdani, M.: TCAM representations of intervals of integers encoded by binary trees. In: Proceedings of IWOCA Conference, Lake Macquarie, Newcastle, Australia (November 2007)
12. Coudert, O.: Two-level logic minimization: an overview. Integr. VLSI J 17(2), 97–140 (1994)

Appendix

Alternative presentation of algorithm *LexTempl*

Assume $\alpha = (a_1, a_2, \ldots a_k)$ and $\beta = (b_1, b_2, \ldots b_k)$ are binary strings of length k. Denote by $[\alpha]_2$ the number corresponding to α in binary. Assume $p = [\alpha]_2 \leq q = [\beta]_2$. The input interval $[p, q]$ is presented in the form (α, β). Denote by $size(\alpha, \beta)$ the length of the interval $[p, q]$.

Denote by $\mathbf{1}_k$, $\mathbf{0}_k$ the sequence of k ones, and k zeros, respectively. It is convenient to assume later that we write $\mathbf{0}$, $\mathbf{1}$ without indices, the length is implied by the other string in a corresponding pair.

The chains correspond to pairs of one of the forms

$$(\alpha, \mathbf{1}), \quad (\mathbf{0}, \beta)$$

The crucial is the notion of complementary interval. The pair $\alpha \leq \beta$ is called **complementary** iff

$$size(\alpha, \mathbf{1}_k) + size(\mathbf{0}_k, \beta) \geq 2^k$$

ALGORITHM *LexTempl'*(α, β)

1. if $[\alpha, \beta]$ is a chain then **return** *ChainTempl*(α, β);

2. if ($\alpha = a \cdot \alpha'$, $\beta = a \cdot \beta'$) for $a \in \{0, 1\}$ then

    return $a \cdot$ *LexTempl'*(α', β');

3. if ($\alpha = 01\alpha'$, $\beta = 10\beta'$) then **return**

    $01 \cdot$ *ChainTempl*$(\alpha', \mathbf{1})$ + $10 \cdot$ *ChainTempl*$(\mathbf{0}, \beta')$;

4. Let $\alpha = a_1 a_2 \cdot \alpha'$, $\beta = b_1 b_2 \cdot \beta'$;

    $\mathcal{R}' := $ *LexTempl'*$(0 \cdot \alpha', 1 \cdot \beta')$; Represent $\mathcal{R}'$ as $0\mathcal{R}'_0 + 1\mathcal{R}'_1 + *\mathcal{R}'_*$;

    if ($a_1 a_2 = 01$, $b_1 b_2 = 11$) then **return**

    $$01 \cdot \mathcal{R}'_0 + 1* \cdot \mathcal{R}'_1 + *1 \cdot \mathcal{R}'_* + 10* \ldots;$$

    if ($a_1 a_2 = 00$, $b_1 b_2 = 10$) then **return**

    $$0* \cdot \mathcal{R}'_0 + 10 \cdot \mathcal{R}'_1 + *0 \cdot \mathcal{R}'_* + 01* \ldots;$$

5. (Now $a_1 a_2 = 00$, $b_1 b_2 = 11$;)

    $\mathcal{R} := 0*\mathcal{R}'_0 + *1\mathcal{R}'_1 + **\mathcal{R}'_* + 10** \cdots *$;

    if α', β' are complementary then **return** $\mathcal{R}$ else **return** $\mathcal{R} + 01** \cdots *$;

The motivation for presenting the algorithm in terms of skeleton trees was to ease the proof of correctness. The skeleton tree is showing much better the structure of the interval from the point of view of constructing a minimal template.

Fast Algorithms for Computing Tree LCS

Shay Mozes[1], Dekel Tsur[2,*], Oren Weimann[3], and Michal Ziv-Ukelson[2,*]

[1] Brown University, Providence, RI 02912-1910, USA
shay@cs.brown.edu
[2] Ben-Gurion University, Beer-Sheva, Israel
{dekelts,michaluz}@cs.bgu.ac.il
[3] Massachusetts Institute of Technology, Cambridge, MA 02139, USA
oweimann@mit.edu

Abstract. The LCS of two rooted, ordered, and labeled trees F and G is the largest forest that can be obtained from both trees by deleting nodes. We present algorithms for computing tree LCS which exploit the *sparsity* inherent to the tree LCS problem. Assuming G is smaller than F, our first algorithm runs in time $O(r \cdot \text{height}(F) \cdot \text{height}(G) \cdot \lg \lg |G|)$, where r is the number of pairs $(v \in F, w \in G)$ such that v and w have the same label. Our second algorithm runs in time $O(Lr \lg r \cdot \lg \lg |G|)$, where L is the size of the LCS of F and G. For this algorithm we present a novel three dimensional alignment graph. Our third algorithm is intended for the constrained variant of the problem in which only nodes with zero or one children can be deleted. For this case we obtain an $O(rh \lg \lg |G|)$ time algorithm, where $h = \text{height}(F) + \text{height}(G)$.

1 Introduction

The *longest common subsequence* (LCS) of two strings is the longest subsequence of symbols that appears in both strings. The *edit distance* of two strings is the minimal number of character deletions insertions and replacements required to transform one string into the other. Computing the LCS or the edit distance can be done using similar dynamic programming algorithms in $O(mn)$ time and space, where m and n ($m \leq n$) are the lengths of the strings [15, 29]. The only known speedups to the edit distance algorithm are by polylogarithmic factors [7, 11, 23]. For the LCS problem however, it is possible to obtain time complexities better than $\widetilde{O}(mn)$ in favorable cases, e.g. [3, 10, 16, 17, 18, 25]. This is achieved by exploiting the sparsity inherent to the LCS problem and measuring the complexity by parameters other than the lengths of the input strings. In this paper, we apply this idea to computing the LCS of rooted, ordered, and labeled trees.

The problem of computing string LCS translates to finding a longest chain of matches in the alignment graph of the two strings. Many string LCS algorithms that construct such chains by exploiting sparsity have their natural predecessors in either Hirschberg [16] or Hunt and Szymanski [18]. Given

---

* The research was supported by the Lynn and William Frankel Center for Computer Sciences at Ben-Gurion University.

P. Ferragina and G. Landau (Eds.): CPM 2008, LNCS 5029, pp. 230–243, 2008.

two strings S and T, let L denote the size of their LCS and let r denote the number of matches in the alignment graph of S and T. Hirschberg's algorithm achieves an $O(nL + n \lg |\Sigma|)$ time complexity by computing chains in succession. The Hunt-Szymanski algorithm achieves an $O(r \lg m)$ time complexity by extending partial chains. The latter can be improved to $O(r \lg \lg m)$ by using the successor data-structure of van Emde Boas [28]. Apostolico and Guerra [21] gave an $O(mL \cdot \min(\lg |\Sigma|, \lg m, \lg \frac{2n}{m}))$ time algorithm, and another algorithm with running time $O(m \lg n + d \lg \frac{nm}{d})$ which can also be implemented to take $O(d \lg \lg \min(d, \frac{nm}{d}))$ time [13]. Here, $d \leq r$ is the number of dominant matches (as defined by Hirschberg [16]). Note that in the worst case both d and r are $\Theta(nm)$, while the parameter L is always bounded by m. When there are $k \geq 2$ input strings, the sparse LCS problem extends to the problem of chaining from fragments in multiple dimensions [1, 24]. Here, the match point arithmetic is extended with range search techniques, yielding a running time of $O(r(\lg n)^{k-2} \lg \lg n)$.

The problem of computing the LCS of two trees was considered by Lozano et al. [22] and Amir et al. [2]. The problem is defined as follows.

Definition 1 (Tree LCS). *The LCS of two rooted, ordered, labeled trees, is the size of the largest forest that can be obtained from both trees by deleting nodes. Deleting a node v means removing v and all edges incident to v. The children of v become children of the parent of v (if it exists) instead of v.*

We also consider the following constrained variant of the problem.

Definition 2 (Homeomorphic Tree LCS). *The Homeomorphic LCS (HLCS) of two rooted, ordered, labeled trees is the size of the largest tree that can be obtained from both trees by deleting nodes, such that in the series of node deletions, a deleted node must have 0 or 1 children at the time the deletion is applied.*

Tree LCS is a popular metric for measuring the similarity of two trees and arises in XML comparisons, computer vision, compiler optimization, natural language processing, and computational biology [6, 8, 20, 26, 31]. To date, computing the LCS of two trees is done by using *tree edit distance* algorithms. Tai [26] gave the first such algorithm with a time complexity of $O(nm \cdot \text{leaves}(F)^2 \cdot \text{leaves}(G)^2)$, where n and m are the sizes of the input trees F and G (with $m \leq n$) and $\text{leaves}(F)$ denotes the number of leaves in F. Zhang and Shasha [31] improved this result to $O(nm \cdot \min\{\text{height}(F), \text{leaves}(F)\} \cdot \min\{\text{height}(G), \text{leaves}(G)\})$, where $\text{height}(F)$ denotes the height of F. In the worst case, their algorithm runs in $O(n^2 m^2) = O(n^4)$ time. Klein [19] improved this result to a worst-case $O(m^2 n \lg n) = O(n^3 \lg n)$ time algorithm and Demaine et al. [12] further improved to $O(nm^2(1 + \lg \frac{n}{m})) = O(n^3)$. Chen [9] gave an $O(nm + n \cdot \text{leaves}(G)^2 + \text{leaves}(F) \cdot M(\text{leaves}(G)))$ time algorithm, where $M(k)$ is the time complexity for computing the distance product of two $k \times k$ matrices. For homeomorphic edit distance (where deletions are restricted to nodes with zero or one child), Zhang et al. [30] gave an $O(mn)$ time algorithm.

Our Results. We modify Zhang and Shasha's algorithms and Klein's algorithm similarly to the modifications of Hunt-Szymanski and Hirschberg to the classical $O(mn)$-time algorithm for string LCS. We present two algorithms for computing the LCS of two rooted, ordered, and labeled trees F and G of sizes n and m. Our first algorithm runs in time $O(r \cdot \text{height}(F) \cdot \text{height}(G) \cdot \lg \lg m)$ where r is the number of pairs $(v \in F, w \in G)$ such that v and w have the same label. Our second algorithm runs in time $O(Lr \lg r \cdot \lg \lg m)$, where $L = |\text{LCS}(F, G)|$. This algorithm is more complicated and requires a novel three dimensional alignment graph. In both these algorithms the $\lg \lg m$ factor can be replaced by $\lg \lg (\min(m, r))$ by noticing that if $r < m$ then there are at least $m - r$ nodes in G that do not match any node in F so we can delete them from G and solve the problem on the new G whose size is now at most r. Finally we consider LCS for the case when only homeomorphic mappings are allowed between the compared trees (i.e. deletions are restricted to nodes with zero or one child). For this case we obtain an $O(rh \lg \lg m)$ time algorithm, where $h = \text{height}(F) + \text{height}(G)$.

Roadmap. The rest of the paper is organized as follows. Preliminaries and definitions are given in Section 2. In Section 3 we present our sparse variant of the Zhang-Shasha algorithm and in sections 4 and 5 we give such variants for Klein's algorithm. Finally, in Section 6 we describe our algorithm for the homeomorphic tree LCS. Due to the space restriction, the proofs are deferred to the full version of this paper.

2   Preliminaries

For a forest F, the node set of F is written simply as F, as when we speak of a node $v \in F$. We denote F_v as the subtree of F that contains the node $v \in F$ and all its descendants. A forest obtained from F by deleting nodes is called a *subforest* of F. For a pair of trees F, G, two nodes $v \in F, w \in G$ with the same label are called a *match pair*. For the tree LCS problem we assume without loss of generality that the roots of the two input trees form a match pair (if this property does not hold for the two input trees, we can add new roots to the trees and solve the tree LCS problem on the new trees).

The *Euler string* of a tree F is the string obtained when performing a left-to-right DFS traversal of F and writing down the label of each node twice: when the DFS traversal first enters the node and when it last leaves the node. We define $e_F(i)$ to be the index such that both the ith and $e_F(i)$th characters of the Euler string of F were generated from the same node of F. Note that $e_F(e_F(i)) = i$.

For $i \leq j$, we denote by $F[i..j]$ the forest induced by all nodes $v \in F$ whose Euler string indices *both* lie between i and j. A *left-to-right postorder traversal* of a tree F whose root v has children $v_1, v_2, \ldots, v_k$ (ordered from left to right) is a traversal which recursively visits $F_{v_1}, F_{v_2}, \ldots, F_{v_k}$, then finally visits v.

For two forests F and G, let $\text{LCS}_R(F, G)$ (resp., $\text{LCS}_L(F, G)$) denote the size of the largest forest that can be obtained from F and G by node deletions without deleting the root of the rightmost (resp., leftmost) tree in F or G. If the roots of the rightmost trees in F and G are not a match pair then we define $\text{LCS}_R(F, G) = 0$. Clearly, $\text{LCS}_R(F, G) \leq \text{LCS}(F, G)$ and $\text{LCS}_L(F, G) \leq \text{LCS}(F, G)$.

Lemma 1. *If F and G are trees whose roots have equal labels then* $\text{LCS}_R(F, G) = \text{LCS}_L(F, G) = \text{LCS}(F, G)$.

A *path decomposition* of a tree F is a set of disjoint paths in F such that (1) each path ends in a leaf, and (2) each node appears in exactly one path. The *main path* of F with respect to a decomposition $\mathcal{P}$ is the path in $\mathcal{P}$ that contains the root of F. A *heavy path decomposition* of a tree F was introduced by Harel and Tarjan [14] and is built as follows. We classify each node of F as either *heavy* or *light*: For each node v we pick the child of v with maximum number of descendants and classify it as *heavy* (ties are resolved arbitrarily). The remaining nodes are classified as *light*. The *main* path p of the heavy path decomposition starts at the root (which is light), and at each step moves from the current node v to its heavy child. We next remove the nodes of p from F, and recursively compute a heavy path decomposition for each of the remaining trees. An important property of this decomposition is that the number of light ancestors of a node $v \in F$ is at most $\lg n + 1$.

A *successor data-structure* is a data-structures that stores a set of elements S with a key for each element and supports the following operations: (1) $\text{insert}(S, x)$: inserts x into S (2) $\text{delete}(S, x)$: removes x from S (3) $\text{pred}(S, k)$: returns the element $x \in S$ with maximal key such that $\text{key}(x) \leq k$ (4) $\text{succ}(S, k)$: returns the element $x \in S$ with minimal key such that $\text{key}(x) \geq k$. Van Emde Boas presented a data structure [28] that supports each of these operations in $O(\lg \lg u)$ time, where the set of legal keys is $\{1, 2, \ldots, u\}$.

3   An $O(r \cdot \text{height}(F) \cdot \text{height}(G) \cdot \lg \lg m)$ Algorithm

In this section we present an $O(r \cdot \text{height}(F) \cdot \text{height}(G) \cdot \lg \lg m)$ time algorithm for computing the LCS of two trees F and G of sizes n and m and heights $\text{height}(F)$ and $\text{height}(G)$ respectively. The relation between this algorithm and Zhang and Shasha's $O(nm \cdot \text{height}(F) \cdot \text{height}(G))$ time algorithm [31] is similar to the relation between Hunt and Szymanski's $O(r \lg \lg m)$ time algorithm [18] and Wagner and Fischer's $O(mn)$ time algorithm [29] in the string LCS world.

We describe an algorithm based on that of Zhang and Shasha using an alignment graph. This approach was also used in [4,5,27]. The *alignment graph* $B_{F,G}$ of F and G is an edge-weighted directed graph defined as follows. The vertices of $B_{F,G}$ are (i, j) for $1 \leq i \leq 2n$ and $1 \leq j \leq 2m$. Intuitively, vertex (i, j) corresponds to $\text{LCS}(F[1..i], G[1..j])$, and edges in the alignment graph correspond to edit operations. The graph has the following edges:

1. Edges $(i-1, j) \rightarrow (i, j)$ and $(i, j-1) \rightarrow (i, j)$ with weight 0 for every i and j. These edges either connect vertices which represent the same pair of forests, or represent deletion of the rightmost root of just one of the forests. Both cases do not change the LCS, hence the zero weight we assign to these edges.

2. An edge for every match pair $v \in F, w \in G$, except for the roots of F and G. Let i and $e_F(i)$ be the two characters of the Euler string of F that correspond to v, where $e_F(i) < i$, and let $e_G(j) < j$ be the two characters of the Euler string of G that correspond to w. We add an edge $(e_F(i), e_G(j)) \to (i,j)$ with weight $\mathrm{LCS}(F_v, G_w)$ to $B_{F,G}$. This edge corresponds to matching the rightmost trees of $F[1..i]$ and $G[1..j]$ and its weight is obtained by recursively applying the algorithm on the trees F_v and G_w. Note that we cannot add an edge of this type for the match pair of the roots of F and G because we cannot compute the weight of such edge by recursion.

3. An edge $(2n-1, 2m-1) \to (2n, 2m)$ with weight 1, which corresponds to the match between the roots of F and G.

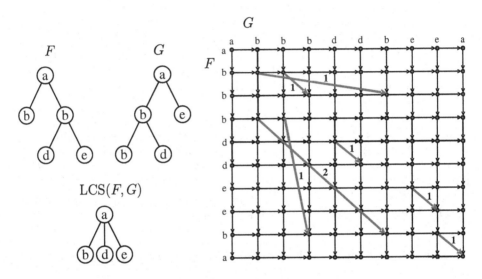

Fig. 1. An alignment graph of two trees F and G. The horizontal and vertical edges are of weight 0. Every diagonal edge e corresponds to a match pair, and e's weight is the LCS of the two Euler substrings between e's endpoints.

See Figure 1 for an example. For an edge $e = (i,j) \to (i',j')$, let $\mathrm{tail}(e) = (i,j)$ and $\mathrm{head}(e) = (i',j')$. The ith coordinate of a vector x is denoted by x_i. For example, for e above, $\mathrm{head}(e)_2 = j'$.

Lemma 2. *The maximum weight of a path in $B_{F,G}$ from vertex $(1,1)$ to vertex (i,j) is equal to $\mathrm{LCS}(F[1..i], G[1..j])$.*

Zhang and Shasha's algorithm computes the maximum weight of a path from $(1,1)$ to (i,j), for every vertex (i,j) of $B_{F,G}$. By Lemma 2, this gives $\mathrm{LCS}(F,G)$ at the vertex $(2n, 2m)$. If there are only few match pairs, we can do better. Denote the set of edges in $B_{F,G}$ with nonzero weights by $E_{F,G}$. Clearly, $|E_{F,G}| = r$. We will exploit the sparsity of the edges $E_{F,G}$ by ignoring the edges with weight 0 and the vertices that are not the endpoint of an edge in $E_{F,G}$. We define the *score* of $e \in E_{F,G}$ as the maximum weight of a path in $B_{F,G}$ from $(1,1)$ to $\text{head}(e)$ that passes through e.

Lemma 3. $\text{score}(e) = \text{LCS}_R(F[1..\text{head}(e)_1], G[1..\text{head}(e)_2])$ *for every edge* $e \in E_{F,G}$.

By Lemmas 1 and 3 we have that $\text{LCS}(F, G) = \text{score}((2n - 1, 2m - 1) \rightarrow (2n, 2m))$. We now describe a procedure that computes $\text{LCS}(F, G)$ in $O(|E_{F,G}| \cdot \lg \lg m)$ time, assuming we have already computed $\text{LCS}(F_v, G_w)$ for every match pair $v \in F, w \in G$ except for the roots of F and G. This procedure computes $\text{score}(e)$ for every $e \in E_{F,G}$. It uses a successor data-structure S that stores edges from $E_{F,G}$, where the key of an edge e is $\text{head}(e)_2$. The pseudocode for the procedure is as follows (we assume that $\text{score}(\text{NULL}) = 0$).

```
1: for i = 1, ..., 2n do
2:     for every e ∈ E_{F,G} with head(e)_1 = i do
3:         j ← head(e)_2
4:         if score(e) > score(pred(S, j)) then
5:             insert(S, e)
6:             while succ(S, j + 1) ≠ NULL and score(succ(S, j + 1)) ≤ score(e) do
7:                 delete(S, succ(S, j + 1))
8:     for every e ∈ E_{F,G} with tail(e)_1 = i do
9:         score(e) ← weight(e) + score(pred(S, tail(e)_2))
```

Let $e_1, e_2, \ldots$ be the edges of $E_{F,G}$ according to the order in which they are processed in line 2. An edge e is t-*relevant* if e is one of the edges $e_1, \ldots, e_t$. We say that a path p is t-relevant if all its nonzero weight edges are t-relevant. Denote by $w(t, i, j)$ the maximum weight of a t-relevant path from $(1,1)$ to (i, j). The correctness of the algorithm follows immediately from the following lemma.

Lemma 4. *For every* t, *the score of* e_t *is computed correctly by the algorithm. Moreover, for every* t, *just after* e_t *is processed in lines 2–7,* $\text{score}(\text{pred}(S, j)) = w(t, \text{head}(e)_1, j)$ *for all* j.

To analyze the running time of the algorithm, let us count the number of times each operation on S is called. Each edge of $E_{F,G}$ is inserted or deleted at most once. The number of successor operations is the same as the number of deletions, and the number of predecessor operations is the same as the number of edges. Hence, the total number of operations on S is $O(|E_{F,G}|)$. Using the successor data-structure of van Emde Boas [28] we can support each operation on S in $O(\lg \lg m)$ time yielding a total running time of $O(|E_{F,G}| \cdot \lg \lg m)$. By running the above procedure recursively on every match pair we get that the total time complexity is bounded by

$$O\left(\sum_{\text{match pair } (v,w)} |E_{F_v, G_w}| \cdot \lg \lg m\right) = O\left(\lg \lg m \cdot \sum_{\text{match pair } (v,w)} \text{depth}(v) \cdot \text{depth}(w)\right)$$

$$= O\left(\lg \lg m \cdot r \cdot \text{height}(F) \cdot \text{height}(G)\right).$$

4   An $O(mr \lg r \cdot \lg \lg m)$ Algorithm

We begin this section by giving an alternative description of Klein's algorithm using an alignment graph. However, as opposed to the alignment graph of [27,4,5] our graph is three dimensional.

Given a tree F and a path decomposition $\mathcal{P}$ of F we define a sequence of subforests of F as follows. $F(n) = F$, and $F(i)$ for $i < n$ is the forest obtained from $F(i+1)$ by deleting one node: If the root of leftmost tree in F is not on the main path of $\mathcal{P}$ then this root is deleted, and otherwise the root of the rightmost tree in F is deleted. Let x_i be the node which is deleted from $F(i)$ when creating $F(i-1)$. Let y_i be the node of G that generates the ith character of the Euler string of G. Let I_{right} be the set of all indices i such that $F(i-1)$ is created from $F(i)$ by deleting the rightmost root of $F(i)$, and $I_{\text{left}} = \{1, \ldots, n\} \setminus I_{\text{right}}$.

The alignment graph $B_{F,G}$ of trees F and G is defined as follows. The vertices of $B_{F,G}$ are (i,j,k) for $0 \le i \le n$, $1 \le j \le 2m$, and $j \le k \le 2m$. Intuitively, vertex (i,j,k) corresponds to $\text{LCS}(F(i), G[j..k])$. For a vertex (i,j,k) with $i \in I_{\text{right}}$ the following edges enter the vertex.

1. If $i \ge 1$, an edge $(i-1,j,k) \to (i,j,k)$ with weight 0. This edge corresponds to deletion of the rightmost root of $F(i)$. This does not increase the LCS hence the zero weight.
2. If $j \le k - 1$, an edge $(i,j,k-1) \to (i,j,k)$ with weight 0. This edge either connects vertices which represent the same pair of forests, or represent deletion of the rightmost root in $G[j..k]$. Both cases do not change the LCS, hence the zero weight.
3. If x_i, y_k is a match pair, $j \le e_G(k) < k$, and x_i is not on the main path of F, an edge $(i - |F_{x_i}|, j, e_G(k)) \to (i,j,k)$ with weight $\text{LCS}(F_{x_i}, G_{y_k})$. This edge correspond to matching the rightmost tree in $F(i)$ to the rightmost tree of $G[j..k]$.
4. If x_i, y_k is a match pair, $j \le e_G(k) < k$, and x_i is on the main path of F, an edge $(i-1, e_G(k), k-1) \to (i,j,k)$ with weight 1. This edge corresponds to matching x_i (the root of $F(i) = F_{x_i}$) to y_k (the rightmost root of $G[j..k]$). If we match these nodes then only descendants of y_k can be matched to the nodes of $F(i-1)$ (since $F(i)$ is a tree). To ensure this, we set the second coordinate of the tail of the edge to $e_G(k)$ (instead of j as in the previous case), since nodes with indices $j' < e_G(k)$ are not descendants of y_k.

Similarly, for $i \in I_{\text{left}}$ the edges that enter (i,j,k) are

1. If $i \ge 1$, an edge $(i-1,j,k) \to (i,j,k)$ with weight 0.
2. If $j \le k - 1$, an edge $(i,j+1,k) \to (i,j,k)$ with weight 0.
3. If x_i, y_j is a match pair, $j < e_G(j) \le k$, and x_i is not on the main path of F, an edge $(i - |F_{x_i}|, e_G(j), k) \to (i,j,k)$ with weight $\text{LCS}(F_{x_i}, G_{y_j})$.
4. If x_i, y_j is a match pair, $j < e_G(j) \le k$, and x_i is on the main path of F, an edge $(i-1, j+1, e_G(j)) \to (i,j,k)$ with weight 1.

The set of all edges in $B_{F,G}$ with nonzero weights is denoted by $E_{F,G}$. In order to build $B_{F,G}$ one needs to know the values of $\text{LCS}(F', G')$ for some pairs of subforests F', G' of F, G. These values are obtained by making recursive calls to Klein's algorithm on the appropriate subforests of F and G.

Lemma 5. *The maximum weight of a path in $B_{F,G}$ from some vertex $(0, l, l)$ to vertex (i, j, k) is equal to $\text{LCS}(F(i), G[j..k])$.*

Klein's algorithm computes the maximum weight path that ends at each vertex in $B_{F,G}$ using dynamic programming, and returns the maximum weight of a path from some vertex $(0, l, l)$ to $(n, 1, 2m)$, which is equal to $\text{LCS}(F, G)$. The path decomposition $\mathcal{P}$ is selected in order to minimize the total size of the alignment graph $B_{F,G}$ and the alignment graphs created by the recursive calls of the algorithm. Using heavy path decomposition [14], the time complexity of Klein's algorithm is $O(n \lg n \cdot m^2)$.

Now, we present an algorithm for computing the LCS based on the sparsity of $E_{F,G}$. Recall that the score of an edge $e \in E_{F,G}$ is the maximum weight of a path in $B_{F,G}$ from some vertex $(0, l, l)$ to $\text{head}(e)$ that passes through e.

Lemma 6. *Let e be an edge in $E_{F,G}$ and denote $\text{head}(e) = (i, j, k)$. If $i \in I_{\text{right}}$ then $\text{score}(e) = \text{LCS}_R(F(i), G[j..k])$, and otherwise $\text{score}(e) = \text{LCS}_L(F(i), G[j..k])$.*

Knowing the scores of the edges gives us $\text{LCS}(F, G)$ as $\text{LCS}(F, G) = \text{score}((n-1, 1, 2m-1) \to (n, 1, 2m))$. In fact, additional LCS values can be obtained from the scores:

Lemma 7. *For every match pair $x \in F, y \in G$ such that x is on the main path of F there is an edge $e \in E_{F,G}$ such that $\text{LCS}(F_x, G_y) = \text{score}(e)$.*

A high-level description of the algorithm for computing the LCS of F and G is:

1: Build a path decomposition $\mathcal{P}$ of F.
2: **for** every node x in F in postorder **do**
3:     **if** x is the first node on some path $P \in \mathcal{P}$ **then**
4:         Build the set $E_{F_x,G}$.
5:         Compute the scores of the edges in $E_{F_x,G}$.
6: Output $\text{score}((n-1, 1, 2m-1) \to (n, 1, 2m))$.

We will explain how to construct the path decomposition $\mathcal{P}$ in step 1 later. For now note just that $\mathcal{P}$ is used when building each of the sets $E_{F_x,G}$ in step 4. To build $E_{F_x,G}$, we need a path decomposition of F_x. We use the path decomposition that is induced from the path decomposition $\mathcal{P}$. In order to build $E_{F_x,G}$ one needs to know the values of $\text{LCS}(F_{x'}, G_y)$ for pairs of nodes x' and y, where x' is a node of F_x that is not on the main path of F_x. By Lemma 7, the value of $\text{LCS}(F_{x'}, G_y)$ is equal to the score of an edge from $E_{F_{x''},G}$ where x'' is the first vertex on the path $p \in \mathcal{P}$ that contains x' (x'' can equal x'). Since the nodes of F are processed in postorder, the scores of the edges in $E_{F_{x''},G}$ are known when building $E_{F_x,G}$.

It remains to show how to compute the scores of the edges in $E_{F,G}$. The algorithm for computing the scores of the edges uses $4m$ successor data-structures

$S_1^{\text{left}}, \ldots, S_{2m}^{\text{left}}$ and $S_1^{\text{right}}, \ldots, S_{2m}^{\text{right}}$. Each of these structures stores a subset of $E_{F,G}$. The key of an edge e in some structure S_i^{right} is $\text{head}(e)_3$, and the key of an edge e in some structure S_i^{left} is $\text{head}(e)_2$.

```
1: for i = 1, . . . , n do
2:    for every e ∈ E_F,G with head(e)_1 = i do
3:       j ← head(e)_2, k ← head(e)_3
4:       if i ∈ I_right and score(e) > score(pred(S_j^right, k)) then
5:          insert(S_j^right, e)
6:          while succ(S_j^right, k + 1) ≠ NULL and score(succ(S_j^right, k + 1)) ≤
             score(e) do
7:             delete(S_j^right, succ(S_j^right, k + 1))
8:       if i ∈ I_left and score(e) > score(succ(S_k^left, j)) then
9:          insert(S_k^left, e)
10:         while pred(S_k^left, j−1) ≠ NULL and score(pred(S_k^left, j−1)) ≤ score(e)
             do
11:            delete(S_k^left, pred(S_k^left, j − 1))
12:   for every e ∈ E_F,G with tail(e)_1 = i do
13:      j ← tail(e)_2, k ← tail(e)_3
14:      score(e) ← weight(e) + max(score(pred(S_j^right, k)), score(succ(S_k^left, j)))
```

We call an edge e with $\text{head}(e)_1 \in I_{\text{right}}$ a *right edge*. Let $e_1, e_2, \ldots$ be the edges of $E_{F,G}$ according to the order in which they are processed in line 2.

Lemma 8. *The scores of all nonzero weight edges are computed correctly by the algorithm. Moreover, for every t, just after e_t is processed in lines 2–11, for all j and k, $\text{score}(\text{pred}(S_j^{\text{right}}, k))$ (resp., $\text{score}(\text{succ}(S_k^{\text{left}}, j)))$ is equal to the maximum weight of a t-relevant path from some vertex $(0, l, l)$ to $(\text{head}(e)_1, j, k)$ whose last nonzero weight edge is a right (resp., left) edge.*

Just as in the previous section, using the successor data-structure of van Emde Boas [28] we have that computing the scores of the edges in $E_{F,G}$ takes $O(|E_{F,G}| \lg \lg m)$ time. The time for computing the LCS between F and G is therefore $O(\sum_{x \in L_\mathcal{P}} |E_{F_x,G}| \lg \lg m)$, where $L_\mathcal{P}$ is the set of the first nodes of the paths in $\mathcal{P}$. In order to minimize $\sum_{x \in L_\mathcal{P}} |E_{F_x,G}|$, we build $\mathcal{P}$ similar to a *heavy path decomposition* but where *heavy* is determined by number of matches and not by size. This is done as follows. We begin building the main path. We start at the root of F and then we repeatedly extend the path by moving to a child w of the current node that maximizes the number of matches between F_w and G (ties are broken arbitrarily). After obtaining the main path, we remove its nodes from F and then recursively build a path decomposition of each of the remaining trees. The decomposition $\mathcal{P}$ that is obtained has the property that for each node $x \in F$, the number of nodes in $L_\mathcal{P}$ that are ancestors of x is at most $\lg r + 1$.

Lemma 9. $\sum_{x \in L_\mathcal{P}} |E_{F_x,G}| \leq 2mr(\lg r + 1)$.

We have therefore shown an algorithm that computes the LCS of two trees in $O(mr \lg r \cdot \lg \lg m)$ time.

5   An $O(Lr \lg r \cdot \lg \lg m)$ Algorithm

In this section we improve the algorithm of the previous section. Notice that in the alignment graph of the previous section each match pair generates up to $O(m)$ edges (while in the alignment graph of Section 3, each match pair generates exactly one edge). Therefore, the time of processing a match pair is $O(m \lg \lg m)$. We will show how to process each group of edges of a match pair in $O(L \lg \lg m)$ time by exploiting additional sparsity properties of the problem.

Formally, we partition the edges of $E_{F,G}$ into groups, where each group is the edges that correspond to some match pair: For $i \in I_{\text{right}}$ let $E_{F,G,i,a} = \{e \in E_{F,G} | \text{head}(e)_1 = i, \text{head}(e)_3 = a\}$, and for $i \in I_{\text{left}}$ let $E_{F,G,i,a} = \{e \in E_{F,G} | \text{head}(e)_1 = i, \text{head}(e)_2 = a\}$. The total number of groups $E_{F',G,i,a}$ for all the alignment graphs $B_{F',G}$ that are built by the algorithm is at most $r(\lg r + 1)$.

Consider some group $E_{F,G,i,k}$ for $i \in I_{\text{right}}$. Let $s = e_G(k)$. We have that $E_{F,G,i,k} = \{e_1, \ldots, e_s\}$ where $\text{head}(e_j) = (i, j, k)$. Denote $l_1 = \text{score}(e_s) = \text{weight}(e_s)$ and $l_2 = \text{score}(e_1)$. By Lemma 6, $\text{score}(e_1) \geq \text{score}(e_2) \geq \cdots \geq \text{score}(e_s)$. Moreover, for all j, $\text{score}(e_j) \in \{0, \ldots, L\}$ and $\text{score}(e_j) - \text{score}(e_{j+1}) \in \{0, 1\}$. Therefore, there are indices $j_{l_1} = s, j_{l_1+1}, \ldots, j_{l_2}$ such that $\text{score}(e_{j_l}) = l$ and $\text{score}(e_{j_{l+1}}) = l - 1$ (if $l \neq l_1$) for all l. These indices are called the *compact representation* of the scores of $E_{F,G,i,k}$.

To improve the algorithm of the previous section, instead of processing individual edges, we will process groups. For each group, we will compute the compact representation of its scores. The time to process each group will be $O(L \lg \lg m)$ so the total time complexity will be $O(Lr \lg r \cdot \lg \lg m)$.

We define two dimensional arrays $A_1, \ldots, A_n$, where $A_i[j, k]$ is the maximum weight of path from some vertex $(0, l, l)$ to (i, j, k). By Lemma 5, every array A_i has the following properties:

1. Each row of A_i is monotonically increasing.
2. Each column of A_i is monotonically decreasing.
3. The difference between two adjacent cells in A_i is either 0 or 1.
4. Each cell of A_i is an integer from $\{0, \ldots, L\}$.

The properties above are the same as the properties of the dynamic programming table for string LCS. Following the approach of [16], we define the *l-contour* of A_i (for $1 \leq l \leq L$) to be the set of all pairs (j, k) such that $A_i[j, k] = l$, $A_i[j+1, k] < l$ (or $j = 2m$), and $A_i[j, k-1] < l$ (or $k = 1$). By properties (1) and (2) of A_i we have that for two pairs (j, k) and (j', k') in the l-contour of A_i we have either $j < j'$ and $k < k'$, or $j > j'$ and $k > k'$.

The algorithm processes each i from 1 to n. Again, iteration i consists of two stages: (1) updating the l-contours according to the groups $E_{F,G,i,a}$ for all a (2) computing the compact representation of the scores for each group $E_{F,G,i',a}$ such that the edges $e \in E_{F,G,i',a}$ satisfy $\text{tail}(e)_1 = i$. We next explain each of the two stages in detail.

Computing the l-contours of A_i for all l is done by updating the l-contours of A_{i-1} that were computed in the previous iteration. The l-contour of A_i for the current value of i is kept using two successor data-structures S_l^1 and S_l^2. The key of a pair (j,k) in S_l^1 is j, while the key of (j,k) in S_l^2 is k.

Suppose that $i \in I_{\text{right}}$ (handling $i \in I_{\text{left}}$ is similar). In order to compute the l-contours of A_i, we process the groups $E_{F,G,i,k}$ for all k. Consider some fixed $E_{F,G,i,k}$, and let $j_{l_1}, j_{l_1+1}, \ldots, j_{l_2}$ be the compact representation of the scores of $E_{F,G,i,k}$ (which was computed in a prior iteration of the algorithm). Updating the l-contours according to the scores of the edges in $E_{F,G,i,k}$ is done by:

```
1: for l = l_1, ..., l_2 do
2:    if pred(S_l^2, k) = NULL or pred(S_l^2, k)_1 < j_l then
3:       p ← (j_l, k)
4:       insert(S_l^1, p)
5:       insert(S_l^2, p)
6:       while succ(S_l^2, k + 1) ≠ NULL and succ(S_l^2, k + 1)_1 ≤ j_l do
7:          p ← succ(S_l^2, k + 1)
8:          delete(S_l^1, p)
9:          delete(S_l^2, p)
```

It remains to describe stage (2), which computes the compact representation of the scores of some group $E_{F,G,i',k'}$ such that the edges $e \in E_{F,G,i',k'}$ satisfy $\text{tail}(e)_1 = i$. Suppose that $i' \in I_{\text{right}}$ and denote $E_{F,G,i',k'} = \{e_1, \ldots, e_s\}$ where $e_j = (i,j,k) \to (i',j,k')$. All the edges in $E_{F,G,i',k'}$ have the same weight w. Suppose that $x_{i'}$ is not on the main path of F. Clearly, $\text{score}(e_j) = w + A_i[j,k]$. Therefore the compact representation of the scores of $E_{F,G,i',k'}$ can be computed using $S_1^2, \ldots, S_L^2$:

```
1: j_w ← s
2: for l = 1, ..., L do
3:    if pred(S_l^2, k) ≠ NULL then j_{l+w} ← pred(S_l^2, k)_1
```

If $x_{i'}$ is on the main path of F then $\text{score}(e_1) = \cdots = \text{score}(e_s) = 1 + A_i[s,k]$, and computing the compact representation of the scores is done similarly. The computation of the compact representation of the scores of a group $E_{F,G,i',j'}$ with $i \in I_{\text{left}}$ is done similarly using the structures $S_1^1, \ldots, S_L^1$.

We have established the following theorem:

Theorem 1. *The tree LCS problem can be solved in time $O(Lr \lg r \cdot \lg \lg m)$.*

6   An $O(rh \lg \lg m)$ Algorithm for Homeomorphic Tree LCS

In this section we address the homeomorphic tree LCS problem. For this problem we obtain an $O(rh \lg \lg m)$ time algorithm, where $h = \text{height}(F) + \text{height}(G)$. We start by describing an $O(nm)$ non-sparse algorithm for the problem, based on the constrained edit distance algorithm of Zhang [30]. Here, the computation of $\text{HLCS}(F,G)$ is done recursively, in a postorder traversal of F and G. For every pair of nodes $v \in F$ and $w \in G$ we compute $\text{score}(v,w)$ which is equal to $\text{HLCS}(F_v, G_w)$. The computation of $\text{score}(v, w)$ is based on the previously computed scores for all children of v and w as follows. Let $c(u)$ denote the number of children of a node u and let $u_1, \ldots, u_{c(u)}$ denote the ordered sequence of the children of u. Then

$$\text{score}(v, w) = \max\left\{ \max_{i \leq c(v)} \{\text{score}(v_i, w)\}, \max_{i \leq c(w)} \{\text{score}(v, w_i)\}, \alpha(v, w) + 1 \right\},$$

where $\alpha(v, w)$ is defined as follows. If (v, w) is not a match pair then $\alpha(v, w) = -1$. Otherwise, $\alpha(v, w)$ is the maximum weight of a non-crossing bipartite matching between the vertices $v_1, \ldots, v_{c(v)}$ and the vertices $w_1, \ldots, w_{c(w)}$, where the weight of matching v_i with w_j is $\text{score}(v_i, w_j)$. Computing $\alpha(v, w)$ takes $O(c(v) \cdot c(w))$ time using dynamic programming on a $c(v) \times c(w)$ table.

In order to obtain a sparse version of this algorithm, there are two goals to be met. First, rather than computing $\text{score}(v, w)$ for all nm node pairs, we will only compute the scores for match pairs. Second, we need to avoid the $O(c(v) \cdot c(w))$ time complexity of the dynamic programming algorithm for computing $\alpha(v, w)$ and replace it with sparse dynamic programming. For every match pair (v, w) we have

$$\text{score}(v, w) = \max\left\{ \max_{v'}\{\text{score}(v', w)\}, \max_{w'}\{\text{score}(v, w')\}, \alpha(v, w) + 1 \right\},$$

where $\max_{v'}$ is maximum over all proper descendants v' of v that have the same label as v, and $\max_{w'}$ is defined similarly. Computing the two maxima above for all match pairs is done as follows. First, we initialize $\text{score}(v, w) = 0$ for all match pairs (v, w). After computing $\text{score}(v, w)$ for some match pair (v, w), we perform $\text{score}(\hat{v}, w) \leftarrow \max\{\text{score}(\hat{v}, w), \text{score}(v, w)\}$ and $\text{score}(v, \hat{w}) \leftarrow \max\{\text{score}(v, \hat{w}), \text{score}(v, w)\}$, where $\hat{v}$ and $\hat{w}$ are the parents of v and w, respectively. Thus, for some match pair (v, w), after processing all match pairs (v', w') where v' is a descendant of v and w' is a descendant of w, we have that $\text{score}(v, w)$ is equal to $\max\{\max_{v'}\{\text{score}(v', w)\}, \max_{w'}\{\text{score}(v, w')\}\}$, so it remains to compute $\alpha(v, w)$.

To compute $\alpha(v, w)$, define $P_{v,w}$ to be the set of all pairs (v_i, w_j) such that $\text{score}(v_i, w_j) > 0$. Applying a sparse dynamic programming approach to the computation of $\alpha(v, w)$ exploit the fact that $P_{v,w}$ can be much smaller than $c(v) \cdot c(w)$. However, note that just querying all pairs of children of v and w to check which ones have a positive score would already consume $O(c(v) \cdot c(w))$ time. But, given the set $P_{v,w}$, the cost of computing $\alpha(v, w)$ is $O(|P_{v,w}| \lg \lg m)$ instead of $O(c(v) \cdot c(w))$. Thus, in the rest of this section we show how to efficiently construct the sets $P_{v,w}$.

Our approach is based on the observation that, even before the scores are computed, a key subset of the match pairs of F_v and G_w can be identified that have the potential to eventually participate in $P_{v,w}$. For every $i \leq c(v)$ and $j \leq c(w)$, let $\hat{S}_{v,w,i,j}$ be the set of all match pairs (x, y) such that x is a descendant of v_i and y is a descendant of w_j, and let $S_{v,w,i,j}$ be the set of all match pairs $(x, y) \in \hat{S}_{v,w,i,j}$ for which there is no match pair $(x', y') \neq (x, y)$ in $S_{v,w,i,j}$ such that x' is an ancestor of x and y' is an ancestor of y.

The following lemma shows that $P_{v,w}$ can be built from the sets $S_{v,w,i,j}$.

Lemma 10. *Let (v, w) be a match pair. Let v_i be a child of v and w_j be a child of w such that (v_i, w_j) is not a match pair. Then, $\text{score}(v_i, w_j)$ is equal to the maximum score of a pair in $S_{v,w,i,j}$, or to 0 if $S_{v,w,i,j} = \emptyset$.*

While it is possible to build the sets $S_{v,w,i,j}$, it is simpler to build sets $S'_{v,w,i,j}$ such that $S_{v,w,i,j} \subseteq S'_{v,w,i,j} \subseteq \hat{S}_{v,w,i,j}$. From the proof of Lemma 10 we have that $\text{score}(v_i, w_j)$ is also equal to the maximum score of a pair in $S'_{v,w,i,j}$, or to 0 if $S'_{v,w,i,j} = \emptyset$. We build the sets $S'_{v,w,i,j}$ as follows. For each match pair (x, y) of F, G we build a list L_x of all proper ancestors v of x such that v is the lowest proper ancestor of x with label equal to $\text{label}(v)$ (the list L_x is generated by traversing the path from x to the root while maintaining a boolean array that stores which characters were already encountered). We also build a list L_y of all proper ancestors w of y such that w is the lowest proper ancestor of y with label equal to $\text{label}(w)$. For every $v \in L_x$ and every proper ancestor w of y with $\text{label}(w) = \text{label}(v)$, we add the pair (x, y) to $S'_{v,w,i,j}$ where v_i is the child of v which is on the path from v to x, and w_j is the child of w which is on the path from w to y. Similarly, for every $w \in L_y$ and every proper ancestor v of x with $\text{label}(v) = \text{label}(w)$, we add the pair (x, y) to $S'_{v,w,i,j}$.

Lemma 11. $S'_{v,w,i,j} \supseteq S_{v,w,i,j}$ *for all match pairs (v, w) and all i and j.*

Theorem 2. *The homeomorphic tree LCS problem can be solved in $O(rh \lg \lg m)$ time, where $h = \text{height}(F) + \text{height}(G)$.*

References

1. Abouelhoda, M.I., Ohlebusch, E.: Chaining algorithms for multiple genome comparison. J. of Discrete Algorithms 3(2-4), 321–341 (2005)
2. Amir, A., Hartman, T., Kapah, O., Shalom, B.R., Tsur, D.: Generalized LCS. In: Ziviani, N., Baeza-Yates, R. (eds.) SPIRE 2007. LNCS, vol. 4726, pp. 50–61. Springer, Heidelberg (2007)
3. Apostolico, A., Guerra, C.: The longest common subsequence problem revisited. Algorithmica 2, 315–336 (1987)
4. Backofen, R., Hermelin, D., Landau, G.M., Weimann, O.: Normalized similarity of RNA sequences. In: Proc. 12th symposium on String Processing and Information Retrieval (SPIRE), pp. 360–369 (2005)
5. Backofen, R., Hermelin, D., Landau, G.M., Weimann, O.: Local alignment of RNA sequences with arbitrary scoring schemes. In: Lewenstein, M., Valiente, G. (eds.) CPM 2006. LNCS, vol. 4009, pp. 246–257. Springer, Heidelberg (2006)
6. Bille, P.: A survey on tree edit distance and related problems. Theoretical computer science 337, 217–239 (2005)
7. Bille, P.: Pattern Matching in Trees and Strings. PhD thesis, ITU University of Copenhagen (2007)
8. Chawathe, S.: Comparing hierarchical data in external memory. In: Proc. 25th International Conference on Very Large Data Bases, Edinburgh, Scotland, U.K, pp. 90–101 (1999)
9. Chen, W.: New algorithm for ordered tree-to-tree correction problem. J. of Algorithms 40, 135–158 (2001)

10. Chin, F.Y.L., Poon, C.K.: A fast algorithm for computing longest common subsequences of small alphabet size. J. of Information Processing 13(4), 463–469 (1990)
11. Crochemore, M., Landau, G.M., Ziv-Ukelson, M.: A subquadratic sequence alignment algorithm for unrestricted scoring matrices. SIAM J. on Computing 32, 1654–1673 (2003)
12. Demaine, E.D., Mozes, S., Rossman, B., Weimann, O.: An optimal decomposition algorithm for tree edit distance. In: Arge, L., Cachin, C., Jurdziński, T., Tarlecki, A. (eds.) ICALP 2007. LNCS, vol. 4596, pp. 146–157. Springer, Heidelberg (2007)
13. Eppstein, D., Galil, Z., Giancarlo, R., Italiano, G.F.: Sparse dynamic programming i: linear cost functions. J. of the ACM 39(3), 519–545 (1992)
14. Harel, D., Tarjan, R.E.: Fast algorithms for finding nearest common ancestors. SIAM J. of Computing 13(2), 338–355 (1984)
15. Hirschberg, D.S.: A linear space algorithm for computing maximal common subsequences. Com. ACM 18(6), 341–343 (1975)
16. Hirschberg, D.S.: Algorithms for the longest common subsequence problem. J. of the ACM 24(4), 664–675 (1977)
17. Hsu, W.J., Du., M.W.: New algorithms for the LCS problem. J. of Computer and System Sciences 29(2), 133–152 (1984)
18. Hunt, J.W., Szymanski, T.G.: A fast algorithm for computing longest common subsequences. Commun. ACM 20(5), 350–353 (1977)
19. Klein, P.N.: Computing the edit-distance between unrooted ordered trees. In: Bilardi, G., Pietracaprina, A., Italiano, G.F., Pucci, G. (eds.) ESA 1998. LNCS, vol. 1461, pp. 91–102. Springer, Heidelberg (1998)
20. Klein, P.N., Tirthapura, S., Sharvit, D., Kimia, B.B.: A tree-edit-distance algorithm for comparing simple, closed shapes. In: Proc. 11th ACM-SIAM Symposium on Discrete Algorithms (SODA), pp. 696–704 (2000)
21. Levenstein, V.I.: Binary codes capable of correcting insetrions and reversals. Sov. Phys. Dokl. 10, 707–719 (1966)
22. Lozano, A., Valiente, G.: On the maximum common embedded subtree problem for ordered trees. In: Iliopoulos, C.S., Lecroq, T. (eds.) String Algorithmics, pp. 155–170. King's College Publications (2004)
23. Masek, W.J., Paterson, M.S.: A faster algorithm computing string edit distances. J. of Computer and System Sciences 20(1), 18–31 (1980)
24. Myers, G., Miller, W.: Chaining multiple-alignment fragments in sub-quadratic time. In: Proc. 6th annual ACM-SIAM symposium on Discrete algorithms (SODA), pp. 38–47 (1995)
25. Rick, C.: Simple and fast linear space computation of longest common subsequences. Information Processing Letters 75(6), 275–281 (2000)
26. Tai, K.: The tree-to-tree correction problem. J. of the ACM 26(3), 422–433 (1979)
27. Touzet, H.: A linear tree edit distance algorithm for similar ordered trees. In: Apostolico, A., Crochemore, M., Park, K. (eds.) CPM 2005. LNCS, vol. 3537, pp. 334–345. Springer, Heidelberg (2005)
28. van Emde Boas, P.: Preserving order in a forest in less than logarithmic time and linear space. Information Processing Letters 6(3), 80–82 (1977)
29. Wagner, R.A., Fischer, M.J.: The string-to-string correction problem. J. of the ACM 21(1), 168–173 (1974)
30. Zhang, K.: Algorithms for the constrained editing distance between ordered labeled trees and related problems. Pattern Recognition 28(3), 463–474 (1995)
31. Zhang, K., Shasha, D.: Simple fast algorithms for the editing distance between trees and related problems. SIAM J. of Computing 18(6), 1245–1262 (1989)

Why Greed Works for Shortest Common Superstring Problem

Bin Ma

Department of Computer Science
University of Western Ontario
London, ON, Canada N6A 5B7
bma@csd.uwo.ca

Abstract. The shortest common superstring problem (SCS) has been widely studied for its applications in string compression and DNA sequence assembly. Although it is known to be Max-SNP hard, the simple greedy algorithm works extremely well in practice. Previous researchers have proved that the greedy algorithm is asymptotically optimal on random instances. Unfortunately, the practical instances in DNA sequence assembly are very different from random instances.

In this paper we explain the good performance of greedy algorithm by using the smoothed analysis. We show that, for *any* given instance I of SCS, the average approximation ratio of the greedy algorithm on a small random perturbation of I is $1 + o(1)$. The perturbation defined in the paper is small and naturally represents the mutations of the DNA sequence during evolution.

Due to the existence of the uncertain nucleotides in the output of a DNA sequencing machine, we also proposed the shortest common superstring with wildcards problem (SCSW). We prove that in worst case SCSW cannot be approximated within ratio $n^{1/7-\epsilon}$, while the greedy algorithm still has $1 + o(1)$ smoothed approximation ratio.

1 Introduction

For n given strings s_1, s_2, ..., s_n, the shortest common superstring (SCS) problem asks for a shortest string s that contains every s_i as a substring. SCS finds applications in data compression [7,8] and in DNA (and other biology sequence) assembly [12,15,13]. Recently SCS has been extensively studied [4,9,19,10,5,11,18,1], largely due to its application in DNA assembly, where many overlapping short segments of DNA "reads" (substrings) need to be put together to construct the original DNA sequence. [1]

SCS is known to be Max-SNP hard, even for binary strings with equal lengths [20]. Therefore, it does not admit a PTAS. The best known approximation algorithm has ratio 2.5 [18]. There is a very simple greedy algorithm that repeatedly

---

[1] SCS can only model the small-scale DNA sequencing. The existence of long repeats in eukaryotic genomes makes SCS an inappropriate model for the whole genome sequencing.

P. Ferragina and G. Landau (Eds.): CPM 2008, LNCS 5029, pp. 244–254, 2008.

merges two maximum overlapping strings into one until there is only one string left. It was conjectured that this simple greedy algorithm has approximation ratio 2 [9], and the ratio was proved to be 4 and 3.5, respectively in [4] and [11]. In practice, this greedy algorithm works extremely well and it was reported that the average approximation ratio is below 1.014 [14]. It was proved that for random instances several greedy algorithms, including the one mentioned above, are asymptotically optimal [6,21]. In fact, because random strings do not overlap very much, the concatenation of the strings is not much longer than the shortest common superstring. As a result, a simple greedy algorithm will perform well on random instances. However, this is not a proper explanation to the good performance of the simple greedy algorithm in practice, because the practical instances arising from DNA assembly are not random and the input strings have significant overlaps.

Here we aim to explain the phenomenon that the greedy algorithm is a good approximation in practical cases, by adopting the smoothed analysis introduced in [17,16]. Average analysis studies the average behavior of an algorithm over all instances of a problem, and therefore the result heavily depends on a probabilistic distribution assumption of the instance space. However, smoothed analysis studies the algorithm's average behavior on each "local region" of the instance space. If the algorithm has good average performance on each local region, then for any reasonable probabilistic distribution on the whole instance space, the algorithm should perform well. For discrete problems, a local region can be viewed as a subset of instances generated by reasonable and small perturbations of a given instance. Clearly, smoothed analysis is in between of the worst case analysis and the average analysis. For a more complete review of smoothed analysis, we refer the readers to [16].

In Section 3, we will introduce a type of small and reasonable perturbations on instances of SCS; and prove that, *for any given instance*, the average approximation ratio of the greedy algorithm on the small perturbations of the instance is better than $1 + o(1)$. The result clearly explain why the greedy algorithm performs well in practical instances.

Because SCS is in Max-SNP hard, in worst case analysis it is not possible to approximate SCS arbitrarily well in polynomial time (unless P=NP). Our result shows the opposite in smoothed analysis. To our knowledge, this is the first time to demonstrate that a problem's lower bound complexity in terms of approximation can be different in worst case analysis and in smoothed analysis.

For DNA assembly, the DNA reads generated by a DNA sequencing machine often contain some undetermined nucleotides, which can be any of the four types A, C, G, and T. The conventional SCS problem does not model these undetermined nucleotides correctly. Therefore, in Section 4 we propose the shortest common superstring with wildcards (SCSW) problem, where the undetermined nucleotides are modeled as wildcards. We will prove that in worst case SCSW cannot be approximated within ratio $n^{1/7-\epsilon}$. However, the smoothed analysis will again show that the simple greedy algorithm has smoothed approximation ratio $1 + o(1)$ for SCSW.

2   Notations

Let s be a string over alphabet Σ. $|s|$ denotes the *length* of s. $s[i]$ denotes the i-th letter of s. Therefore, $s = s[1]s[2]\ldots s[|s|]$. Let $s[i..j]$ denote the substring $s[i]s[i+1]\ldots s[j]$.

A *string with wildcards* is a string over alphabet $\Sigma^* = \Sigma \cup \{*\}$, where $*$ indicates a wildcard. Given two strings s and t with or without wildcards, s *matches* t if (1) $|s| = |t|$, and (2) $\{s[i], t[i]\} \subset \Sigma \Rightarrow t[i] = s[i]$ for $i = 1, \ldots, |s|$.

Let s and t be two strings. If there is a suffix of s that is equal to a prefix of t, we say that s *overlaps* with t, or equivalently, there is an overlap between s and t. Notice that under this definition the fact that s overlaps t does not necessarily mean that t also overlaps s. Let s and t be two strings with wildcards, s *overlaps* with t if there are a suffix of s and a prefix of t that match each other.

Let s be a string. Let $s_1 = s[j_1..j_1']$, ..., $s_n = s[j_n..j_n']$ be substrings of s. s is called an original string of the SCS instance $I = \{s_1, s_2, \ldots, s_n\}$. If s_i and s_k are such that $j_i \leq j_k \leq j_i' \leq j_k'$, we say that s_i overlaps s_k in the original string s. $O_{ik} = s[j_k..j_i']$ is called the *original overlap* between s_i and s_k. $|O_{ik}|$ is called the length of the original overlap. Notice that under this definition, unless $j_i = j_k$ and $j_i' = j_k'$, at most one of $|O_{ik}|$ and $|O_{ki}|$ can be greater than zero.

3   Smoothed Analysis of SCS

3.1   The Practical SCS Instances

For DNA assembly, all the short DNA reads are substrings from the original DNA. An instance of SCS is therefore generated as follows: *Given a string s, select n substrings $s_1, s_2, \ldots, s_n$ as the instance of SCS.* We further require that $s_1, s_2, \ldots, s_n$ cover all positions of s. Otherwise, s can be replaced by deleting the uncovered positions.

For clarity of the presentation we assume all substrings s_i have the same length m. A discussion of instances with substrings with different lengths can be found in Section 5.

We denote such an instance as $I = I(s, m, (j_1, \ldots, j_n)))$, where j_i is the starting position of s_i in s. That is,

$$I = \{s_1 = s[j_1..j_1 + m - 1], \ldots, s_n = s[j_n..j_n + m - 1]\}.$$

A *perturbed instance* of I is defined to be $I' = I(s', m, (j_1, \ldots, j_n))$, where m and j_i $(i = 1, \ldots, n)$ remain unchanged and s' is obtained by uniform-randomly mutate each letter of s with a small probability $p > 0$. Therefore

$$I' = \{s_1' = s'[j_1..j_1 + m - 1], \ldots, s_n' = s'[j_n..j_n + m - 1]\}.$$

Figure 1 illustrates an example. If the original instance I represents a DNA sequencing experiment, then the perturbed instance I' represents the same experiment on a mutated DNA sequence, assuming all of the reads are taken from the same locations.

Fig. 1. An illustration of an SCS instance. The input strings s_1, s_2 and s_3 are all substrings of the original string s. An exemplary perturbation changed the dot positions of s. As a result, all of the corresponding positions in s_1, s_2 and s_3 are changed together.

In the rest of the paper we assume that $m = \Omega(\log n)$. We note that the proof of Max-SNP hardness of SCS in [20] was based on instances with this restriction. We will prove that even for a very small $p = \frac{2 \log(nm)}{\epsilon m}$, for *any* given original instance I, the average approximation ratio of the greedy algorithm on the perturbed instances I' is at most $1 + 3\epsilon$.

Because of the Max-SNP hardness of SCS, consider I to be a hard instance of SCS where the greedy algorithm does not approximate well. Our result indicates that the hardness of the instance can be destroyed by a very small perturbation. As today's natural DNA sequences are all evolved from their ancestral sequences by random mutations, our result explains why the greedy algorithm works well in practical instances of SCS.

3.2   Smoothed Analysis of the Greedy Algorithm

The simplest greedy algorithm for closest superstring problem is to repeatedly merge two strings with the longest overlap, until there is only one string left. In this section we provide the smoothed analysis of this simple greedy algorithm.

Let $I = I(s, m, (j_1, \ldots, j_n))$ be an instance of SCS and $I' = I(s', m, (j_1, \ldots, j_n))$ be the perturbed instance as described above. Let S'_o denote the optimal solution of the perturbed instance I'. Let S'_g denote the solution of I' computed by the greedy algorithm. Our task is to upper bound the weighted average of $|S'_g|/|S'_o|$ over all perturbations of the given instance.

Let $s_i = s[j_i..j_i + m - 1]$ and $s'_i = s'[j_i..j_i + m - 1]$. From the definition of the perturbation, for any non-empty original overlap between s_i and s_j in s, there is an overlap between s'_i and s'_j with the same length $|O_{ij}|$. This overlap is called a *consistent* overlap. All the other overlaps between s'_i and s'_j are then called *inconsistent* overlaps.

A major difference between the consistent overlaps and inconsistent overlaps is that consistent overlaps are guaranteed by our perturbation, whereas the following lemma shows that an inconsistent overlap is rarely long.

Lemma 1. *Let $p < \frac{1}{2}$ be the mutation probability in the perturbation. For any i, j, the probability that s'_i and s'_j has an inconsistent overlap of length k is no more than $(1 - p)^k$.*

Proof. First of all, the probability of having a length-k inconsistent overlap is

$$P = Pr(s'_i[m - k + l] = s'_j[l] \text{ for } 1 \leq l \leq k).$$

Because of the perturbation and $p \leq \frac{1}{2}$, for each l,

$$Pr(s_i'[m - k + l] = s_j'[l]) \leq 1 - p.$$

If the events $s_i'[m - k + l] = s_j'[l]$ are independent for different l, then the lemma is proved.

The trouble is that for different l and l', $s_j'[l]$ and $s_i'[m - k + l']$ can be from the same position of the original string s. Hence the independence does not hold. However, for an inconsistent overlap, it is easy to see that at least one of the following two facts is true:

1. the mutation at $s_j'[1]$ is independent from $s_i'[m - k + 1..m]$;
2. the mutation at $s_j'[k]$ is independent from $s_i'[m - k + 1..m]$.

Without loss of generality, we assume the second fact is true. Then

$$Pr(s_i'[m - k + l] = s_j'[l] \text{ for } 1 \leq l \leq k)$$
$$\leq (1 - p) \times Pr(s_i'[m - k + l] = s_j'[l] \text{ for } 1 \leq l \leq k - 1)$$

By applying the above reasoning recursively, we can prove that $Pr(s_i'[m-k+l] = s_j'[l]$ for $1 \leq l \leq k) \leq (1 - p)^k$. $\square$

Let P_{ij} be the maximum inconsistent overlap length between a suffix of s_i' and a prefix of s_j'.

Lemma 2. *Let $\epsilon > 0$ be a small number and $p \geq \frac{2 \log(nm)}{\epsilon m}$. Then when m is a sufficiently large number,*

$$Pr(P_{ij} < \epsilon m \text{ for all } i, j) \geq 1 - \frac{\epsilon}{m \log(nm)}.$$

Proof. From Lemma 1, for any given i and j,

$$Pr(P_{ij} \geq \epsilon m) \leq \sum_{k=\lceil \epsilon m \rceil}^{m} Pr(\text{There is a length } k \text{ inconsistent overlap})$$

$$\leq \sum_{k=\lceil \epsilon m \rceil}^{m} (1 - p)^k$$

$$\leq p^{-1} \times (1 - p)^{\lceil \epsilon m \rceil}$$

$$\leq p^{-1} \times \left(1 - \frac{2 \log(nm)}{\epsilon m}\right)^{\lceil \epsilon m \rceil}$$

$$\leq p^{-1} \times 2 \times e^{-2 \log(nm)} \tag{1}$$

$$\leq \frac{\epsilon}{n^2 m \log(nm)}.$$

Here Inequality (1) is because $(1 - x)^{1/x} \rightarrow e^{-1}$ when $x \rightarrow 0$.

Therefore,

$$Pr\left(P_{ij} \geq \epsilon m \text{ for some } i, j\right) \leq n^2 \times \frac{\epsilon}{n^2 m \log(nm)} = \frac{\epsilon}{m \log(nm)}.$$

The lemma is proved.   □

Lemma 3. *If $P_{ij} < \epsilon m$ for all $i \neq j$, then the approximation ratio of the greedy algorithm is at most $\frac{1}{(1-\epsilon)^2}$.*

Proof. Let S'_o and S'_g be defined as before. Because of the definition of perturbation, s' is a solution of the perturbed instance. Therefore, $|S'_o| \leq |s'| = |s|$. Next, let us lower bound $|S'_o|$.

In S'_o, denote the non-empty overlap between the suffix of s'_i and the prefix of s'_j by O'_{ij}. O'_{ij} is either consistent or inconsistent. Clearly, if both O'_{ij} and O'_{jk} are consistent, and O'_{ik} is non-empty, then O'_{ik} is also consistent. Define $i \prec j$ if and only if there is a sequence $i_1 = i, i_2, \ldots, i_k = j$ such that $O'_{i_l i_{l+1}}$ is consistent for $l = 1, \ldots, k-1$. Define $i \equiv j$ if either $i \prec j$ or $j \prec i$. Then clearly $\equiv$ is an equivalence relation.

Therefore, $s'_1, \ldots, s'_n$ are classified into k groups of equivalence classes T_1, $T_2, \ldots, T_k$, under the $\equiv$ relation. By overlapping the strings in T_i together using the consistent overlaps, each T_i naturally defines a string t_i, which is a substring of both s' and S'_o. Because the length of inconsistent overlap, $P_{ij} < \epsilon m$ for all $i \neq j$, the overlap length of a pair of t_i and t_j is also less than ϵm. Consequently,

$$|S'_o| \geq \sum_{i=1}^{k} |t_i| - (k-1)\epsilon m \geq (1-\epsilon) \sum_{i=1}^{k} |t_i| \geq (1-\epsilon)|s'| = (1-\epsilon)|s|. \quad (2)$$

Next let us examine the relationship between $|s|$ and $|S'_g|$. The perturbation does not destroy the original overlap between s_i and s_j in s. Therefore, the maximum overlap length between s'_i and s'_j is at least $|O_{ij}|$. On the other hand, $P_{ij} < \epsilon m$ for all $i \neq j$. As a result, for every i and j such that $O_{ij} \geq \epsilon m$, a greedy algorithm will always assemble s'_i and s'_j in the same way as s_i and s_j being assembled in s.

Assume assembling all pairs of s_i and s_j for $O_{ij} \geq \epsilon m$ gives us several longer strings $t_1, t_2, \ldots, t_k$. Then

$$|S'_g| \leq \sum_{i=1}^{k} |t_i|.$$

Without loss of generality, assume that t_i is before t_{i+1} in S'_g. Because all $O_{ij} \geq \epsilon m$ have been assembled, the overlap between t_i and t_{i+1} is shorter than ϵm. Therefore,

$$|s| \geq \sum_{i=1}^{k} |t_i| - (k-1)\epsilon m > (1-\epsilon) \sum_{i=1}^{k} |t_i| \geq (1-\epsilon)|S'_g|. \quad (3)$$

The theorem is the direct consequence of (2) and (3).   □

The following is our main theorem.

Theorem 1. *For any given small $\epsilon > 0$, let the perturbation probability be $p \geq \frac{2\log(nm)}{\epsilon m}$. Then for sufficiently large m, the expected ratio of the greedy algorithm on the perturbed instances is $(1 + 3\epsilon)$.*

Proof. Let $p_{good} = Pr(P_{ij} < \epsilon m$ for all $i, j)$. Then with probability p_{good}, Lemma 3 is true.

Lemma 2 says that $p_{good} \geq 1 - \frac{\epsilon}{m\log(nm)}$. Moreover, in [11], the greedy algorithm was proved to have worst-case approximation ratio 3.5. Therefore,

$$E(ratio) \leq (1 - p_{good}) \times 3.5 + p_{good} \times \frac{1}{(1-\epsilon)^2}$$

$$\leq \frac{\epsilon}{m\log(nm)} \times 3.5 + 1 \times \frac{1}{(1-\epsilon)^2}$$

$$\leq (1 + 3\epsilon). \qquad \square$$

Remark. Our proof can actually allow $\epsilon = o(1)$. This shows that a problem in Max-SNP hard can have arbitrarily good smoothed approximation ratio.

Remark. Our perturbation is very small comparing to the instance size. With perturbation probability $p = \frac{2\log(nm)}{\epsilon m}$, each length-$m$ substring is only expected to change $O(\log(nm))$ letters.

4   Shortest Common Superstring with Wildcards

For DNA assembly, the DNA reads (substrings) are generated with a sequencing machine. Very often, these reads contain undetermined nucleotides because the sequencing machine has very low confidence to determine the type of nucleotides at those locations. For example, in read "...`TAAAACAANNANTTCCGAAGA`..." each letter N indicates an uncertain nucleotide which can be any of A, C, G, or T. When these reads are put together, those uncertain letters should behave like wildcards that match any letters. Hence, we propose the following variant of SCS.

Shortest Common Superstring with Wildcards (SCSW). Given n strings $s_1, s_2, \ldots, s_n$ over alphabet $\Sigma^* = \Sigma \cup \{*\}$, where $*$ is a wildcard that matches any single letter, find the shortest string s such that for each i, s_i matches a substring of s.

Clearly, the SCS problem is a special case of SCSW. Because SCS belongs to Max-SNP hard, so does SCSW. In this section, we first prove a much stronger hardness result for SCSW. Then we demonstrate that when the wildcards (the sequencing errors) are distributed randomly, the greedy algorithm still has the smoothed approximation ratio $1 + o(1)$.

To prove the inapproximability, we reduce the minimum chromatic number problem to SCSW.

Minimum Chromatic Number Problem. Given an undirected graph $G = \langle V, E \rangle$, find a partition of V into disjoint sets $V_1, V_2, \ldots, V_k$ such that each V_i is an independent set and k is minimized.

The following property about the minimum chromatic number problem was proved in [3].

Lemma 4 ([3]). *The minimum chromatic number problem cannot be approximated in polynomial time within ratio* $|V|^{1/7-\epsilon}$ *unless P=NP.*

Theorem 2. *The shortest common superstring with wildcards problem cannot be approximated in polynomial time within ratio* $n^{1/7-\epsilon}$ *unless P=NP.*

Proof. Suppose $G = \langle V, E \rangle$ is an instance of the minimum chromatic number problem. Let $V = \{v_1, v_2, \ldots, v_n\}$ and $E = \{e_1, e_2, \ldots, e_m\}$. We construct an instance of SCSW as follows.

Let the alphabet $\Sigma = \{\mathtt{A}, \mathtt{T}, \mathtt{G}, \mathtt{C}\}$ and the wildcard be $\mathtt{N}$. Each v_i corresponds to a string t_i with length m. For each $k = 1, 2, \ldots, m$,

$$
t_i[k] = \begin{cases} \mathtt{A}, & \text{if } e_k = (v_i, v_j) \text{ and } i < j, \\ \mathtt{T}, & \text{if } e_k = (v_j, v_i) \text{ and } j < i, \\ \mathtt{N}, & \text{otherwise.} \end{cases}
$$

For example, if $m = 7$ and v_3 has three adjacent edges $e_2 = (v_1, v_3)$, $e_4 = (v_3, v_5)$, and $e_5 = (v_3, v_6)$, then $t_3 = \mathtt{NTNAANN}$. Such a construction ensures that t_i and t_j match each other if and only if $(v_i, v_j) \notin E$.

Let $X = \mathtt{G}^{mn}\mathtt{C}^{mn}$ be a length $2mn$ string. Let $s_i = Xt_iX$ $(i = 1, \ldots, n)$ be the instance of the SCSW.

Suppose the minimum chromatic number instance has the optimal solution $V = V_1 \cup V_2 \cup \ldots \cup V_K$. We construct a solution of SCSW as follows. For each $j = 1, \ldots, K$, suppose $V_j = \{v_{i_1}, \ldots, v_{i_l}\}$. Let T_j be a length-m string obtained by "fusing" $t_{i_1}, \ldots, t_{i_l}$ together. More specifically, put $t_{i_1}, \ldots, t_{i_l}$ into different rows as follows

$$
\begin{aligned}
t_{i_1} &: \ldots\mathtt{NTNANNN}\ldots \\
t_{i_2} &: \ldots\mathtt{NNTNNNN}\ldots \\
&\ldots \\
t_{i_l} &: \ldots\mathtt{NNNNANT}\ldots
\end{aligned}
$$

Because V_j is an independent set, from the construction of each t_i, one can easily see that each column has at most one letter that is not the wildcard $\mathtt{N}$. If there is such a letter at the k-th column, let $T_j[k]$ be that letter. Otherwise, let $T_j[k]$ be any letter. Then T_j is a length-m string that is matched by each of $t_{i_1}, \ldots, t_{i_l}$.

Because $V = V_1 \cup V_2 \cup \ldots \cup V_K$, it is easy to see that $S = XT_1XT_2X \ldots T_KX$ is a solution of the SCSW with length $2mn + m(2n+1)K$. Therefore, the length of the shortest common superstring is at most $2mn + m(2n+1)K$.

On the other hand, suppose SCSW has a solution (not necessarily optimal). Because of the existence of $X = \mathtt{G}^{mn}\mathtt{C}^{mn}$ in each s_i, the solution must have the form $S = XT_1XT_2X \ldots T_kX$, where each T_j is a length-m string that is matched by some t_i, and each t_i must match one of the T_j $(j = 1, \ldots, k)$. Because of the construction of t_i, if both t_i and $t_{i'}$ are matched by T_j, then there is no edge connecting v_i and $v_{i'}$ in graph G. Therefore, by letting $V'_j = \{v_i | t_i \text{ matches } T_j\}$,

V_j' is an independent set and $V = V_1' \cup V_2' \cup \ldots \cup V_k'$. Let $V_j = V_j' \setminus \cup_{i=1}^{j-1} V_i'$. We get a solution for the minimum chromatic number problem.

Therefore, from a solution of the constructed instance with length $2mn+m(2n+1)k$, we can also construct in polynomial time a solution of the original instance with chromatic number k. From the above discussion, it is easy to verify that the reduction is an L-reduction. Because of Lemma 4, the theorem is proved.   □

Although the worst case analysis shows that SCSW has very high complexity in terms of approximation. We show that the greedy algorithm works well on average case using smoothed analysis. Here we assume an instance is generated by first generating an SCS instance $I = I(s, m, (j_1, \ldots, j_n)) = \{s_1, \ldots, s_n\}$, and then turn some positions of each s_i to be the wildcard letter.

The greedy algorithm for SCS can be straightforwardly adopted to find a solution of SCSW. The only difference is that the definitions of overlap are different in SCS and SCSW (see Section 2).

A perturbation to this instance I is done in three steps:

1. For each letter in s, change it with a small probability p to get s'. This step represents the DNA sequence mutation during evolution;
2. Generate $I' = I(s', m, (j_1, \ldots, j_n)) = \{s_1', \ldots, s_n'\}$; and
3. For each s_i', each letter is changed to the wildcard letter with probability q, independently. This step represents the sequencing error in the DNA sequencing machine.

Theorem 3. *For any small number $\epsilon > 0$, for $p \geq \frac{2\log(nm)}{\epsilon m}$ and a constant $0 \leq q < 1$, the expected ratio of the greedy algorithm on the perturbed instances is $1 + \frac{3\epsilon}{1-q}$.*

Proof. The proof is very similar to the proof of Theorem 1. We similarly define the concepts of consistence and inconsistence. Because the wildcards are independently assigned with probability q, for each position of string $s_i'[j]$, the probability of that it is neither a wildcard nor $s_i[j]$ is $p(1-q)$. Similarly to Lemma 1, we can show that an inconsistent overlap of length k happens with probability no more than $(1 - p(1-q))^k$. The rest of the proof just follows all the proofs after Lemma 1 in Section 3.2.   □

5   Discussion

We proposed a natural perturbation model for the shortest common superstring problem (SCS), and proved that the greedy algorithm has average ratio $1 + o(1)$ over the perturbations of any given instance. Because the average is taken over the perturbations of any given instance, the result is stronger than showing the average ratio is $1 + o(1)$ over all the instances. This smoothed analysis explains why the greedy algorithm performs well in practice, regardless the Max-SNP hardness of SCS. This shows that the hard instances of SCS are very "unstable", a small perturbation on the instance will destroy the hardness.

Due to the uncertain letters in the output of a DNA sequencing machine, we proposed the shortest common superstring with wildcards problem (SCSW). We proved that this variant is much harder than SCS in terms of approximation. However, we showed that when the uncertain letters are drawn randomly, the smoothed analysis for SCS still works. As a result, the simple greedy algorithm still has $1 + o(1)$ smoothed approximation ratio.

Another variant of SCS studied in the literature is the shortest approximate common super string problem, where the common superstring need to contain a substring within certain Hamming distance to each input string [21,2]. We claim that our smoothed analysis still works when the allowed errors are bounded by $p'm$ for p' much smaller than the perturbation probability p. This is because a similar result as Lemma 1 still holds. Here we omit the proof.

It is noteworthy that our algorithmic results do not need the alphabet to be finite; whereas the hardness result only needs an alphabet of size 4.

All the proofs in the paper assumed that the input strings s_i have the same length $m = \Omega(\log n)$. When the input strings have different lengths, one can easily see that by changing m to be the minimum length of all input strings, all the results still hold when $m = \Omega(\log n)$. In fact, as long as most of the input strings have length $\Omega(\log n)$, the contribution of the very short strings to the length is negligible and our results still hold. Because in practice very short strings are not a problem, the exact bound for the allowed number of short strings is omitted here.

Acknowledgment

The work was supported in part by NSERC, Canada Research Chair, China NSF 60553001, National Basic Research Program of China 2007CB807900, 2007CB807901, and was partially done when he visited Prof. Andrew Yao at ITCS at Tsinghua University. The author thanks Dr. Shanghua Teng for valuable discussions.

References

1. Armen, C., Stein, C.: A 2 2/3-approximation algorithm for the shortest superstring problem. In: Hirschberg, D.S., Meyers, G. (eds.) CPM 1996. LNCS, vol. 1075, pp. 87–101. Springer, Heidelberg (1996)
2. Rebaï, A.S., Elloumi, M.: Approximation algorithm for the shortest approximate common superstring problem. In: Proc. 12th Word Academy of Science, Engineering and Technology, pp. 302–307 (2006)
3. Bellare, M., Goldreich, O., Sudan, M.: Free bits, pcps and non-approximability - towards tight results. SIAM Journal on Computing 27, 804–915 (1998)
4. Blum, A., Jiang, T., Li, M., Tromp, J., Yannakakis, M.: Linear Approximation of Shortest Superstrings. Journal of the Association for Computer Machinery 41(4), 630–647 (1994)
5. Breslauer, D., Jiang, T., Jiang, Z.: Rotations of periodic strings and short superstrings. Journal of Algorithms 24(2), 340–353 (1997)

6. Frieze, A.M., Szpankowski, W.: Greedy algorithms for the shortest common superstring that are asymptotically optimal. Algorithmica 21(1), 21–36 (1998)
7. Gallant, J., Maier, D., Storer, J.: On finding minimal length superstrings. Journal of Computer and System Sciences 20, 50–58 (1980)
8. Storer, J.: Data Compression: Methods and Theory. Addison-Wesley, Reading (1988)
9. Tarhio, J., Ukkonen, E.: A greedy approximation algorithm for constructing shortest common superstrings. Theoretical Computer Science 57, 131–145 (1988)
10. Turner, J.: Approximation algorithms for the shortest common superstring problem. Information and Computation 83, 1–20 (1989)
11. Kaplan, H., Shafrir, N.: The greedy algorithm for shortest superstrings. Information Processing Letters 93, 13–17 (2005)
12. Li, M.: Towards a DNA sequencing theory. In: Proc. of the 31st IEEE Symposium on Foundations of Computer Science, pp. 125–134 (1990)
13. Waterman, M.S.: Introduction to Computational Biology: Maps, Sequences, and Genomes. Chapman and Hall, Boca Raton (1995)
14. Romero, H.J., Brizuela, C.A., Tchernykh, A.: An experimental comparison of approximation algorithms for the shortest common superstring problem. In: Proc. Fifth Mexican International Conference in Computer Science (ENC 2004), pp. 27–34 (2004)
15. Shapiro, M.B.: An algorithm for reconstructing protein and RNA sequences. Journal of ACM 14(4), 720–731 (1967)
16. Spielman, D.A., Teng, S.-H.: Smoothed analysis: Motivation and discrete models. In: Dehne, F., Sack, J.-R., Smid, M. (eds.) WADS 2003. LNCS, vol. 2748, pp. 256–270. Springer, Heidelberg (2003)
17. Spielman, D.A., Teng, S.-H.: Smoothed analysis of algorithms: Why the simplex algorithm usually takes polynomial time. Journal of ACM 51(3), 385–463 (2004)
18. Sweedyk, Z.: 2.5-approximation algorithm for shortest superstring. SIAM Journal on Computing 29(3), 954–986 (2000)
19. Teng, S.H., Yao, F.: Approximating shortest superstrings. In: Proc. 34th IEEE Symposium on Foundations of Computer Science, pp. 158–165 (1993)
20. Vassilevska, V.: Explicit inapproximability bounds for the shortest superstring problem. In: Jedrzejowicz, J., Szepietowski, A. (eds.) MFCS 2005. LNCS, vol. 3618, pp. 793–800. Springer, Heidelberg (2005)
21. Yang, E.H., Zhang, Z.: Shortest common superstring problem: average case analysis for both exact and approximate matching. IEEE Transactions on Information Theory 45(6), 1867–1886 (1999)

Constrained LCS: Hardness and Approximation

Zvi Gotthilf[1], Danny Hermelin[2], and Moshe Lewenstein[1]

[1] Department of Computer Science, Bar-Ilan University, Ramat Gan 52900, Israel
{gotthiz,moshe}@cs.biu.ac.il
[2] Department of Computer Science, University of Haifa,
Mount Carmel, Haifa 31905, Israel
danny@cri.haifa.ac.il

Abstract. The problem of finding the longest common subsequence (LCS) of two given strings A_1 and A_2 is a well-studied problem. The constrained longest common subsequence (C-LCS) for three strings A_1, A_2 and B_1 is the longest common subsequence of A_1 and A_2 that contains B_1 as a subsequence. The fastest algorithm solving the C-LCS problem has a time complexity of $O(m_1 m_2 n_1)$ where m_1, m_2 and n_1 are the lengths of A_1, A_2 and B_1 respectively. In this paper we consider two general variants of the C-LCS problem. First we show that in case of two input strings and an arbitrary number of constraint strings, it is NP-hard to approximate the C-LCS problem. Moreover, it is easy to see that in case of an arbitrary number of input strings and a single constraint, the problem of finding the constrained longest common subsequence is NP-hard. Therefore, we propose a linear time approximation algorithm for this variant, our algorithm yields a $1/\sqrt{m_{min}|\Sigma|}$ approximation factor, where m_{min} is the length of the shortest input string and $|\Sigma|$ is the size of the alphabet.

1 Introduction

The problem of finding the longest common subsequence (LCS) of two given strings A_1 and A_2 is a well-studied problem, see [3,6,7,1]. The *constrained* longest common subsequence (C-LCS) for three strings A_1, A_2 and B_1 is the longest common subsequence of A_1 and A_2 that contains B_1 as a subsequence. Tsai [10] gave a dynamic programming algorithm for the problem which runs in $O(n^2 m^2 k)$ where m, n and k are the lengths of A_1, A_2 and B_1 respectively. Improved dynamic programming algorithms were proposed in [2,4] which run in time $O(nmk)$. Approximated results for this C-LCS variant presented in [5].

Many problems in pattern matching are solved with dynamic programming solutions. Among the most prominent of these is the LCS problem. These solutions are elegant and simple, yet usually their running times are quadratic or more, i.e. they are not effective in the case of multiple strings. It is a desirable goal to find algorithms which offer faster running times. One slight improvement, a reduction of a log factor, is the classical Four-Russians trick, see [9]. However, in general, faster algorithms have proven to be rather elusive over the years (and perhaps it is indeed impossible).

P. Ferragina and G. Landau (Eds.): CPM 2008, LNCS 5029, pp. 255–262, 2008.

The classical LCS problem has many applications in various fields. Among them applications in string comparison, pattern recognition and data compression. Another application, motivated from computational biology, is finding the commonality of two DNA molecules. Closely related, Tsai [10] gave a natural application for the C-LCS problem: in the computation of the commonality of two biological sequences it may be important to take into account a common specific structure.

1.1 Our Contribution

We propose to consider two general variants of the C-LCS problem. First, we prove that in case of two input strings and an arbitrary number of constraint strings, it is NP-hard to approximate the C-LCS problem. In addition, we obtain the first approximation algorithm for the case of many input strings and a single constraint. Our algorithm yields a $1/\sqrt{m_{min}|\Sigma|}$ approximation factor, where m_{min} is the length of the shortest input string and $|\Sigma|$ is the size of the alphabet. The running time of our algorithm is linear.

2   Preliminaries

Let $A_1 = \langle a_{1_1}, a_{1_2}, \ldots, a_{1_{m1}} \rangle$, $A_2 = \langle a_{2_1}, a_{2_2}, \ldots, a_{2_{m2}} \rangle$, ..., $A_k = \langle a_{k_1}, a_{k_2}, \ldots, a_{k_{mk}} \rangle$ and $B_1 = \langle b_{1_1}, b_{1_2}, \ldots, b_{1_{n1}} \rangle$, $B_2 = \langle b_{2_1}, b_{2_2}, \ldots, b_{2_{n2}} \rangle$, ..., $B_l = \langle b_{n_1}, b_{n_2}, \ldots, b_{1_{nl}} \rangle$ be an input of the C-LCS problem. The longest constrained subsequence (C-LCS, for short) of A_1, A_2, ..., A_k and B_1, B_2, ..., B_l is the longest common subsequence of A_1, A_2, ..., A_k that contains each of B_1, B_2, ..., B_l as a subsequence. The approximation version of the C-LCS problem is defined as follows. Let OPT_{clcs} be the optimal solution for the C-LCS problem and APP_{clcs} the result of the approximation algorithm APP such that:

- APP_{clcs} is a common subsequence of A_1, A_2, ..., A_k.
- B_1, B_2, ..., and B_l are subsequences of APP_{clcs}.

The approximation ratio of the APP algorithm will be the smallest ratio between $|APP_{clcs}|$ and $|OPT_{clcs}|$ over all possible input strings A_1, A_2, ..., A_k and B_1, B_2, ..., B_l.

Clearly, not every instance of the C-LCS problem must have a feasible solution, i.e. there is no common subsequence of all input strings that contains every constraint string as a subsequence. It can be seen in figure 1 that the left instance is an example of a non-feasible C-LCS instance, while for the right instance "*bcabcab*" is a feasible constrained common subsequence.

3   Arbitrary Number of Constraints

In this section we prove that given two input strings and an arbitrary number of constrains the problem of finding the C-LCS is NP-hard. In addition, we show that it is NP-hard to approximate C-LCS for such instances.

A1 [a][b][c][a][b][c][a][b][a]

A2 [b][c][a][a][b][c][c][a][b]

B1 [a][a][c]

B2 [c][a][b][b]

B3 [c][b][a]

A1 [a][b][c][a][b][c][a][b][a]

A2 [b][c][a][a][b][c][c][a][b]

B1 [c][b][c][b]

B2 [b][c][a][a]

Fig. 1. Non feasible and feasible C-LCS instances

Theorem 1. *The C-LCS problem in case of an arbitrary number of constraints is NP-complete.*

Proof: We prove the hardness of the problem by a reduction from 3-SAT.

Given a 3-SAT instance with variables $x_1, x_2, \ldots, x_k$ and clauses $c_1, c_2, \ldots, c_l$, we construct an instance of C-LCS with two input strings and $k + l - 1$ constraints.

The alphabet of A_1 and A_2 is the set of clauses $c_1, c_2, \ldots, c_l$ and a set of separators $\{s_1, s_2, \ldots, s_{k-1}\}$ separating between the variables.

We construct A_1 as follows. For each variable x_i we create a substring X_i by setting all the clauses satisfied with $x_i = true$ followed by all the clauses satisfied with $x_i = false$ (we set the clauses in a sorted order). We then set A_1 to be $X_1 s_1 X_2 s_2 \ldots s_{k-1} X_k$, the X_i substrings separated by the appropriate separators.

We similarly construct A_2. We create a substring X_i' by setting all the clauses satisfied with $x_i = false$ followed by all the clauses satisfied with $x_i = true$ (we set the clauses in a sorted order). We then set A_2 to be $X_1' s_1 X_2' s_2 \ldots s_{k-1} X_k'$, the X_i' substrings separated by the appropriate separators.

Let $c_1, c_2, \ldots, c_l$ and $s_1, s_2, \ldots, s_{k-1}$ be the group of constraints. Note that, all of them are of length one.

See figure 2 as an example of our constriction from the following 3-SAT instance to a C-LCS instance that contains two input strings and $k + l - 1$ constraints (all of length one):

$$(x_1 \vee x_2 \vee x_3) \wedge (\bar{x}_1 \vee \bar{x}_2 \vee x_4) \wedge (x_2 \vee \bar{x}_3 \vee \bar{x}_4) \wedge (x_1 \vee \bar{x}_3 \vee x_4)$$

Lemma 1. *A 3-SAT instance can be satisfied iff there exists a C-LCS of length $\geq k + l - 1$.*

Proof: For simplicity, we assume that there are no clauses that contains both x_i and $\bar{x}_i$.

($\Rightarrow$) Suppose a 3-SAT instance can be satisfied.

Let X be an assignment on the variables satisfying the 3-SAT instance. Let Y be the variables assigned *true* values of X and Z be the variables assigned

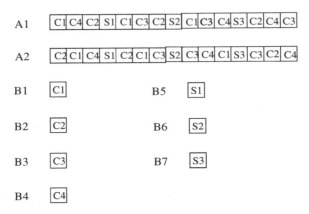

Fig. 2. Construction example

false values. For each variable $x_i \in Y$, let $\{c_{i_j}, \ldots, c_{i_r}\}$ be the clauses which are satisfied by setting x_i to *true*.

We construct a valid C-LCS as follows. Add to the C-LCS the $c'_{i_j}s$ from X_i and X'_i. Clearly they cannot cross each other as they are ordered. Likewise for $x'_i \in Z$ we do the same. Moreover, we select $s_1, s_2, \ldots, s_{k-1}$. Note that, since x_i is either *true* or *false* we will have:

1. No internal crossings within X_i and X'_i.
2. No crossing over the separators.

Obviously, since all clauses are satisfied (by some variable) they appear within the LCS. Since also $s_1, s_2, \ldots, s_{k-1}$ appear, all the C-LCS constraints are satisfied. Therefore, $|C - LCS| \geq l + k - 1$.

($\Leftarrow$) Note that all constraints must be satisfied. Hence, $s_1, s_2, \ldots, s_{k-1}$ appears in the C-LCS. Therefore, any clause c_i appearing in the C-LCS must be within a given X_i, X'_i. Thus, there cannot be an inconsistency of the x_i assignments. Because the clauses $c_1, c_2, \ldots, c_l$ are constraints, they must appear in the C-LCS.

Therefore, the assignment of $x_1, x_2, \ldots, x_k$ must satisfy the 3-SAT instance, since every clause must be satisfied in the C-LCS instance.   □

The following theorem derived from our reduction.

Theorem 2. *The C-LCS problem in case of an arbitrary number of constraints cannot be approximated.*

Proof: By the C-LCS definition and according to Lemma 1, any valid solution for the C-LCS must satisfy all the constraints (and must be of length $\geq k + l - 1$). Therefore, any approximation algorithm must yield an appropriate solution to the 3-SAT problem. In case that an approximation algorithm fails to find a C-LCS, we can conclude that the corresponding 3-SAT instance could not be satisfied.   □

Note that, our reduction is based on a C-LCS instance in which all the constraints are of length one.

4   Single Constraint

In this section we consider the case of an arbitrary number of input strings and a single constraint. It is easy to see that the problem of finding the constrained longest common subsequence is NP-hard. Therefore, we present an approximation algorithm for this case. Our algorithm yields a $1/\sqrt{m_{min}|\Sigma|}$ approximation factor within a linear running time (while m_{min} is the length of the shortest input string). Let $A_1, A_2, \ldots, A_k$ be the input strings. Throughout this section we assume a single constraint string exists, denote it by $B = \langle b_1, b_2, \ldots, b_n \rangle$.

The following result follows from the NP-hardness of the LCS [8] and by setting $B = \epsilon$.

Observation 1. *Given an arbitrary number of input strings and a single constraint, the problem of finding the C-LCS of such instances is NP-hard.*

4.1   Approximation Algorithm

Now we present a linear time approximation algorithm. First we give some useful notations that will be used throughout this subsection.

Let $A_i = \langle A_{i_1}, A_{i_2}, \ldots, A_{i_{m_i}} \rangle$ be an input string of length m_i. Denote with $A_i[s, e]$ the substring of A_i that starts at location s and ends at location e. Denote by $start(A_i, j)$ the leftmost location in A_i such that $b_1, b_2, \ldots, b_j$ is a subsequence of $A_i[1, start(A_i, j)]$. Symmetrically, denote by $end(A_i, j)$ the rightmost location in A_i such that $b_j, b_{j+1}, \ldots, b_n$ is a subsequence of $A_i[end(A_i, j), m_i]$. See Figure 3 as an example of $start(A_i, j)$ and $end(A_i, j)$. For the simplicity of the analysis assume that $start(A_i, 0) + 1 = A_{i_1}$ and $end(A_i, n + 1) - 1 = A_{i_{m_i}}$.

Let OPT_{clcs} be an optimal C-LCS solution. By definition, B must be a subsequence of OPT_{clcs} and a subsequence of every input string A_i $(1 \leq i \leq k)$.

Choose an arbitrary embedding of B over OPT_{clcs} (as a subsequence) and denote with $p_1, p_2, \ldots, p_n$ the positions of $b_1, b_2, \ldots, b_n$ in OPT_{clcs}. For simplicity assume $p_0 + 1$ and $p_{n+1} - 1$ are the positions of the first and the last characters of OPT_{clcs} respectively. Note that there may be many possible embeddings of B over OPT_{clcs}.

The following lemma and corollaries are instrumental in achieving the desirable approximation ratio.

Lemma 2. *Let $B = \langle b_1, b_2, \ldots, b_n \rangle$ be the constraint string and OPT_{clcs} be an optimal C-LCS, then for any assignment of B over OPT_{clcs} and for every $0 \leq i \leq n$ the following statement holds:*

$$|LCS(A_1[start(A_1, i)+1, end(A_1, i+1)-1], A_2[start(A_2, i)+1, end(A_2, i+1)-1],$$
$$\ldots, A_m[start(A_m, i) + 1, end(A_m, i + 1) - 1])| \geq |OPT_{clcs}[p_i + 1, p_{i+1} - 1]|.$$

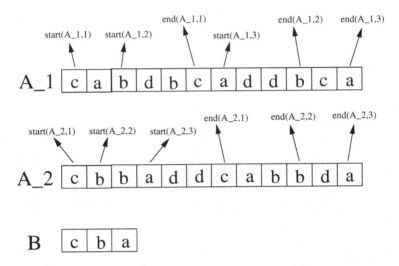

Fig. 3. An example of $start(A_i, j)$ and $end(A_i, j)$

Proof: Let us assume that there is an assignment of B over OPT_{clcs} such that:
$|LCS(A_1[start(A_1, i)+1, end(A_1, i+1)-1], A_2[start(A_2, i)+1, end(A_2, i+1)-1],$
$\ldots, A_m[start(A_m, i)+1, end(A_m, i+1)-1])| < |OPT_{clcs}[p_i + 1, p_{i+1} - 1]|.$

Note that, $OPT_{clcs}[p_i + 1, p_{i+1} - 1]$ must be a common subsequence of substrings of A_1, A_2, ..., A_m. For every $j \leq m$, those substrings must start at a location $\geq start(A_j, i) + 1$ and end at a location $\leq end(A_j, i + 1) - 1$. This contradicts the fact that the LCS of the substrings cannot be longer than the LCS of the original complete strings. □

The next two corollaries follows from Lemma 2.

Corollary 1. *Let* $B = \langle b_1, b_2, \ldots, b_n \rangle$ *be the constraint string and* OPT_{clcs} *be an optimal C-LCS. If we can find the LCS of* $A_1, A_2, \ldots, A_m$, *then we can approximate the C-LCS with a* $\frac{1}{n+1}$-*approximation ratio.*

Proof: Choosing the maximal LCS of $A_1[start(A_1, i) + 1, end(A_1, i + 1) - 1]$, $A_1[start(A_1, i) + 1, end(A_1, i + 1) - 1]$, ..., $A_m[start(A_m, i) + 1, end(A_m, i + 1) - 1]$ (over $0 \leq i \leq n$). W.L.O.G. let LCS_j be the maximal LCS and let j be the corresponding index. By Lemma 2 we get that $\langle b_1, b_2, \ldots, b_j \rangle \cdot LCS_j \cdot \langle b_{j+1}, b_{j+2}, \ldots, b_n \rangle \geq \frac{|OPT_{clcs}|}{(n+1)}$, where '·' denotes string concatenation. □

Corollary 2. *Let* $B = \langle b_1, b_2, \ldots, b_n \rangle$ *be the constraint string and* OPT_{clcs} *be an optimal C-LCS. If we can find an approximate LCS of* $A_1, A_2, \ldots, A_m$, *within an approximation ratio* $\frac{1}{r}$, *then we can approximate the C-LCS with a* $\frac{1}{r(n+1)}$-*approximation ratio.*

Proof: Using similar arguments to Corollary 1 and according to Lemma 2. □

Now, we give a short description of our algorithm (see Algorithm 1 for details). The structure of our algorithm is derived from Corollary 2. For every $i \leq n$, we simply compute an approximated LCS between $A_1[start(A_1, i) + 1, end(A_1, i + 1) - 1]$, $A_1[start(A_1, i)+1, end(A_1, i+1)-1], \ldots, A_m[start(A_m, i)+1, end(A_m, i+1) - 1]$. We find the approximate LCS as follows:

For every $\sigma \in \Sigma$ and for every input string, denote with $C_{A_i}(\sigma, e, f)$ the number of $\sigma's$ in $A_i[e, f]$. For every $i \leq n$, let $C[\sigma, e_i, f_i] = \min(C_{A_i}(\sigma, e_i, f_i))$ and let $C^*(e_i, f_i) = \max C[\sigma, e_i, f_i]$ over all $\sigma \in \Sigma$.

With the use of $C[\sigma, e_i, f_i]$ and some additional arrays, the following lemma can be straightforwardly be seen to be true.

Lemma 3. $C^*(e_i + 1, f_i)$ and $C^*(e_i, f_i + 1)$ can be computed from $C^*(e_i, f_i)$ in $O(k)$ time, given $O(\sum_{i=1}^{k} m_i)$ space.

Our algorithm, perform one scan of A_i $(1 \leq i \leq k)$, from left to right. We can use two pointers for every string in order to scan it appropriately.

Algorithm 1. Linear Time Approximation Algorithm

---
1 $Occ \leftarrow 0$;
2 $bLoc \leftarrow 0$;
3 **for** $j \leftarrow 0$ **to** n **do**
    /* $1 \leq i \leq k$ */
4   **if** $|C^*[start(A_i, j) + 1, end(A_i, j + 1) - 1]| > Occ$ **then**
5          $Symbol \leftarrow$ The corresponding symbol of the above C^* ;
6          $Occ \leftarrow |C^*[start(A_i, j) + 1, end(A_i, j + 1) - 1]|$;
7          $bLoc \leftarrow j$;
8 **return** $B[1, bLoc] \cdot \langle Symbol^{Occ} \rangle \cdot B[bLoc + 1, n]$;

Time and Correctness Analysis:
Let C_{out} be the output string of the Algorithm 1, note that:

1) C_{out} is common subsequence of $A_1, A_2, \ldots, A_m$.
2) C_{out} contains B as a subsequence.

Thus, C_{out} is a feasible solution.

The running time is linear. The computation of $C^*[start(A_i, j)+1, end(A_i, j+1)-1]$ is a process of $2(\Sigma_{i=1}^{k}|m_i|)$ updates operations (we insert and delete every character of the input strings exactly once). Moreover, according to Lemma 3, we can perform k update operations in $O(k)$ time. Thus, the total running time remains linear.

Lemma 4. Algorithm 1 yields an approximation ratio of $\frac{1}{\sqrt{m_{min}|\Sigma|}}$.

Proof: We divide the proof into three cases. If $n \leq \sqrt{\frac{m_{min}}{|\Sigma|}} - 1$, then according to Lemma 2 and since the approximate LCS provide a $1/\Sigma$ approximation ratio, the

length of the C-LCS returned by Algorithm 1 is at least $|OPT_{clcs}|/\sqrt{m_{min}|\Sigma|}$.
Therefore, it is sufficient to prove that Algorithm 1 also yields an approximation
ratio of $\frac{1}{\sqrt{m_{min}|\Sigma|}}$ in case that $n > \sqrt{\frac{m_{min}}{|\Sigma|}} - 1$.

Note that, if $n \geq \sqrt{\frac{m_{min}}{|\Sigma|}}$ any valid solution for the C-LCS must also provide
an approximation ratio of $\frac{1}{\sqrt{m_{min}|\Sigma|}}$. Moreover, if $OPT_{clcs} > n$, we can see that
Algorithm 1 returns at least one extra character over B. Thus, in case that
$\sqrt{\frac{m_{min}}{|\Sigma|}} - 1 \leq n < \sqrt{\frac{m_{min}}{|\Sigma|}}$, our algorithm also yields an approximation ratio of
$\frac{1}{\sqrt{m_{min}|\Sigma|}}$. □

5   Open Questions

A natural open question is whether there are better approximation algorithms for
the single constraint C-LCS problem, which improves the above approximation
factor ? Another interesting question is regarding the existence of a lower bound
for this C-LCS variant.

References

1. Aho, A.V., Hirschberg, D.S., Ullman, J.D.: Bounds on the Complexity of the Longest Common Subsequence Problem. Journal of the ACM 23(1), 1–12 (1976)
2. Arslan, A.N., Egecioglu, Ö.: Algorithms For The Constrained Longest Common Subsequence Problems. International Journal of Foundations of Computer Science 16(6), 1099–1109 (2005)
3. Bergroth, L., Hakonen, H., Raita, T.: A Survey of Longest Common Subsequence Algorithms. In: Proc. SPIRE 2000, pp. 39–48 (2000)
4. Chin, F.Y.L., De Santis, A., Ferrara, A.L., Ho, N.L., Kim, S.K.: A simple algorithm for the constrained sequence problems. Information Processing Letters 90(4), 175–179 (2004)
5. Gotthilf, Z., Lewenstein, M.: Approximating Constrained LCS. In: Ziviani, N., Baeza-Yates, R. (eds.) SPIRE 2007. LNCS, vol. 4726, pp. 164–172. Springer, Heidelberg (2007)
6. Hirschberg, D.S.: A Linear Space Algorithm for Computing Maximal Common Subsequences. Communications of the ACM 18(6), 341–343 (1975)
7. Hirschberg, D.S.: Algorithms for the Longest Common Subsequence Problem. Journal of the ACM 24(4), 664–675 (1977)
8. Maier, D.: The Complexity of Some Problems on Subsequences and Supersequences. Journal of the ACM 25(2), 322–336 (1978)
9. Masek, W.J., Paterson, M.: A Faster Algorithm Computing String Edit Distances. Journal of Computer and System Sciences 20(1), 18–31 (1980)
10. Tsai, Y.-T.: The constrained longest common subsequence problem. Information Processing Letters 88(4), 173–176 (2003)

Finding Additive Biclusters with Random Background

(Extended Abstract)

Jing Xiao[1], Lusheng Wang[2], Xiaowen Liu[3], and Tao Jiang[4]

[1] Department of Computer Science and Technology, Tsinghua University
xiaojing00@mails.tsinghua.edu.cn
[2] Department of Computer Science, City University of Hong Kong, Hong Kong
lwang@cs.cityu.edu.hk
[3] Department of Computer Science, University of Western Ontario, London, Ontario, Canada
N6A 5B7
liuxiaowencs@gmail.com
[4] Department of Computer Science and Engineering, University of California, Riverside
jiang@cs.ucr.edu

Abstract. The biclustering problem has been extensively studied in many areas including e-commerce, data mining, machine learning, pattern recognition, statistics, and more recently in computational biology. Given an $n \times m$ matrix A ($n \geq m$), the main goal of biclustering is to identify a subset of rows (called objects) and a subset of columns (called properties) such that some objective function that specifies the quality of the found bicluster (formed by the subsets of rows and of columns of A) is optimized. The problem has been proved or conjectured to be NP-hard under various mathematical models. In this paper, we study a probabilistic model of the implanted additive bicluster problem, where each element in the $n \times m$ background matrix is a random number from $[0, L-1]$, and a $k \times k$ implanted additive bicluster is obtained from an error-free additive bicluster by randomly changing each element to a number in $[0, L-1]$ with probability θ. We propose an $O(n^2 m)$ time voting algorithm to solve the problem. We show that for any constant δ such that $(1-\delta)(1-\theta)^2 - \frac{1}{L} > 0$, when $k \geq \max\left\{\frac{8}{\alpha}\sqrt{n\log n}, \frac{8\log n}{c} + \log(2L)\right\}$, where c is a constant number, the voting algorithm can correctly find the implanted bicluster with probability at least $1 - \frac{9}{n^2}$. We also implement our algorithm as a software tool for finding novel biclusters in microarray gene expression data, called VOTE. The implementation incorporates several nontrivial ideas for estimating the size of an implanted bicluster, adjusting the threshold in voting, dealing with small biclusters, and dealing with multiple (and overlapping) implanted biclusters. Our experimental results on both simulated and real datasets show that VOTE can find biclusters with a high accuracy and speed.

Keywords: bicluster, Chernoff bound, polynomial-time algorithm, probability model, computational biology, gene expression data analysis.

1 Introduction

Biclustering has proved extremely useful for exploratory data analysis. It has important applications in many fields, *e.g.*, e-commerce, data mining, machine learning, pattern

P. Ferragina and G. Landau (Eds.): CPM 2008, LNCS 5029, pp. 263–276, 2008.
© Springer-Verlag Berlin Heidelberg 2008

recognition, statistics, and computational biology [24]. Data arising from text analysis, market-basket data analysis, web logs, microarray experiments *etc.* are usually arranged in a co-occurrence table or a matrix, such as word-document table, product-user table, cpu-job table, or webpage-user table. Discovering a large bicluster in a product-user matrix indicates, for example, which users share the same preferences. Biclustering has therefore applications in recommender systems and collaborative filtering, identifying web communities, load balancing, discovering association rules, *etc.*

Recently, biclustering becomes an important approach to microarray gene expression data analysis [5]. The underlying bases for using biclustering in the analysis of gene expression data are (i) similar genes may exhibit similar behaviors only under a subset of conditions, not all conditions, and (ii) genes may participate in more than one function, resulting in a regulation pattern in one context and a different pattern in another. Using biclustering algorithms, one may obtain subsets of genes that are co-regulated under certain subsets of conditions.

Given an $n \times m$ matrix A, the main goal of biclustering is to identify a subset of rows (called *objects*) and a subset of columns (called *properties*) such that a pre-determined objective function which specifies the quality of the bicluster (consisting of the found subsets of rows and columns) is optimized.

Biclustering is also known under several different names, *e.g.*, "co-clustering", "two-way clustering", and "direct clustering". The problem was first introduced by Hartigan in the 70's [8]. Since then, it has been extensively studied in many areas. Several objective functions have also been proposed for measuring the quality of a bicluster. Almost all of them have been proved or conjectured to be NP-hard [16,19].

Let $A(I, J)$ be an $n \times m(n \geq m)$ matrix, where $I = \{1, 2, \ldots, n\}$ is the set of rows and $J = \{1, 2, \ldots, m\}$ is the set of columns. Each element $a_{i,j}$ of $A(I, J)$ is an integer in $[0, L - 1]$ indicating the weight of the relationship between object i and property j. For subset $I' \subseteq I$ and subset $J' \subseteq J$, $A(I', J')$ denotes the bicluster of $A(I, J)$ that contains only the elements $a_{i,j}$ satisfying $i \in I'$ and $j \in J'$. When a bicluster contains only a single row i and a column set J', we simply use $A(i, J')$ to represent it. Similarly, we use $A(I', j)$ to represent the bicluster with a row set I' and a single column j. There are several ways to model the relationship between objectives (or genes) [24].

Constant model: A bicluster $A(I', J')$ is an *error-free constant* bicluster if for each column $j \in J'$, for all $i \in I'$, $a_{i,j} = c_j$, where c_j is a constant for any column j.

Additive model: A bicluster $A(I', J')$ is an *error-free additive* bicluster if for any pair of rows i_1 and i_2 in $A(I', J')$, $a_{i_1,j} - a_{i_2,j} = c_{i_1,i_2}$, where c_{i_1,i_2} is a constant for any pair of rows i_1 and i_2.

The additive model is a general model of biclusters that covers several other popular models as its special cases. See [17] for a detailed discussion on various models of biclusters. This model has many applications and has been extensively studied [2,11,13,15,16,17,19,20,21,24]. In this paper, we will focus on the additive model. In particular, we study a probabilistic model of implanted additive biclusters that has recently been used in the literature for evaluating biclustering algorithms [15,20].

The probabilistic additive model: Our probabilistic model for generating the implanted bicluster and background matrix is as follows. Let $A(I, J)$ be an $n \times m$ matrix, where each element $a_{i,j}$ is a random number in $[0, L-1]$ generated independently. Let B be an error-free $k \times k$ additive bicluster. The additive bicluster B' with noise is generated from B by changing each element $b_{i,j}$, with probability θ, into a random number in $[0, L-1]$. We then implant B' into the background matrix $A(I, J)$ and randomly shuffle its rows and columns to obtain a new matrix $A'(I, J)$. For convenience, we will still denote the elements of $A'(I, J)$ as $a_{i,j}$'s.

From now on, we will consider matrix $A'(I, J)$ as the input matrix. Let $I_B \subseteq I$ and $J_B \subseteq J$ be the row and column sets of the implanced bicluster in A'. The implanted bicluster is denoted as $A'(I_B, J_B)$.

The implanted additive bicluster problem: Given the $n \times m$ matrix $A'(I, J)$ with an implanted additive bicluster as described above, find the implanted additive bicluster B'.

Based on the above probabilistic model, we propose an $O(n^2 m)$ time voting algorithm for finding the implanted bicluster. We show that for any constant δ such that $(1 - \delta)(1 - \theta)^2 - \frac{1}{L} > 0$, when $n < m^3$ and $k \geq \max\left\{\frac{8}{\alpha}\sqrt{m \log m}, \frac{8 \log m}{c} + \log(2L)\right\}$, where $c = \min\{\frac{(1-\theta)\delta^2 k}{2L}, \frac{(1-2\theta)^2}{8L}, \frac{(L-2)^2}{12L^3}\}$, the voting algorithm can correctly find the implanted bicluster with probability at least $1 - 9m^{-2}$. We also implement our algorithm into a software tool, called VOTE. In order to make tool applicable in a real setting, the implementation has to incorporate several nontrivial ideas for estimating the size of an implanted bicluster, adjusting the threshold in voting, dealing with small biclusters, and dealing with multiple and overlapping biclusters. Our extensive experiments on both simulated and real datasets show that VOTE can find implanted additive biclusters with high accuracy and efficiency. More specifically, VOTE has a comparable performance/accuracy as the best programs compared in [20,15], but much faster speed.

We note in passing that a closely related problem of finding an implanted clique/distribution in a random graph has been studied in the graph theory community [1,6,12]. In [12], Kucera claimed that when the size of the implanted clique is at least $\Omega(\sqrt{m \log m})$, where m is the number of vertices in the input random graph, a simple approach based counting the degrees of vertices can find the clique with a high probability. Alon *et al.* gave an improved algorithm that can find implanted cliques of sizes at least $\Omega(\sqrt{m})$ with a high probability [1]. Feige and Krauthgamer gave an algorithm that can find implanted cliques of similar sizes in semi-random graphs [6]. It is easy to see that this problem of finding implanted cliques is a special case of our implanted bicluster problem, where the input matrix is binary and all the elements in the bicluster matrix are 1's. We observe that while it may be easy to modify Kucera's simple degree-based method to work for implanted constant biclusters under our probabilistic model, it is not obvious that the above results would directly imply our results on implanted additive biclusters.

In the rest of the paper, we first present the voting algorithm and analyze its theoretical performance on the above probabilistic model. We then describe the implementation of VOTE, and the experimental results. Due to the page limit, the proofs will be omitted in this extended abstract but will be provided in the full paper.

2   The Three Phase Voting Algorithm

We start the construction of the algorithm with some interesting observations. Recall that B is an error-free $k \times k$ additive bicluster and A' is the random input matrix with a noisy additive bicluster B' implanted.

Observation 1. *Consider the k rows in B. There are at least $\frac{k}{L}$ rows that are identifical. That is, there exists a row set $I_C \subseteq I_B$ with $|I_C| \geq \frac{k}{L}$ such that $A'(I_C, J_B)$ is a constant bicluster with noise.*

Consider a row $i_1 \in I_B$ and a column $j_1 \in J_B$. For each row $i_2 \in I_B$, $c_{i_1,i_2} = a_{i_1,j_1} - a_{i_2,j_1}$ is an integer in $[a_{i_1,j_1} - L + 1, a_{i_1,j_1}]$. Based on the value c_{i_1,i_2}, we can partition I_B into L different row sets $I_B^d = \{i_2 | i_2 \in I_B \ \& \ c_{i_1,i_2} = d\}$, $d = a_{i_1,j_1} - L + 1, \ldots, a_{i_1,j_1}$. Let I_C be one of the row sets with the maximum cardinality, $|I_C| = \max_d |I_B^d|$. Then, $A(I_C, J_B)$ is a constant bicluster and $|I_C| \geq \frac{k}{L}$. Let $|I_C| = l$.

Our algorithm has three phases. In the first phase of the algorithm, we want to find the row set I_C in $A'(I, J)$. In order to vote, we first convert the matrix $A'(I, J)$ into a *distance* matrix $D(I, J)$ containing the same sets of rows and columns, and then focus on $D(I, J)$.

Distance matrix: Given an $n \times m$ matrix $A'(I, J)$, we can convert it into a distance matrix based on a row in the matrix. Let $i^* \in I$ be any row in the matrix A. We refer to row i^* as the *reference* row. Define $d_{i,j} = a_{i,j} - a_{i^*,j}$. In the transformation, we subtract the reference row i^* from every row in $A'(I, J)$. We use $D(I, J)$ to denote the $n \times m$ distance matrix containing the set of rows I and the set of columns J with every element $d_{i,j}$. For a row $i \in I$ and a column set $J' \subseteq J$, the number of occurrences of u, $u \in [-L+1, L-1]$, in $D(i, J')$ is the number of elements with value u in $D(i, J')$, denoted by $f(i, J', u) = |\{d_{i,j} | d_{i,j} = u \ \& \ j \in J'\}|$. The number of occurrences of the element that appears the most in $D(i, J')$ is $f^*(i, J') = \max_u f(i, J', u)$. Similarly, for a row set $I' \subseteq I$ and a column $j \in J$, the number of occurrences of u in $D(I', j)$ is the number of elements with value u in $D(I', j)$, denoted by $f(I', j, u)$. The number of occurrences of the element that appears the most in $D(I', j)$ is $f^*(I', j) = \max_u f(I', j, u)$.

Observation 2. *Suppose that we use a row $i^* \in I_C$ as the reference row. For each row i_1 in I_C, the expectation of the number of 0's in row i_1 of $D(I, J)$ is at least $\frac{m-k}{L} + (1 - \theta)^2 k$. For each row i_2 in $I_B - I_C$, the expectation of the number of 0's in row i_2 of $D(I, J)$ is at most $\frac{m-k}{L} + \frac{2\theta k}{L}$. For each row i_3 in $I - I_B$, the expectation of the number of 0's in row i_3 of $D(I, J)$ is at most $\frac{m-k}{L} + \frac{k}{L}$.*

Based on the observation, if the reference row i^* is in I_C, we can find the rows with the most 0's in the distance matrix to obtain a row set I_0 by using the following voting method.

The first phase voting

1. **for** $i = 1$ to n **do**
2.   compute $f(i, J, 0)$.
3. select rows i such that $f(i, J, 0) > \frac{m}{L} + 4\sqrt{m \log m}$ to form I_0.

When m and k are sufficiently large and θ is sufficiently small, we can prove that, with a high probability, the row set I_0 is equal to I_C. The proof will be given in the next section.

In the second phase voting of the algorithm, we attempt to find locate the column set J_B of the implanted bicluster. It is based on the following observation.

Observation 3. *For a column j_1 in J_B, the expectation of the number of occurrences of the element that appears the most in $D(I_C, j_1)$ is $(1 - \theta)|I_C|$. For a column j_2 in $J - J_B$, the expectation of the number of occurrences of an element u in $D(I_C, j_1)$ is $\frac{1}{L}|I_C|$.*

With a high probability (and again assuming that θ is sufficiently small), the number of occurrences of the element that appears the most in the columns of J_B is greater than the number of occurrences of the element that appears the most in the columns of $J - J_B$. That is, for two columns $j_1 \in J_B$ and $j_2 \notin J_B$, with a high probability, $f^*(I_0, j_1) > \frac{|I_0|}{2} > f^*(I_0, j_2)$. Based on the property, we can use voting to find a column set J_1.

The second phase voting
1. **for** $j = 1$ to m **do**
2.   compute $f^*(I_0, j)$.
3.   select columns j such that $f^*(I_0, j) > \frac{|I_0|}{2}$ to form J_1.

We can prove (in the next section) that, with a high probability, J_1 is equal to the implanted column set J_B.

Similarly, the third phase voting of the algorithm is designed to locate the row set I_B of the implanted bicluster. But, before the voting, we need correct corrupted columns of the distance matrix $D(I, J)$ caused by the elements of the reference row i^* that were changed during the generation of B'. Recall that $f^*(I_0, j) = \max_u f(I_0, j, u)$. Let $f(I_0, j, u_j) = f^*(I_0, j)$. For every $j \in J_1$, if $u_j \neq 0$, then the element $a_{i*,j}$ was changed when B' was generated (assuming $J_1 = J_B$), and we can thus correct each element $d_{i,j}$ in the jth column of the matrix $D(I, J)$ by subtracting u_j from it.

In the following, let us assume that the entries in the submatrix $D(I, J_B)$ have been adjusted according to the correct reference row i^* as described above. The following observation holds.

Observation 4. *For a row i_1 in I_B, the expectation of the number of occurrences of the element that appears the most in $D(i_1, J_B)$ is at least $(1 - \theta)k$. For a row i_2 in $I - I_B$, the expectation of the number of occurrences of the element that appears the most in $D(i_2, j_B)$ is $\frac{k}{L}$.*

We can thus find a row set I_1 in $A'(I, J_1)$ as follows.

The third phase voting
1. **for** $i = 1$ to n **do**
2.   compute $f^*(i, J_1)$.
3.   select rows i such that $f^*(i, J_1) > \frac{|J_1|}{2}$ to form I_1.

We can prove (in the next section) that, if $|I_1| \geq k$, with a high probability, I_1 is equal to the implanted column set I_B. Therefore, a voting algorithm based on the above

The Three Phase Voting Algorithm
Input: An $n \times m$ matrix $A'(I, J)$, an integer k, noise level θ, and L.
Output: A bicluster $A'(I_1, J_1)$.
1. **for** each row $i^* \in I$, **do**
2.   construct the $n \times m$ distance matrix $D(I, J)$ from $A'(I, J)$ with reference row i^*.
3.   find a row set I_0 by using the first phase voting.
4.   **if** $\|I_0\| \geq \frac{k}{L}$, **then**
5.   find a column set J_1 by using the second phase voting.
6.   correct the corrupted columns in submatrix $D(I, J_1)$.
7.   find a row set I_1 by using the third phase voting.
8.   **if** $\|I_1\| \geq k$ and $\|J_1\| \geq k$, output $A'(I_1, J_1)$ and return.

Fig. 1. The three phase voting algorithm

procedures, as given in Figure 1, can be used to find the implanted bicluster with a high probability. Since the time complexity of the steps 2 - 7 of the algorithm is $O(nm)$ and these steps are repeated n times, the time complexity of the algorithm is $O(n^2m)$.

3 Analysis of the Algorithm

In this section, we will prove that, with a high probability, the above voting algorithm correctly outputs the implanted bicluster.

Recall that in the submatrix $A'(I_B, J_B)$, each element was changed with probability θ to generate B' from B. We will show that, with a high probability, there exists a row $i \in I_C$ such that row i has at least $(1 - \delta)(1 - \theta)k$ unchanged elements in $A'(i, J_B)$ for any $0 < \delta < 1$.

In the analysis, we need the following two lemmas from [18,14].

Lemma 1. [18] *Let $X_1, X_2, \ldots, X_n$ be n independent random binary (0 or 1) variables, where X_i takes on the value of 1 with probability p_i, $0 < p_i < 1$. Let $X = \sum_{i=1}^n X_i$ and $\mu = E[X]$. Then for any $0 < \delta < 1$,*

(1) $\mathbf{Pr}(X > (1 + \delta)\mu) < \left[\frac{e^\delta}{(1+\delta)^{(1+\delta)}} \right]^\mu$,

(2) $\mathbf{Pr}(X < (1 - \delta)\mu) \leq e^{-\frac{1}{2}\mu\delta^2}$.

Lemma 2. [14] *Let $X_i, 1 \leq i \leq n$, X and μ be defined as in Lemma 1. Then for any $0 < \epsilon < 1$,*

(1) $\mathbf{Pr}(X > \mu + \epsilon n) \leq e^{-\frac{1}{3}n\epsilon^2}$,

(2) $\mathbf{Pr}(X < \mu - \epsilon n) \leq e^{-\frac{1}{2}n\epsilon^2}$.

These two lemmas will be used to establish the next lemma.

Lemma 3. *For any $0 < \delta < 1$, with probability at least $1 - e^{-\frac{1}{2L}(1-\theta)k^2\delta^2}$, there exists a row $i \in I_C$ that has at least $(1 - \delta)(1 - \theta)k$ unchanged elements in $A'(i, J_B)$.*

Suppose that there is a row $i^* \in I_C$ with $(1 - \delta)(1 - \theta)k$ unchanged elements in $A'(i, J_B)$. Now, let us consider the distance matrix $D(I, J)$ with the reference row i^*. We now show that, with a high probability, the rows in I_C have more 0's than those in $I - I_C$ in matrix $D(I, J)$. That is, with a high probability, our algorithm will find the row set I_C in the first phase voting.

Lemma 4. *Let $i^* \in I_C$ be the reference row with $(1 - \delta)(1 - \theta)k$ unchanged elements in $A'(i^*, J_B)$, and $D(I, J)$ the distance matrix as described above. When $\alpha = (1-\delta)(1-\theta)^2 - \frac{1}{L} > 0$ and $k \geq \frac{8}{\alpha} \sqrt{m \log m}$, with probability at least $1 - m^{-7} - nm^{-5}$, $f(i, J, 0) > \frac{m}{L} + \frac{\alpha}{2}k$ for all $i \in I_C$, and $f(i, J, 0) < \frac{m}{L} + \frac{\alpha}{2}k$ for all $i \in I - I_C$.*

The above lemma shows that, when a row i^* with $(1 - \delta)(1 - \theta)k$ unchanged elements in $A'(i, J_B)$ is selected as the reference row, and m and k are large enough, $I_0 = I_C$ with a high probability. Next, we prove that, with a high probability, our algorithm will find the implanted column set J_B.

Lemma 5. *Suppose that the row set I_0 found in the first phase voting of Algorithm 1 is indeed equal to I_C. With probability at least $1 - ke^{-\frac{(1-2\theta)^2}{8L}k} - L(m-k)e^{-\frac{(L-2)^2}{12L^3}k}$, the column set J_1 found in the second phase voting of Algorithm 1 is equal to J_B.*

Similarly, we can prove that, with a high probability, our algorithm will find the implanted row set I_B.

Lemma 6. *Suppose that the column set J_1 found in the second phase voting of Algorithm 1 is indeed equal to J_B. With probability at least $1 - ke^{-\frac{(1-2\theta)^2}{8}k} - 2L(n - k)e^{-\frac{(L-2)^2}{12L^2}k}$, the row set I_1 found in the third phase voting of Algorithm 1 is equal to I_B.*

Finally, we can prove that, with a high probability, no columns or rows other than those in the implanted bicluster will be output by the voting algorithm.

Lemma 7. *With probability at least $1 - Ln(m-k)e^{-\frac{(L-2)^2}{12L^3}k} - 2Ln(n-k)e^{-\frac{(L-2)^2}{12L^2}k}$, no columns or rows of $A'(I, J)$ other than those in $A'(I_B, J_B)$ will be output by the Algorithm 1.*

Based on Lemmas 3, 4, 5, 6 and 7, we can show that, when m and k are large enough, the three phase voting algorithm can find the implanted bicluster with a high probability. Let c be a constant such that $c < \min\{\frac{(1-\theta)\delta^2 k}{2L}, \frac{(1-2\theta)^2}{8L}, \frac{(L-2)^2}{12L^3}\}$. In most applications, we may assume that $n < m^3$. Then, we have the following theorem.

Theorem 1. *When $n < m^3$, $\alpha = (1-\delta)(1-\theta)^2 - \frac{1}{L} > 0$ and $k \geq \max\{\frac{8}{\alpha}\sqrt{m \log m}, \frac{8 \log m}{c} + \log(2L)\}$, the voting algorithm correctly outputs the implanted bicluster with probability at least $1 - 9m^{-2}$.*

If we replace m by n in the above analysis, the same proof shows that

Corollary 1. *When $\alpha = (1 - \delta)(1 - \theta)^2 - \frac{1}{L} > 0$ and $k \geq \max\{\frac{8}{\alpha}\sqrt{n \log n}, \frac{8 \log n}{c} + \log(2L)\}$, the voting algorithm correctly outputs the implanted bicluster with probability at least $1 - 9n^{-2}$.*

In the practice of microarray data analysis, the number of conditions m is much smaller than the number of genes n. Thus, Theorem 1 allows the parameter k to be smaller (*i.e.*

it works for smaller implanted biclusters) than Corollary 1, although it assumes a slightly more complicated condition ($n < m^3$) and has a slightly worse success probability.

4 The Implementation of the Voting Algorithm

The voting algorithm described in Section 2 is originally based on the probabilistic model for generating the implanted additive bicluster. Many assumptions have been used to prove its correctness. To deal with real data, we have to carefully resolve the following issues.

Estimation of the bicluster size. In the voting algorithm, we assume that the size k of the implanted bicluster is part of the input. However, in practice, the size of the implanted bicluster is unknown. Here we develop a method to estimate the size of the bicluster. We first set k to be a large number such that $k \geq |J_B|$. Let q be the maximum number of rows such that $f(i, J, u) > (m-k)Pr(d_{i,j} = u)+k$ among all $u \in [-L+1, L-1]$. Our key observation here is that if k is greater than $|J_B|$, then q will be smaller than $|I_B|$. If k is smaller than $|J_B|$, then q will be greater than $|I_B|$. Thus, we can gradually decrease the value of k while observing that the value of q increases accordingly. The process stops when $q \geq 2k$.

To set the initial value of k such that $k \geq |J_B|$, we set $k = 3 \cdot max_u(Pr(d_{i,j} = u)) \cdot m$. This worked very well in our experiments.

Dealing with retangular biclusters. Many interesting biclusters in the practice of microarray gene expression data are non-square. To deal with such rectangular biclusters, where $|I_B| \neq |J_B|$, we first try to obtain a square bicluster in the first phase voting (assuming $|I_B| \geq |J_B|$) and then use the k rows in I_0 for the second phase voting. The third phase voting may in fact generate a rectangular bicluster with unequal numbers of rows and columns.

Adjusting the threshold used in the first phase voting for a real input matrix. In Step 3 of the first phase voting, we use the threshold $f(i, J, 0) > \frac{m}{L} + 4\sqrt{m \log m}$ to select rows to form I_0. This is based on the assumption that in the random background matrix, $d_{i,j} = 0$ with probability $\frac{1}{L}$. In order for the algorithm to work for any input data, we consider the distribution of numbers in the whole input matrix. We calculate the probability $Pr(d_{i,j} = l)$ for each $l \in [-L + 1, L + 1]$ in the input matrix. In Step 3 of the first phase voting, we choose all the rows such that $f(i, J, u) > (m-k)Pr(d_{i,j} = u)+k$. In this way, we were able to make our algorithm to work well for real microarray data where the background did not seem to follow some simple uniform/normal distribution.

When $|I_c|$ is too small for voting. Recall that I_c is the set of the rows identical to the reference row I^* in the implanted bicluster. In other words, the set I_c contains all the rows i with $d_{i,j} = 0$ for $j \in J_B$. The expectation of $|I_c|$ is $\frac{k}{L}$. When k is small and L is large, $|I_c|$ (and thus I_0) could be too small for the voting in the second phase to be effective. To enhance the performance of the algorithm, we consider the set I_B^u for each $u \in [-L + 1, L - 1]$ as defined in the beginning of Section 2, and approximate it using a set I_0^u in the algorithm just like how we approximated the set $I_C = I_B^0$ by the set I_0 in the first phase voting. Thus, the second phase voting becomes:

The second phase voting
1. **for** $j = 1$ to m **do**
2.   compute $f(I_0^u, j, u)$ for each $u \in [-L + 1, L - 1]$.
3.   select columns j such that $\sum_{u=-L+1}^{L-1} f(I_0^u, j, u) > (\sum_{u=-L+1}^{L-1} |I_0^u|)/2$ to form J_1.

Dealing with multiple and overlapping biclusters. In microarray gene expression analysis, a real input matrix may contain multiple biclusters of interest, some of which could overlap. We could easily modify the voting algorithm to find multiple implanted biclusters by forcing it to go through all the n rounds (*i.e.* considering each of the n rows as the reference row) and recording all the biclusters found. If the two biclusters found in two different rounds overlap (in terms of the area) by more than 25% of the area of the smaller biclcuster, then we consider them as the same bicluster.

5   Experimental Results

We have implemented the above voting algorithm in C++ and produced a software, named VOTE. In this section, we will compare VOTE with some well-known biclustering algorithms in the literature on both simulated and real microarray datasets. The tests were performed on a desktop PC with P4 3.0G CPU and 512M memory running Windows operating system.

To evaluate the performance of different methods, we use a measure (called *match score*) similar to the score introduced in Prelić *et al.* [20]. Let M_1, M_2 be two sets of biclusters. The match score of M_1 with respect to M_2 is given by

$$S(M_1, M_2) = \frac{1}{|M_1|} \sum_{A(I_1, J_1) \in M_1} \max_{A(I_2, J_2) \in M_2} \frac{|I_1 \cap I_2| + |J_1 \cap J_2|}{|I_1 \cup I_2| + |J_1 \cup J_2|}.$$

Let M_{opt} denote the set of implanted biclusters and M the set of the output biclusters of a biclustering algorithm. $S(M_{opt}, M)$ represents how well each of the true biclusters is discovered by a biclustering algorithm.

5.1   Simulated Datasets

Following the method in [15,20], we consider an $n \times m$ background matrix A. Let $L = 30$. We generate the elements in the background matrix A such that the data fits the standard normal distribution with the mean of 0 and the standard deviation of 1. To generate an additive $b \times c$ bicluster, we first randomly generate the expression values in a reference row $(a_1, a_2, \ldots, a_c)$ according to the standard normal distribution. To obtain a row $(a_{i1}, a_{i2}, \ldots, a_{ic})$ in the additive bicluster, we randomly generate a distance d_i (based on the standard normal distribution) and set $a_{i,j} = a_j + d_i$ for $j = 1, 2, \ldots, c$. After we obtain the $b \times c$ additive bicluster, we add some noise by

Table 1. Parameter settings for different biclustering methods

Method	Type of Bicluster	Parameter Setting
BiMax	Constant	minimum number of genes and chips: 4
ISA	Constant/Additive	$t_g = 2.0, t_c = 2.0, seeds = 500$
CC	Constant	$\delta = 0.5, \alpha = 1.2$
CC	Additive	$\delta = 0.002, \alpha = 1.2$
RMSBE	Constant/Additive	$\alpha = 0.4, \beta = 0.5, \gamma = \gamma_e = 1.2$
OPSM	Additive	$l = 100$

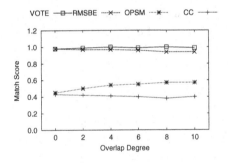

Fig. 2. Performance on small additive biclusters

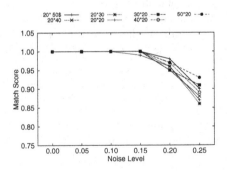

Fig. 3. Performance on biclusters of different sizes

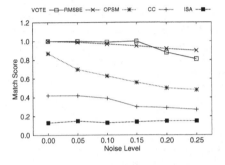

Fig. 4. Performance on overlapping biclusters

Fig. 5. Performance on rectangular biclusters

randomly selecting $\theta \cdot b \cdot c$ elements in the bicluster and changing their values to a random number (according to the standard normal distribution). Finally, we insert the obtained bicluster into the background matrix A and shuffle the rows and columns. We compare our program, VOTE, with several well-known programs for biclustering from the literature including ISA, CC, OPSM, and RMSBE [3,5,9,10,15]. The parameter settings of different methods are listed in Table 1.

Testing the performance on small biclusters. First, we test how well the programs are able to find small implanted additive biclusters. Let $n = m = 100$ and $b = c = 15 \times 15$,

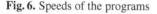

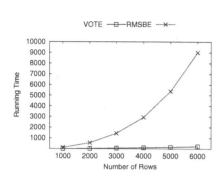

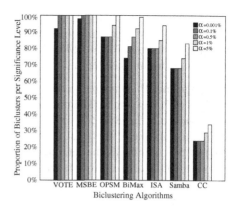

Fig. 6. Speeds of the programs

Fig. 7. Proportion of biclusters significantly enriched by a GO category. Here, α is the adjusted significance score of a bicluster.

and consider implanted biclusters generated with different noise levels θ in the range of $[0, 0.25]$. Figure 2 shows that VOTE and RMSBE outperform CC, OPSM and ISA with on all noise levels.

Testing the performance on biclusters of different sizes. Since RMSBE has the best performance among the existing programs considered here, we compare VOTE with RMSBE on different bicluster sizes. In this test, the noise level is set as $\theta = 0.2$. The sizes of the implanted (square) biclusters vary from 30×30 to 100×100 and the background matrix is of size 500×500. As illustrated in Figure 3, VOTE outperforms RMSBE when the size of the square bicluster is greater than 40, while RMSBE is more powerful in finding small biclusters.

Finding multiple biclusters. To test the ability of finding multiple biclusters, we first generate two $b \times b$ additive biclusters with o overlapped rows and columns. The parameter o is called the *overlap degree*. The background matrix size is fixed as 100×100. Both the background matrix and the biclusters are generated as before. To find multiple biclusters in a given matrix, some methods, *e.g.*, CC, needs to mask the previously discovered biclusters with random values. One of the advantages of the approaches based on a reference row, *e.g.*, VOTE and RMSBE, is that it is unnecessary to mask previously discovered biclusters. We test the performance of VOTE, RMSBE, CC and OPSM on overlapping biclusters by using 20×20 additive biclusters with noise level $\theta = 0.1$ and overlap degree o ranging from 0 to 10. The results are shown in Figure 4. We can see that both VOTE and RMSBE are only marginally affected by the overlap degree of the implanted biclusters. VOTE is slightly better than RMSBE, especially when o increases.

Finding rectangular biclusters. We generate rectangular additive biclusters with different sizes and noise levels. The row and column sizes of the implanted biclusters range from 20 to 50. The noise level θ is from the range $[0, 0.25]$. The background matrix is of size 100×100. The results are shown in Figure 5. We can see that the performance of VOTE is not affected by the shapes of the rectangular biclusters. Since RMSBE can only find near square biclusters, we compare the performance of VOTE with that of an extension of RMSBE. Comparing Figure 5 with the test results given in [15], our algorithm is better in finding rectangular biclusters.

Running time. To compare the speeds of VOTE and RMSBE, we consider background matrices of 200 columns. The number of rows ranges from 1000 to 6000. The size of the implanted bicluster is 50×50. The running time of VOTE and RMSBE is shown in Figure 6. In the test, we let RMSBE randomly select 10% rows as the reference row and 50 columns as the reference column. We can see that VOTE is much faster than RMSBE. Moreover, for the real gene expression data of *S. cerevisiae* provided by Gasch *et al.* [7], our algorithm runs in 66 seconds and RMSBE (randomly selecting 300 genes as the reference row and 40 conditions as the reference column) runs in 1230 seconds.

5.2 Real Dataset

Similar to the method used by Tanay et al. [22] and Prelić *et al* [20], we investigate whether the set of genes discovered by a biclustering method shows significant enrichment with respect to a specific GO annotation provided by the Gene Ontology Consortium [7]. We use the web tool funcAssociate of Berriz *et al.* [4] to evaluate the discovered biclusters. FuncAssociate first uses Fisher's exact test to compute the hypergeometric functional score of a gene set, then it uses the Westfall and Young procedure [23] to compute the adjusted significance score of the gene set. The analysis is performed on the gene expression data of *S. cerevisiae* provided by Gasch *et al.* [7]. The dataset contains 2993 genes and 173 conditions. We set $L = 30$, filter out the biclusters with over 25% overlapped elements, and output the largest 100 biclusters. The running time of VOTE on this dataset is 66 seconds. The adjusted significance scores (adjusted p-values) of the 100 biclusters are computed by using FuncAssociate. Here, we compare the significance scores for RMSBE, OPSM, BiMax [20], ISA, Samba [22], and CC obtained from Figure 7 in Liu *et al.* [15]. The result is summarized in Figure 7. We can see that 92% of discovered biclusters by VOTE are statistically significant, *i.e.* with $\alpha \leq 5\%$. Moreover, the performance of VOTE in this regard is comparable to that of RMSBE and is better than those of the other programs compared in [15].

6 Conclusion

Based on a simple probabilistic model, we have designed a three phase voting algorithm to find implanted additive biclusters. We proved that when the size of the implanted bicluster is $\Omega(\sqrt{m \log m})$, the voting algorithm can correctly find the implanted bicluster with a high probability. We have also implemented the voting algorithm as a software tool, VOTE, for finding novel biclsuters in real microarray gene expression data. Our extensive experiments on simulated datasets demonstrate that VOTE performs very well in terms of both accuracy and speed. Future work includes testing VOTE on more real datasets, which could be a bit challenging since true biclusters for most gene expression datasets are unknown.

Acknowledgments

JX's research is supported in part by the National Natural Science Foundation of China Grant 60553001, and the National Basic Research Program of China Grant 2007CB807900,2007CB807901, LW's research is supported by a grant from City University of Hong Kong [Project No. 7001996], and TJ's research is supported by NSF grant IIS-0711129, NIH grant LM008991-01, National Natural Science Foundation of China grant 60528001, and a Changjiang Visiting Professorship at Tsinghua University.

References

1. Alon, N., Krivelevich, M., Sudakov, B.: Finding a Large Hidden Clique in a Random Graph. Random Structures and Algorithms 13(3-4), 457–466 (1998)
2. Barkow, S., Bleuler, S., Prelić, A., Zimmermann, P., Zitzler, E.: BicAT: a biclustering analysis toolbox. Bioinformatics 22(10), 1282–1283 (2006)
3. Ben-Dor, A., Chor, B., Karp, R., Yakhini, Z.: Discovering local structure in gene expression data: the order-preserving submatrix problem. In: Proceedings of Sixth International Conference on Computational Molecular Biology (RECOMB), pp. 45–55. ACM Press, New York (2002)
4. Berriz, G.F., King, O.D., Bryant, B., Sander, C., Roth, F.P.: Charactering gene sets with FuncAssociate. Bioinformatics 19, 2502–2504 (2003)
5. Cheng, Y., Church, G.M.: Biclustering of expression data. In: Proceedings of the 8th International Conference on Intelligent Systems for Molecular (ISMB 2000), pp. 93–103. AAAI Press, Menlo Park (2000)
6. Feige, U., Krauthgamer, R.: Finding and certifying a large hidden clique in a semirandom graph. Random Structures and Algorithms 16(2), 195–208 (2000)
7. Gasch, A.P., Spellman, P.T., Kao, C.M., Carmel-Harel, O., Eisen, M.B., Storz, G., Botstein, D., Brown, P.O.: Genomic expression programs in the response of yeast cells to enviormental changes. Molecular Biology of the Cell 11, 4241–4257 (2000)
8. Hartigan, J.A.: Direct clustering of a data matrix. J. of the American Statistical Association 67, 123–129 (1972)
9. Ihmels, J., Friedlander, G., Bergmann, S., Sarig, O., Ziv, Y., Barkai, N.: Revealing modular organization in the yeast transcriptional network. Nature Genetics 31, 370–377 (2002)
10. Ihmels, J., Bergmann, S., Barkai, N.: Defining transcription modules using large-scale gene expression data. Bioinformatics 20(13), 1993–2003 (2004)
11. Kluger, Y., Basri, R., Chang, J., Gerstein, M.: Spectral biclustering of microarray data: coclustering genes and conditions. Genome Research 13, 703–716 (2003)
12. Kucera, L.: Expected complexity of graph partitioning problems. Disc. Appl. Math. 57, 193–212 (1995)
13. Li, H., Chen, X., Zhang, K., Jiang, T.: A general framework for biclustering gene expression data. Journal of Bioinformatics and Computational Biology 4(4), 911–933 (2006)
14. Li, M., Ma, B., Wang, L.: On the closest string and substring problems. J. ACM 49(2), 157–171 (2002)
15. Liu, X., Wang, L.: Computing the maximum similarity biclusters of gene expression data. Bioinformatics 23(1), 50–56 (2007)
16. Lonardi, S., Szpankowski, W., Yang, Q.: Finding biclusters by random projections. In: Proceedings of the Fifteenth Annual Symposium on Combinatorial Pattern Matching, pp. 102–116 (2004)

17. Madeira, S.C., Oliveira, A.L.: Biclustering algorithms for biological data analysis: a survey. IEEE/ACM Transactions on Computational Biology and Bioinformatics 1(1), 24–45 (2004)
18. Motwani, R., Raghavan, P.: Randomized algorithms. Cambridge University Press, Cambridge (1995)
19. Peeters, R.: The maximum edge biclique problem is NP-complete. Disc. Appl. Math. 131(3), 651–654 (2003)
20. Prelić, A., Bleuler, S., Zimmermann, P., Wille, A., Bühlmann, P., Gruissem, W., Hennig, L., Thiele, L., Zitzler, E.: A systematic comparison and evaluation of biclustering methods for gene expression data. Bioinformatics 22(9), 1122–1129 (2006)
21. Shamir, R., Maron-Katz, A., Tanay, A., Linhart, C., Steinfeld, I., Sharan, R., Shiloh, Y., Elkon, R.: EXPANDER - an integrative program suite for microarray data analysis. BMC Bioinformatics 6, 232 (2005)
22. Tanay, A., Sharan, R., Shamir, R.: Discovering statistically significant biclusters in gene expression data. Bioinformatics 18, suppl. 1, 136–144 (2002)
23. Westfall, P.H., Young, S.S.: Resampling-based multiple testing. Wiley, New York (1993)
24. Yang, J., Wang, W., Wang, H., Yu, P.: δ-clusters: capturing subspace correlation in a large data set. In: Proceedings of the 18th International Conference on Data Engineering, pp. 517–528 (2002)

An Improved Succinct Representation for Dynamic k-ary Trees

Diego Arroyuelo[*]

Dept. of Computer Science, University of Chile
darroyue@dcc.uchile.cl

Abstract. *k-ary trees* are a fundamental data structure in many text-processing algorithms (e.g., text searching). The traditional pointer-based representation of trees is space consuming, and hence only relatively small trees can be kept in main memory. Nowadays, however, many applications need to store a huge amount of information. In this paper we present a *succinct* representation for dynamic k-ary trees of n nodes, requiring $2n + n \log k + o(n \log k)$ bits of space, which is close to the *information-theoretic lower bound*. Unlike alternative representations where the operations on the tree can be usually computed in $O(\log n)$ time, our data structure is able to take advantage of asymptotically smaller values of k, supporting the basic operations parent and child in $O(\log k + \log \log n)$ time, which is $o(\log n)$ time whenever $\log k = o(\log n)$. Insertions and deletions of leaves in the tree are supported in $O((\log k + \log \log n)(1 + \frac{\log k}{\log (\log k + \log \log n)}))$ amortized time. Our representation also supports more specialized operations (like subtreesize, depth, etc.), and provides a new trade-off when $k = O(1)$ allowing faster updates (in $O(\log \log n)$ amortized time, versus the amortized time of $O((\log \log n)^{1+\epsilon})$, for $\epsilon > 0$, from Raman and Rao [21]), at the cost of slower basic operations (in $O(\log \log n)$ time, versus $O(1)$ time of [21]).

1   Introduction and Previous Works

In this paper we study the problem of the succinct representation of *dynamic k-ary trees*, or *cardinal* trees, or simply *tries*, i.e. trees such that the children of a node are sorted and labeled with a symbol drawn from the *alphabet* $\{1, \ldots, k\}$. We assume that k is fixed, yet we note it as a variable in our analysis: think for example of a trie representing information about a text on an (large) alphabet.

A *succinct data structure* requires space close to the *information-theoretic lower bound* (besides lower-order additive terms). Since the number of different k-ary trees with n nodes is $\frac{1}{kn+1}\binom{kn+1}{n}$, the information-theoretical lower bound for the number of bits to represent a k-ary tree is $C(n,k) = \log\left(\frac{1}{kn+1}\binom{kn+1}{n}\right)$ ($\log x$ means $\lceil \log_2 x \rceil$ in this paper) which, assuming that k is a function of n, is $C(n,k) \approx 2n + n \log k - o(n + \log k)$ bits. In most succinct representations of k-ary trees, the first term stands for the encoding of the tree structure, and the second term for the space required to code the labels of the edges.

---

[*] Supported in part by *Yahoo! Research Latin America* and Fondecyt Grant 1-080019.

P. Ferragina and G. Landau (Eds.): CPM 2008, LNCS 5029, pp. 277–289, 2008.

Besides requiring little space, succinct data structures in general support operations as efficiently as their non-space-efficient counterparts. In the case of trees, we are interested in succinct representations that can be navigated in the usual way, as many compact representations cannot be navigated [15]. We are interested in the following operations: parent(x), which gets the parent of node x; child(x, i), which gets the i-th child of node x; child(x, α), which gets the child of node x labeled by symbol $\alpha \in \{1, \ldots, k\}$; depth($x$), which gets the depth of node x in the tree; degree(x), which gets the number of children of node x; subtreesize(x), which gets the size of the subtree of node x; preorder(x), which gets the preorder number of node x; and isancestor(x, y), which tells us whether node x is an ancestor of node y. In the context of dynamic trees, the representation should support operations insert and delete as well, which respectively allow us to add new nodes and delete existing nodes from the tree.

We consider the *standard word* RAM model of computation, in which every word of size $w = \Theta(\log n)$ bits can be accessed in constant time. Basic arithmetic and logical operations can be computed in constant time. Unless stated, we will assume that all navigations start from the root of the tree, and that insertions and deletions occur at leaves, which is usual in many applications [19,21].

The pointer-based representation of a tree requires $O(n \log n)$ bits for the tree structure, which is space consuming and hence only small trees can be kept in a fast memory. Typical examples where this matters are that of DOM trees for XML documents, and suffix trees [1] for full-text search applications.

Starting with the work of Jacobson [13], a number of succinct representations have been defined for static trees [18,5,20,10,14,8], each providing a different set of operations and complexities. The case of succinct representation of dynamic trees has been studied only for binary trees by Munro et al.[19] and Raman and Rao [21]; the latter work provides a representation requiring $2n + o(n)$ bits, and supports basic navigation operations in worst-case constant time, while updates in the tree can be performed in $O((\log \log n)^{1+\epsilon})$ amortized time, for any constant $\epsilon > 0$. The efficient representation of succinct dynamic k-ary trees was posed as an open problem by Munro et al. [19], as adapting these representations for dynamic k-ary trees by transforming the tree into binary gives poor results: basic navigation operations like parent and child now become $O(k)$ in the worst case, which is not so advantageous if k is not a constant, as in many applications.

Another alternative to represent a dynamic k-ary tree succinctly is to use the dynamic data structure for balanced parentheses of Chan et al. [7] to represent the *Depth-First Unary Degree Sequence* (DFUDS) [5] of the tree. Thus, the basic update and navigation operations are supported basically in $O(\log n)$ time; however, the time complexity of those operations depends on n, the number of nodes in the tree, rather than on k, the alphabet size. Then, this structure cannot take advantage of asymptotically smaller values of k.

In this paper we present an improved succinct representation for dynamic k-ary trees, which is faster than both Raman and Rao [21] and Chan et al. [7]'s solutions. Our data structure requires $2n + n \log k + o(n \log k)$ bits of space, and supports operations parent(x) and child(x, i) in $O(\log k + \log \log n)$ worst-case time, which is $o(\log n)$ whenever $\log k = o(\log n)$ holds. Updates are supported in $O((\log k + \log \log n)(1 + \frac{\log k}{\log (\log k + \log \log n)}))$ amortized time. We are also able to compute operation child(x, α) in $O((\log k + \log \log n)(1 + \frac{\log k}{\log (\log k + \log \log n)}))$ worst-case time. For the particular case where $k = O(1)$ (for example, binary trees), our representation provides a new trade-off, supporting basic operations in $O(\log \log n)$ time (versus $O(1)$ time of [21]) and updates in $O(\log \log n)$ amortized time (versus $O((\log \log n)^{1+\epsilon})$ amortized time of [21], for any constant $\epsilon > 0$).

Our basic approach to solve the problem is similar to existing approaches [19,21]: we divide the tree into small blocks which are easier to update upon tree modifications. However, in the case of k-ary trees we have to face additional problems: structuring inside blocks, defining the adequate block size in order to require $o(n)$ bits for the inter-block pointers, supporting block overflows (which cannot be carried out by using table lookups [19,21], since our blocks are large enough so as to be used to index a table), etc.

Besides the fact that dynamic k-ary trees is a fundamental data structure, our work is also motivated by previous works [3,2] on compressed full-text indexes, in particular Lempel-Ziv compressed indexes. The results of this paper will help us to improve the construction time of Lempel-Ziv indexes [3], and will be a base to define a dynamic compressed full-text index on that of [2].

2 Preliminary Concepts

Data Structures for rank and select Queries. Given a sequence $S[1..n]$ over an alphabet $\{1, \ldots, k\}$ and given any $c \in \{1, \ldots, k\}$, we define operation rank$_c(S, i)$ as the number of cs up to position i of S. Operation select$_c(S, j)$ yields the position of the j-th c in S. In the dynamic case we also want to insert/delete symbols into/from the sequence. The data structure of González and Navarro [11] supports all the operations (including insert and delete) in $O(\log n(1 + \frac{\log k}{\log \log n}))$ worst-case time, requiring $nH_0(S) + o(n \log k)$ bits of space, where $H_0(S)$ denotes the 0-th order empirical entropy of S [17].

Data Structures for Searchable Partial Sums. Throughout this paper we will need data structures for searchable partial sums [12]. Given an array $A[1..n']$ of n' integers, these data structures allow one to retrieve $A[i]$ and support operations Sum(A, i), which computes $\sum_{j=1}^{i} A[j]$; Search(A, i), which finds the smallest j' such that Sum$(A, j') \geqslant i$; Update(A, i, δ), which sets $A[i] \leftarrow A[i] + \delta$; Insert$(A, i, e)$, which adds a new element e to the set between elements $A[i-1]$ and $A[i]$; and Delete(A, j), which deletes element $A[j]$.

Such an ADT is traditionally implemented through a red-black tree T_A such that the leaves of the tree are the elements of A. Each internal node of T_A stores a parent pointer, one bit indicating whether the node is the left or right child of its parent, the number nl of elements (leaves) in the left subtree, and the sum sl of the values in the leaves of its left subtree. The space required is $O(n' \log n')$ bits. To compute the operations we navigate the tree using the information stored in each node. The tree can be also navigated from leaves to root, in order to compute the sum up to a given position in A (leaf of T_A) without necessarily knowing the position j', but just the corresponding leaf of T_A. Thus, all the operations can be supported in $O(\log n')$ time.

Data Structures for Balanced Parentheses. Given a sequence P of $2n$ balanced parentheses, we want to support the following operations. findclose(P, i): given an opening parenthesis at position i, finds the matching closing parenthesis; findopen(P, j): given a closing parenthesis at position j, finds the matching opening parenthesis; excess(P, i): yields the difference between the number of opening and closing parenthesis up to position i in P; enclose(P, i): given a parenthesis pair whose opening parenthesis is at position i, yields the opening parenthesis corresponding to the closest matching pair enclosing i.

Munro and Raman [18] showed how to implement all these operations in constant time and using $2n + o(n)$ bits. They also showed one of the applications of sequences of balanced parentheses: the succinct representation of general trees.

In the dynamic case, the parentheses sequence can change over time, by inserting/deleting a new pair of *matching* parenthesis into/from the sequence. The data structure of Chan et al. [7] supports all of the operations, including insertions and deletions, in $O(\log n)$ time and requires $O(n)$ bits of space. The space can be dropped to $2n + o(n)$ bits of space if we represent the parentheses as a binary string of length $2n$, which is then represented using the dynamic data structure of Mäkinen and Navarro [16], requiring overall $2n + o(n)$ bits and supporting rank and select on the parentheses in $O(\log n)$ worst-case time, as well as the insertion and deletion of new elements in $O(\log n)$ worst-case time. Since both solutions are similar, we add the extra index of [7] to this data structure in order to be able to compute the parenthesis operations, including updates, in $O(\log n)$ worst-case time and using $o(n)$ extra bits of space.

Lemma 1. *There exists a representation for a dynamic sequence of $2N$ balanced parentheses using $2N + o(N)$ bits of space and supporting operations* findclose, findopen, excess, enclose, rank$_)$, select$_)$, insert, *and* delete, *all of them in* $O(\log N)$ *worst-case time.*

Succinct Representation of Trees. There are a number of succinct representations of static trees, such as LOUDS [13], *balanced parentheses* [18], DFUDS [5], *ordinal trees* [10], and the *ultra succinct representation* of [14], requiring $2n + o(n)$ bits and allowing different sets of operations.

In particular, the DFUDS [5] representation supports all of the static-tree operations, including child(x, i), child(x, α), degree, and depth (the latter using the approach of Jansson et al. [14]) in constant time. To get this representation [5] we perform a preorder traversal on the tree, and for every node reached we write its degree in unary using parentheses. What we get is almost a balanced parentheses representation: we only need to add a fictitious '(' at the beginning of the sequence. A node of degree d is identified by the position of the first of the $d + 1$ parentheses representing the node.

In order to support child(x, α) on the DFUDS representation we store the sequence S of edge labels according to a DFUDS traversal of the tree. In this way, the labels of the children of a given node are all stored contiguously in S. We represent S with a data structure for rank and select [9]. Let p be the position of node x within the DFUDS sequence D, and let $p' = \mathrm{rank}_((D, p)$ be the position in S of the symbol for the first child of x. Let $n_\alpha = \mathrm{rank}_\alpha(S, p' - 1)$ and $i = \mathrm{select}_\alpha(S, n_\alpha + 1)$. If i lies within positions p' and $p' + \mathrm{degree}(x) - 1$, the child we are looking for is the $(i - p' + 1)$-th child of x, which is computed in constant time as $\mathrm{child}(x, i - p' + 1)$; otherwise x has not a child labeled α. This approach, which is also used by Barbay et al. [4], is different to the original one [5], yet ours is easier to be dynamized by using the data structures of [11].

For succinct dynamic *binary* trees, the representation of Munro et. al. [19] requires $2n + o(n)$ bits, and they allow updates in $O(\log^2 n)$ amortized time, operations parent and child in $O(1)$ time, and subtreesize in $O(\log^2 n)$ time. Raman and Rao [21] improve the update time to $O((\log \log n)^{1+\epsilon})$ amortized. More specialized operations are also supported in constant time.

3   Improved Succinct Dynamic k-ary Trees

Given a static succinct representation of a tree, as for example DFUDS, if we want to insert a new node at any position in the tree, we must rebuild the corresponding sequence from scratch. The methods for succinct representation of dynamic binary trees [19,21] can be adapted to represent dynamic k-ary trees by transforming the tree into a binary one. However, the cost of basic navigation operations such as parent and child now becomes $O(k)$ in the worst case, while the insertion cost remains the same, $O((\log \log n)^{1+\epsilon})$ amortized time.

Another alternative is to represent the DFUDS of the k-ary tree with the data structure of Lemma 1, and the edge labels by a dynamic data structure for rank and select [11]. Thus, the basic navigation operations are supported in $O(\log n)$ time ($\mathrm{child}(x, \alpha)$ is supported in $O(\log n(1 + \frac{\log k}{\log \log n}))$ time if we use [11]). The overall space requirement is $2n + n \log k + o(n \log k)$ bits. Notice that the time complexity of these navigation operations is related to n, the number of nodes in the tree, rather than to k, the alphabet size. Thus, this data structure cannot take advantage of asymptotically smaller values of k, e.g. $k = O(\mathrm{polylog}(n))$.

In this section we present a succinct representation for dynamic k-ary trees, which is faster both than representing the k-ary tree using a data structure for dynamic binary trees [19,21] and than representing DFUDS with the data structure of Lemma 1 whenever $\log k = o(\log n)$ holds.

3.1   Basic Tree Representation

To allow efficient navigation and modification of the tree, we incrementally divide it into disjoint *blocks*, as in previous approaches [3,19,21]. Every block represents a connected component of N nodes of the whole tree, such that $N_{min} \leqslant N \leqslant N_{max}$, for given minimum and maximum block sizes N_{min} and N_{max}, respectively. We arrange these blocks in a tree by adding inter-block pointers, and thus the entire tree is represented by a tree of connected components.

For each internal node x of the tree there are two cases: node x is internal to a block p or x is a leaf of block p (but not a leaf of the whole tree). The latter nodes form what we call the *frontier* of a block, and every such node x stores an inter-block pointer to a child block q where the representation of the subtree of node x starts. We duplicate node x by storing it as a fictitious root of q, such that *every* block is a tree by itself. Therefore, every such node x in the frontier of a block p has two representations: (1) as a leaf in block p; (2) as the root node of the child block q. Note that the former point enforces that sibling nodes are all stored in the same block, and the latter point enforces that every node is stored in the same block as its children. This property will be useful to simplify the navigation on the tree, as well as ensuring that a block can always be partitioned in the right way upon block overflows.

We have reduced the size of the problem, as insertions (or deletions) only need to update the block where the insertion is carried out, and not the whole tree. As every block is a tree by itself, we can represent them by using any succinct tree representation, which makes this representation very flexible. We can use DFUDS [5] to get constant-time navigation inside a block. Yet, this is a static representation, and so the update time would be linear in the block size.

Defining Block Sizes. Let us now define the values N_{min} and N_{max}, the minimum and maximun block size respectively. Since inter-block pointers should require $o(n)$ bits overall, a pointer of $O(\log n)$ bits must point to a block of size $\Omega(\log^2 n)$ nodes, and therefore $N_{min} = \Theta(\log^2 n)$. On the other hand, a block p should have room to store at least the potential k children of the root of the block (recall that sibling nodes must be stored all in the same block). Also, we must define N_{max} in such a way that when we insert a node in a block of maximal size N_{max} (i.e., the block overflows), we can split the block into two blocks, each of size at least N_{min}. By defining $N_{max} = \Theta(k \log^2 n)$, in the worst case (i.e., the case where the new created block has the smallest possible size) the root of the block has its k possible children, the subtree of each such child having $\Theta(\log^2 n)$ nodes. Thus, upon an overflow, any of such subtrees of size at least N_{min} can be copied to a new block, requiring overall $o(n)$ bits for the pointers.

Existing related works [3,19,21] use a static block representation, which is rebuilt from scratch upon insertions or deletions. In the case of binary trees [19,21], these total reconstructions are carried out in constant time by using precomputed tables. However, for k-ary trees the blocks are large enough so as to be used to index a table. Thus, we use a different approach: we first reduce the size of the problem by updating smaller subtrees (the blocks), and then we make these smaller subtrees dynamic to avoid the linear update time.

Let us study now the block layout. Every block p of N nodes, having N_c child blocks and root node r_p is represented by: the tree topology of the block; a set of N_c pointers PTR_p to child blocks; a set of flags F_p indicating the nodes in the frontier of p; and a pointer to the representation of r_p in the parent block.

Representing the Tree Topology of Blocks. We represent the tree structure T_p of block p plus the edge labels S_p by using suitable dynamic data structures, to avoid rebuilding them from scratch upon updates.

The tree structure T_p of each block p is represented by the following data structure, where operation selectnode$_p(j)$ yields the DFUDS position of the node with preorder j inside block p. (From now on we use the subscript p to indicate operations local to a block p, i.e., disregarding the inter-block structure.)

Lemma 2. *There exists a dynamic* DFUDS *representation for a tree T_p of N nodes requiring $2N + o(N)$ bits of space and allowing us to compute operations* parent$_p$, child$_p(x, i)$, degree$_p$, subtreesize$_p$, preorder$_p$, selectnode$_p$, insert$_p$, *and* delete$_p$, *all of them in $O(\log N)$ worst-case time.*

Proof. We represent the DFUDS sequence of T_p using the data structure of Lemma 1, requiring $2N + o(N)$ bits of space. Except for insert$_p$ and delete$_p$, operations on T_p can be computed as defined originally in [5] for the static case, using the operations provided in Lemma 1 as a base, which take $O(\log N)$ time.

For operation insert$_p$, notice that the insertion of a new leaf x will increase the degree of its parent node y in the tree; this increase is carried out by adding a new opening parenthesis at the corresponding position within the representation of y. To represent the new leaf node, on the other hand, we must add a new closing parenthesis at the corresponding position within T_p. We can show that this new pair of opening and closing parentheses is a matching pair, and hence the insertion can be handled by the data structure of Lemma 1. Deletions are handled in a similar way. (Further details are deferred to the full paper.) □

The overall space requirement of the tree structure for all the blocks is $2n + o(n)$ bits. In our case $N = O(k \log^2 n)$, so the operations on T_p can be supported in $O(\log k + \log \log n)$ worst-case time.

We store the symbols labeling the edges of T_p in array S_p, sorted according to DFUDS as in Section 2. We preprocess S_p with a dynamic data structure for rank$_\alpha(S_p, i)$ and select$_\alpha(S_p, j)$ queries [11]. We can now compute operation child$_p(x, \alpha)$ inside block p using the operations provided by the representation of T_p (i.e., child$_p(x, i)$) and that of S_p (i.e., rank$_\alpha$ and select$_\alpha$), as in Section 2, in $O(\log N(1 + \frac{\log k}{\log \log N})) = O((\log k + \log \log n)(1 + \frac{\log k}{\log (\log k + \log \log n)}))$ worst-case time. The insertion/deletion of a symbol to/from S_p can be carried out within the same time complexity [11]. The space requirement is $N \log k + O(N \frac{\log k}{\sqrt{\log N}})$ bits of space per block. In the worst case every block has size N_{min}, and therefore the overall space is $n \log k + O(\frac{n \log k}{\sqrt{\log \log n}}) = n \log k + o(n \log k)$ bits of space.

Representing the Frontier of a Block. We could use a bit vector supporting rank and select to represent F_p, indicating with a 1 the nodes in the frontier. However, this would require $n + o(n)$ extra bits, exceeding our space limitation.

The frontier is instead represented by a *conceptual* increasingly-sorted array $Pre_p[0..N_c]$ storing the preorders (within block p) of the nodes in the frontier of p (i.e., those nodes having an inter-block pointer), except for $Pre_p[0] = 0$. Since the preorder of a node can change upon updates in T_p, we avoid the linear-time reconstruction of Pre_p by defining array $F_p[1..N_c]$, which stores the difference between consecutive preorders in Pre_p, i.e. $F_p[i] = Pre_p[i] - Pre_p[i-1]$, for $i = 1, \ldots, N_c$. Array F_p is preprocessed with a data structure for searchable partial sums (see Section 2), denoting with T_{F_p} the balanced tree representing the searchable partial sums. Since the total number of entries in arrays F_p equals the number of blocks, the overall extra space requirement is $o(n)$ bits.

The preorder represented by a given $F_p[j]$ (i.e., the conceptual value $Pre_p[j]$), is computed in $O(\log N) = O(\log k + \log \log n)$ time by $\mathsf{Sum}(F_p, j)$. Then, by using $\mathsf{selectnode}_p(\mathsf{Sum}(F_p, j))$ we can get the DFUDS position (and hence the representation) for that node in $O(\log k + \log \log n)$ time.

Representing Inter-block Pointers. In block p we store the pointers to child blocks in the *conceptual* array $PTR_p[1..N_c]$, increasingly sorted according to the preorders of the nodes in the frontier of p. Since every pointer is associated to a node in the frontier of p, we store $PTR_p[i]$ along with $F_p[i]$ in the leaves of T_{F_p}.

For parent pointers, we store in each block p a pointer to the representation of the root r_p in the parent block q. As the DFUDS position of a node can change upon tree updates, we cannot store absolute parent pointers, which must be updated in linear worst-case time. Since r_p lies within the frontier of q, we store in p a pointer to the representation of r_p in F_q (i.e., a pointer to the leaf of T_{F_q} corresponding to r_p). As a result, parent pointers are easily updated as needed, and we get the absolute parent pointer for block p by first following the pointer to the leaf of T_{F_q} representing r_p, and then computing Sum in F_q up to that leaf.

The overall number of pointers equals the number of blocks in the structure, which is $O(n/\log^2 n)$. Thus, the overall space for pointers is $o(n)$ bits.

3.2   Supporting Basic Operations

We define the basic navigation operations for our dynamic data structure.

Operation child. To compute $\mathsf{child}(x, i)$, if node x is not a leaf, we use operation $\mathsf{child}_p(x, i)$ inside block p, since each node is stored in the same block as its children. Operation $\mathsf{child}(x, \alpha)$ is computed similarly using $\mathsf{child}_p(x, \alpha)$. If, on the other hand, node x is a leaf, we check whether x is a leaf of the whole tree (in whose case operation child gets undefined), or just a leaf of block p (in whose case we have to follow a pointer to a child block).

We carry out that checking by computing the position $j = \mathsf{Search}(F_p, \text{preorder} (x))$ in F_p for the greatest preorder which is smaller or equal than the preorder of node x. Then we check whether the preorder represented by $F_p[j]$ (i.e., the value $Pre_p[j] = \mathsf{Sum}(F_p, j)$) equals the preorder of node x. In such a case, x is not a leaf of the whole tree, and then we have to follow the pointer $PTR_p[j]$ to get the child block p', to finally apply the corresponding $\mathsf{child}_{p'}$ operation on the root of block p'. Hence, operation $\mathsf{child}(x, i)$ is computed in $O(\log N) = O(\log k + \log \log n)$ time, and operation $\mathsf{child}(x, \alpha)$ takes $O(\log N(1 + \frac{\log k}{\log \log N})) = O((\log k + \log \log n)(1 + \frac{\log k}{\log (\log k + \log \log n)}))$ worst-case time.

Operation parent(x). If x is not the root of block p storing it, the operation is computed locally by using operation $\mathsf{parent}_p(x)$, since every non-root node is stored in the same block as its parent. Otherwise, we first follow the pointer to the parent block q, and then we compute the position of the representation of x in the parent block q as $\mathsf{selectnode}_q(\mathsf{Sum}(F_q, j))$, assuming that block p is the j-th child of block q. (As the parent pointer points to a leaf in T_{F_q}, we do not need to know j, but just to use the parent pointers in the searchable partial sum data structure for F_q.) Finally we apply parent_q on the representation of x in block q, as we are sure that the parent of node x is stored in q (i.e., because node x cannot be the root of block q, given the properties of our data structure). Operation parent is therefore computed in $O(\log N) = O(\log k + \log \log n)$ time.

Operation insert. Since we insert a new node x in block p, we have to update the block accordingly. We first insert node x in T_p, using operation insert_p of Lemma 2. Then, we insert in S_p the new symbol s labeling the new edge. Since the new node increases the preorders of some nodes in block p, every preorder in F_p whose value is greater or equal to the preorder of x must be increased: we look for position $j = \mathsf{Search}(F_p, \mathsf{preorder}(x))$ in F_p from where the preorders must be increased, and then we increase $F[j]$ by using operation Update. This automatically updates all preorders that have changed after the insertion of the new node. Notice that we are also automatically updating the parent pointers for the child blocks of p. In this way we avoid the (worst-case) linear update time of the frontier F_p. The insertion cost according to this procedure is $O((\log k + \log \log n)(1 + \frac{\log k}{\log (\log k + \log \log n)}))$ time, because of the time to update S_p.

Block Overflows. When inserting in a block p of maximal size N_{max}, we first divide the block p into two blocks, both of size between N_{min} and N_{max}, by selecting a node z in block p whose subtree will be reinserted in a new child block p' (including z itself) and then will be deleted from p (leaving node z still in p). In this way z is duplicated, since it is stored along with its children in p', and along with its siblings and parent in p, thus maintaining the properties of our data structure. Then, the insertion of node x is carried out in the adequate block, either p or p', without a new overflow since there is room for a new node in any of these blocks. When the subtree of node z is reinserted in block p', we copy to p' the portions of arrays F_p and PTR_p corresponding to node z, via insertions in $F_{p'}$ and $PTR_{p'}$ and the corresponding deletions in F_p and PTR_p.

After splitting p, we insert a new inter-block pointer in PTR_p pointing to block p', and we add the preorder of node z in F_p, at the corresponding position (marking that z lies now within the frontier of p). We add also a parent pointer in p', pointing to the leaf corresponding to z in the tree T_{F_p} representing F_p. In this simple way we keep up-to-date all of the parent pointers for the children of p, since the other pointers do not change after adding a new child block to p.

In order to amortize the insertion cost, the overall reinsertion process must be carried out in time proportional to the size of the reinserted tree. The work on T_p, S_p, F_p, and pointers can be done in this time, by using the corresponding insert and delete operations on them. We must be careful, however, with the selection of node z, since naively this would take linear time. Thus, we define a list of *candidate nodes* C_p for every block p, storing the local preorders of candidate nodes to be reinserted upon overflow. We represent C_p in the same way as F_p, in differential form and using a searchable partial sum data structure.

To maintain C_p we must dynamically sample some nodes of T_p such that, every time we need to split p, there is *at least* a candidate subtree to be reinserted in the new child block. We must also ensure that the overall space for the C_p data structures is $o(n)$ bits, so we cannot maintain too many candidates.

Thus, every time we descend in the tree we maintain the last node z in block p such that $\mathsf{subtreesize}_p(z) \geqslant N_{min}$ holds. When we find the insertion point of the new node x, say at block p, before adding z to C_p we first perform $p_1 = \mathsf{Search}(C_p, \mathsf{preorder}_p(z))$, and then $p_2 = \mathsf{Search}(C_p, \mathsf{preorder}_p(z) + \mathsf{subtreesize}_p(z))$. Then, we add z to C_p whenever: (1) z is not the root of block p; and (2) it holds that $p_1 = p_2$, which means that there is no other candidate in the subtree of z. If in the descent we find a candidate node z' which is an ancestor of z, then after inserting z to C_p we delete z' from C_p. In this way we keep the lowest possible candidates, avoiding that the subtree of a candidate becomes so large, which would not guarantee a fair partition into two blocks of size between N_{min} and N_{max} upon overflow.

As a result we ensure that the local subtree size of every candidate is at least N_{min}, and also that given a candidate node z, there are no candidate nodes in the subtree of z. Thus, we have a candidate node out of (at least) N_{min} nodes, and hence the total space to manage the candidates is $o(n)$. We are also ensuring that every time a block becomes full we have at least one candite node in C_p to be reinserted, because there were sufficient insertions in p (to become full) in order to find at least a candidate in it. This is because of the maximum block size $N_{max} = \Theta(k \log^2 n)$ that we have chosen: this ensures that whenever a block becomes full, at least one of the children of the block root has size at least N_{min}.

The reinsertion cost is proportional to the size of the reinserted subtree; since we have already paid to insert these nodes for the first time, the insertion cost is $O(\log N(1 + \frac{\log k}{\log \log N})) = O((\log k + \log \log n)(1 + \frac{\log k}{\log (\log k + \log \log n)}))$ amortized.

Operation delete. To delete a node x in block p we update the data structure by using operation delete_p. After deleting x, we check whether there is a candidate node z in C_p which is ancestor of x and whose subtree becomes smaller than N_{min} after deleting x. As there is at most one ancestor of x in C_p, z can be found as the node represented by $C_p[\mathsf{Search}(C_p, \mathsf{preorder}_p(x)) - 1]$; the subtraction comes from the fact that with the search in C_p we find a candidate which is next (in preorder) to z in C_p. After deleting z from C_p, we try to insert in C_p the last node z' found in the descent (before the deletion) whose subtree size is greater or equal to N_{min}, following the same policies as for operation insert.

If we delete x from a block p of size N_{min}, then a *block underflow* occurs. In such a case, we find the representation of the block root r_p in the parent block q, by using the corresponding parent pointer. From that node we reinsert in q all of the nodes of block p. Note that in the worst case there will be only one block overflow in q when reinserting, since block p has less than N_{min} nodes, and after an overflow in q there will be room for *at least* N_{min} new nodes. If p is not a leaf in the tree of blocks, we reinsert the frontier of p within the frontier of q.

Managing Dynamic Memory. The model of memory allocation is a fundamental issue of succinct dynamic data structures, since we must be able to manage the dynamic memory fast and without requiring so much memory space due to memory fragmentation. We assume a standard model where the memory is regarded as an array, with words numbered 0 to $2^w - 1$. The space usage of an algorithm at a given time is the highest memory word currently in use by the algorithm. This corresponds to the so-called $\mathcal{M}_B$ memory model [21].

We manage the memory of every tree block separately, each in a "contiguous" memory space. However, tree blocks are dynamic and therefore this memory space must grow and shrink accordingly. If we use an *Extendible Array* (EA) [6] to manage the memory of a given block, we end up with a collection of at most $O(n/\log^2 n)$ EAs, which must be maintained under the operations: create, which creates a new empty EA in the collection; destroy, which destroys an EA from the collection; grow(A), which increases the size of array A by one; shrink(A), which shrinks the size of array A by one; and access(A, i), which access the i-th item in array A. Raman and Rao [21] show how operation access can be supported in $O(1)$ worst-case time, create, grow and shrink in $O(1)$ amortized time, and destroy in $O(s'/w)$ time, where s' is the nominal size (in bits) of array A to be destroyed. The space requirement for the whole collection is $s + O(a^*w + \sqrt{sa^*w})$ bits, where a^* is the maximum number of EAs that ever existed simultaneously in the collection, and s is the nominal size of the collection.

To simplify the analysis we store every part of a block in different collection of EAs (i.e., we have a collection for T_ps, a collection for S_ps, and so on). The memory for S_p and T_p inside the corresponding EA is managed as in the original works [11,16]. For the case of F_p, C_p, etc., we manage the corresponding EA by using standard techniques to allocate and free dynamic memory. Thus, we use operation grow on the corresponding EA every time we insert a node in the tree, operation shrink when we delete a node, and operation create upon block overflows, all of them in $O(1)$ amortized time. Operation destroy, on the other hand, is used upon block underflows. Consider the EA collection storing S_p for every block p of the tree. The block p' which underflows has size less than $\Theta(\log^2 n)$, and thus the nominal size for the EA storing $S_{p'}$ is less than $\Theta(\log^2 n \log k)$ bits. Therefore operation destroy takes less than $\Theta(\log n \log k)$ time, which is negligible since we have to reinsert all of the nodes of p' in the parent block, at a higher cost. The EAs storing the remaining parts of p' can be destroyed even faster.

For the space analysis, it is important to note that every time $\log n$ changes, the tree must be rebuilt from scratch to adapt these changes. This also involves rebuilding the data structures needed to maintain the collections of EAs. The amortized cost of update operations over the tree still remains the same. Let n' be the maximum number of nodes that ever existed in the tree since the last reconstruction (i.e., the last change of $\log n$). As reconstructions occur when n

is a power of two, then both n and n' lie between (the same) two consecutive powers of two, and thus we can prove that $n \leqslant n' \leqslant 2n$ holds, which means $n' = \Theta(n)$. Thus, we can conclude that the maximum number of EAs that we can have between reconstructions is $a^* = O(n/\log^2 n)$.

The nominal size of the EA collection for T_ps is $2n + o(n)$ bits. Then, this collection requires $2n + o(n) + O(\frac{n}{\log n} + \frac{n}{\sqrt{\log n}}) = 2n + o(n)$ bits of space [21]. The nominal size of the collection for S_ps is $n \log k + o(n \log k)$, and thus we have $n \log k + o(n \log k) + O(\frac{n}{\log n} + n\sqrt{\log k/\log n}) = n \log k + o(n \log k)$ bits overall.

Therefore we have proved:

Theorem 1. *There exists a representation for dynamic k-ary trees using $2n + n \log k + O(\frac{n \log k}{\sqrt{\log \log n}})$ bits of space supporting operations* parent *and* child(x, i) *in $O(\log k + \log \log n)$ worst-case time, operation* child(x, α) *in $O((\log k + \log \log n)(1 + \frac{\log k}{\log (\log k + \log \log n)}))$ worst-case time, and operations* insert *and* delete *in $O((\log k + \log \log n)(1 + \frac{\log k}{\log (\log k + \log \log n)}))$ amortized time.*

In the case of binary non-labeled trees [19,21] we have:

Corollary 1. *There exists a representation for dynamic binary trees using $2n + o(n)$ bits of space and supporting operations* parent *and* child(x, i) *in $O(\log \log n)$ worst-case time, and operations* insert *and* delete *in $O(\log \log n)$ amortized time.*

Thus we improve the $O((\log \log n)^{1+\epsilon})$ update time of [21], but at the price of more expensive navigations ([21] provides $O(1)$ time for these operations).

We leave the definition of more involved operations (like subtreesize, depth, etc.) for the full paper. Most of them are supported in $O(\log k + \log \log n)$ time.

4   Conclusions and Further Works

We have defined a succinct representation for dynamic k-ary trees (or tries) of n nodes, requiring $2n + n \log k + o(n \log k)$ bits of space and supporting navigation operations in $O(\log k + \log \log n)$ time, as well as insertion and deletion of leaves in $O((\log k + \log \log n)(1 + \frac{\log k}{\log (\log k + \log \log n)}))$ amortized time. Our representation is able to take advantage of asymptotically smaller values of k, thus improving the $O(\log n)$ time achieved by alternative representation of Lemma 2 whenever $\log k = o(\log n)$, which covers many interesting applications in practice.

An interesting future work is to reduce the extra space of $O(\frac{n \log k}{\sqrt{\log \log n}}) = o(n \log k)$ bits needed by our representation. This comes from the data structure of [11] used to represent S_p in each block. Also, it would be interesting to reduce the time of the operations (e.g., to $O(\frac{\log \log n}{\log \log \log n})$ time) in the case of small alphabets, e.g. $k = O(\text{polylog}(n))$ (a particular case is that of binary trees).

Acknowledgments. We thank Jérémy Barbay for proofreading this paper.

References

1. Apostolico, A.: The myriad virtues of subword trees. In: Combinatorial Algorithms on Words. NATO ISI Series, pp. 85–96. Springer, Heidelberg (1985)
2. Arroyuelo, D., Navarro, G.: A Lempel-Ziv text index on secondary storage. In: Ma, B., Zhang, K. (eds.) CPM 2007. LNCS, vol. 4580, pp. 83–94. Springer, Heidelberg (2007)
3. Arroyuelo, D., Navarro, G.: Space-efficient construction of LZ-index. In: Deng, X., Du, D.-Z. (eds.) ISAAC 2005. LNCS, vol. 3827, pp. 1143–1152. Springer, Heidelberg (2005)
4. Barbay, J., He, M., Munro, J.I., Rao, S.S.: Succinct indexes for strings, binary relations and multi-labeled trees. In: Proc. SODA, pp. 680–689 (2007)
5. Benoit, D., Demaine, E., Munro, J.I., Raman, R., Raman, V., Rao, S.S.: Representing trees of higher degree. Algorithmica 43(4), 275–292 (2005)
6. Brodnik, A., Carlsson, S., Demaine, E., Munro, J.I., Sedgewick, R.: Resizable arrays in optimal time and space. In: Dehne, F., Gupta, A., Sack, J.-R., Tamassia, R. (eds.) WADS 1999. LNCS, vol. 1663, pp. 37–48. Springer, Heidelberg (1999)
7. Chan, H.L., Hon, W.K., Lam, T.W., Sadakane, K.: Compressed indexes for dynamic text collections. ACM TALG 3(2) (article 21) (2007)
8. Ferragina, P., Luccio, F., Manzini, G., Muthukrishnan, S.: Structuring labeled trees for optimal succinctness, and beyond. In: Proc. FOCS, pp. 184–196 (2005)
9. Ferragina, P., Manzini, G., Mäkinen, V., Navarro, G.: Compressed representations of sequences and full-text indexes. ACM TALG 3(2) (article 20) (2007)
10. Geary, R., Raman, R., Raman, V.: Succinct ordinal trees with level-ancestor queries. In: Proc. SODA, pp. 1–10 (2004)
11. González, R., Navarro, G.: Improved dynamic rank-select entropy-bound structures. In: Proc. LATIN (to appear, 2008)
12. Hon, W.K., Sadakane, K., Sung, W.K.: Succinct data structures for searchable partial sums. In: Ibaraki, T., Katoh, N., Ono, H. (eds.) ISAAC 2003. LNCS, vol. 2906, pp. 505–516. Springer, Heidelberg (2003)
13. Jacobson, G.: Space-efficient static trees and graphs. In: Proc. FOCS, pp. 549–554 (1989)
14. Jansson, J., Sadakane, K., Sung, W.K.: Ultra-succinct representation of ordered trees. In: Proc. SODA, pp. 575–584 (2007)
15. Katajainen, J., Mäkinen, E.: Tree compression and optimization with applications. Int. J. Found. Comput. Sci. 1(4), 425–448 (1990)
16. Mäkinen, V., Navarro, G.: Dynamic entropy-compressed sequences and full-text indexes. ACM TALG (to appear, 2007)
17. Manzini, G.: An analysis of the Burrows-Wheeler transform. Journal of the ACM 48(3), 407–430 (2001)
18. Munro, J.I., Raman, V.: Succinct representation of balanced parentheses and static trees. SIAM Journal on Computing 31(3), 762–776 (2001)
19. Munro, J.I., Raman, V., Storm, A.: Representing dynamic binary trees succinctly. In: Proc. SODA, pp. 529–536 (2001)
20. Raman, R., Raman, V., Rao, S.S.: Succinct indexable dictionaries with applications to encoding k-ary trees and multisets. In: Proc. SODA, pp. 233–242 (2002)
21. Raman, R., Rao, S.S.: Succinct dynamic dictionaries and trees. In: Baeten, J.C.M., Lenstra, J.K., Parrow, J., Woeginger, G.J. (eds.) ICALP 2003. LNCS, vol. 2719, pp. 357–368. Springer, Heidelberg (2003)

Towards a Solution to the "Runs" Conjecture

Maxime Crochemore[1,2,*], Lucian Ilie[3,**,***], and Liviu Tinta[3]

[1] Department of Computer Science, King's College London, London WC2R 2LS, UK
[2] Institut Gaspard-Monge, Université Paris-Est, F-77454 Marne-la-Vallée, France
maxime.crochemore@kcl.ac.uk
[3] Department of Computer Science, University of Western Ontario
London, Ontario, N6A 5B7, Canada
{ilie,ltinta}@csd.uwo.ca

Abstract. The "runs" conjecture, proposed by [Kolpakov and Kucherov, 1999], states that the number of occurrences of maximal repetitions (runs) in a string of length n is at most n. The best bound to date, due to [Crochemore and Ilie, 2007], is $1.6n$. Here we improve very much this bound using a combination of theory and computer verification. Our best bound is $1.048n$ but actually solving the conjecture seems to be now only a matter of time.

1 The Conjecture

Repetitions in strings constitute one of the most fundamental areas of string combinatorics with very important applications to text algorithms, data compression, or analysis of biological sequences. The result of a two-decade effort in the stringology community to find an algorithm to compute all repetitions in a string in linear time resulted in the paper of Kolpakov and Kucherov [7] that (i) used previous techniques of Crochemore [1], Main and Lorentz [9], and Main [8] to construct an algorithm that computes all maximal repetitions (or runs, see the next section for precise definition) in time proportional to the size of the output and (ii) proved that the maximum number of runs in a string of length n, RUNS(n), is linear, i.e., RUNS(n) $\leqslant cn$, where c is a constant. Therefore, the crucial contribution of [7] was (ii). However, they could not provide any bound on the constant c but, based on numerical evidence, stated the following conjecture, for *binary alphabets*:

Conjecture 1 (The "runs" conjecture) *For any $n \geqslant 1$, RUNS(n) $\leqslant n$.*

Several bounds were proved later, all for *arbitrary alphabets*, as follows. The first bound for the number of runs was given by Rytter [11] and is $5n$. A more careful analysis of [11] was done by Puglisi, Simpson, and Smyth [10] to improve the bound to $3.48n$ and by Rytter himself for $3.44n$. All these papers counted each

---

* Research supported in part by CNRS.
** Corresponding author.
*** Research supported in part by NSERC.

P. Ferragina and G. Landau (Eds.): CPM 2008, LNCS 5029, pp. 290–302, 2008.
© Springer-Verlag Berlin Heidelberg 2008

run at the position where they start. A different approach was considered by Crochemore and Ilie [2] where runs are counted at their center (beginning of the second period, see later for precise definition). This latter approach is somewhat counterintuitive as linearly many runs can share the same center as opposed to logarithmically many with the same beginning. However, a much better bound, $1.6n$, was obtained.

Information about the history of the problem can be found in the introduction of [2] and, more generally, on current problems in string repetitions in the coming survey [4]. Here we mention only that the only known lower bound is $\text{RUNS}(n) \geqslant 0.927..n$ due to Franek, Simpson, and Smyth [5].

Besides the obvious mathematical importance of obtaining better bounds, they also provide more accurate analysis of the computational complexity of the algorithms that compute repetitions. In this paper we continue the work in [2] and improve significantly the bound with a combination of theory and computer checking. A common feature of the approaches in Crochemore and Ilie [2] and Rytter [11] is the distinction made between runs with short period (microruns) and long ones. As mentioned in [2], the procedure that leads to bounding the number of microruns can be automatized. Using this idea, Giraud [6] communicated to us that he improved the bound to $1.5n$ by bounding the number of microruns with period up to 9 by $0.924n$. (Our bound in Table 1 for this case is $0.85n$.)

The rather ad-hoc approach for microruns in [2] was good enough for hand computation of all the possibilities up to period 9. When attempting to prove the conjecture using the above idea, a rigorous approach is needed. After the basic definitions in the next section, the idea from [2] is described in detail in Section 3. What we need is an algorithm for verifying (using a computer) improved bounds on the number of microruns. Such an algorithm is given in Section 4 but, however, the computational task is totally infeasible. We develop several powerful heuristics in Section 5 which reduce very much the number of cases that need to be investigated. The results obtained so far are shown in Section 6. The paper concludes with a brief discussion in Section 7. Some proofs are omitted due to limited space.

2   Runs

We denote the length of a string w by $|w|$, its ith letter by $w[i]$, and the factor $w[i]w[i+1]\cdots w[j]$ by $w[i \mathinner{..} j]$. We index w from 0 to $|w|-1$, that is, $w = w[0 \mathinner{..} |w|-1]$, unless otherwise specified. The string w has period p if $w[i] = w[i+p]$ whenever both are defined. We say that there is *period p at i in w* if $w[i-p \mathinner{..} i-1] = w[i \mathinner{..} i+p-1]$, that is, the factor $w[i-p \mathinner{..} i+p-1]$ exists and has period p.

A *run* is a maximal (non-extendable) occurrence of a repetition of exponent at least two. That means, the interval $[i \mathinner{..} j]$ is a run if

(i) $w[i \mathinner{..} j]$ has period p,
(ii) $j - i + 1 \geqslant 2p$,

(iii) $w[i-1] \neq w[i+p-1]$ (if $w[i-1]$ is defined), $w[j+1] \neq w[j-p+1]$ (if $w[j+1]$ is defined) and

(iv) $w[i \mathinner{..} i+p-1]$ is primitive, that is, it is not a proper integer power (2 or larger) of another string.

In such a case we say that there is a run with period p at $i+p$ in w or, briefly, *run p at $i+p$ in w* ($i+p$ is the *center* of the run; see below). Note that "run p at i" implies "period p at i" but not viceversa. The latter needs only (i) and (ii) above. For period p at i to be run p at i it requires the primitivity condition at (iv) and *only* the first part of (iii), that is, the non-left-extendability: $w[i-1] \neq w[i+p-1]$. Right-extendability, that is, when $w[j+1] = w[j-p+1]$, extends the run to the right but does not move its center! In fact, we shall always use only the initial *square*, $w[i-p \mathinner{..} i+p-1]$, of a run, as this is the only part we can always count on and, as explained above, it is enough to define a run p at i.

Here is an example: the string $w[0 \mathinner{..} 10] = \mathsf{abbababbaba}$ has a run $[2 \mathinner{..} 6]$ with period 2 and exponent 2.5, that is, $w[2 \mathinner{..} 6] = \mathsf{babab} = (\mathsf{ba})^{2.5}$. We say that there is run 2 at 4. For instance, there is period 2 at 5 but it is not a run because it can be extended by one position to the left. Other runs are $[1 \mathinner{..} 2]$, $[6 \mathinner{..} 7]$, $[7 \mathinner{..} 10]$, $[4 \mathinner{..} 9]$ and $[0 \mathinner{..} 10]$. For a run $[i \mathinner{..} j]$ of period p, the positions i, $i+p$, and j are its *beginning*, *center*, and *end* respectively.

3   The Idea for Better Bounds

The idea used by Crochemore and Ilie [2] is to count the runs at their centers (the starting position of the second period) as well as count separately *microruns* (runs with "short" period). The following proposition is used in [2] to bound the number of runs with period p or larger; $\mathrm{RUNS}_{\geqslant p}(n)$ denotes the maximum number of runs with period p or larger in a string of length n.

Proposition 1. *For any n and p, $\mathrm{RUNS}_{\geqslant p}(n) \leqslant \frac{6}{p}n$.*

The runs with short periods — microruns — are counted as follows. For a given bound b and maximum period p, the centers of the runs are non-uniformly distributed and we try to amortize their number as follows. Given a position i, we try all possible combinations of periods for the runs with centers to the left of i, until a position $i-j$ is found such that the ratio between the number of centers inside the interval $[i-j \mathinner{..} i]$ and its length, $j+1$, falls below b. Also, at any moment, the number of centers of runs that have both the center and the beginning inside the interval should satisfy the corresponding amortizing condition. In [2], we did this for $p=9, b=1$ but mentioned that it can be done for any p and b (assuming the bound holds). When successful, this procedure proves that $\mathrm{RUNS}_{\leqslant p}(n) \leqslant bn$. Putting these two bounds together for $p=9$ and $b=1$, we obtained in [2] the bound $\mathrm{RUNS}(n) \leqslant (6/10)n + n = 1.6n$. Better bounds can be obtained by increasing p, and/or decreasing b, however, the computation may become very demanding. Our main goal in this paper is to decrease as much as possible the amount of computation required so that it becomes doable.

4 An Algorithm for Microruns

Recall that microruns are runs with period bounded by a fixed value, which we henceforth denote by max_per. We need to consider all possible combinations of periods (up to max_per) of runs in a string to the left of a given position until the total number of centers divided by the length of the factor falls below a given bound b. In [2], for max_per $= 9$ and $b = 1$, we found 61 possible cases. One such case looks like this: $(\emptyset, \{2\}, \{8\}, \{1, 3\})$, where each set contains the periods of runs having their center at that position. In order to amortize the two centers with periods 1 and 3, we go, in this case, three positions to the left. The same combination can be amortized within 2 positions in the case $(\emptyset, \{5\}, \{1, 3\})$ or only 1 position in the case $(\emptyset, \{1, 3\})$.

Such arrays of sets of positive integers will be called histories. Precisely, given a string $s = s[m \ldots o]$ and $m \leqslant i \leqslant j \leqslant o$, we denote $\text{history}_s[i \ldots j] = (H_i, H_{i+1}, \ldots, H_j)$, where $H_g = \{k \mid \text{run } k \text{ at } g \text{ in } s\}$. As an example, if $s[0 \ldots 6] = $ aaabaab, then $\text{history}_s[2 \ldots 5] = (\emptyset, \emptyset, \{3\}, \{1\})$. An array h of sets of positive integers is called a *history* if $h = \text{history}_s[i \ldots j]$, for a string $s[m \ldots o]$ and $m \leqslant i \leqslant j \leqslant o$. In order to amortize the number of microruns, we need to be able to detect (efficiently) the histories.

We see next how this can be done for an arbitrary array of sets of positive integers, say $h = (H_0, H_1, \ldots, H_{k-1})$. (We call the elements of the sets H_i runs.) Construct first the leftmost and rightmost position, respectively, where a run from h can reach, that is,

$$\ell = \min(\{i - p \mid 0 \leqslant i \leqslant k - 1, p \in H_i\} \cup \{0\})),$$
$$r = \max(\{i + p - 1 \mid 0 \leqslant i \leqslant k - 1, p \in H_i\} \cup \{k - 1\}).$$

For example, consider $h = (\emptyset, \emptyset, \{3\}, \{1\})$. Then $\ell = -1$ and $r = 4$, see Fig. 1.

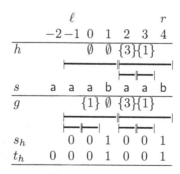

Fig. 1. For the history $h = (\emptyset, \emptyset, \{3\}, \{1\}) = \text{history}_s[0 \ldots 3]$, the string s_h, the history $g = \text{history}_{s_h}[0 \ldots k - 1]$, and the string t_h are shown. The initial squares of the runs are shown as segments.

Consider the set $S = \{\ell, \ell+1, \ldots, r\}$ which is $\{-1, 0, 1, 2, 3, 4\}$ for our example. Let $\equiv_s$ be the smallest equivalence relation that contains the relation $R = \{(j-p, j) \mid i \leqslant j \leqslant i+p-1,$ for some $0 \leqslant i \leqslant k - 1$ and $p \in H_i\}$. That is, $i \equiv_s j$ means that positions i and j should contain the same letters in any string whose history is h. This is implied by the initial squares of the existing runs in h. For our example, $R = \{(2, 3), (-1, 2), (0, 3), (1, 4)\}$ and $S|_{\equiv_s} = \{\{-1, 0, 2, 3\}, \{1, 4\}\}$. Consider also a naming function f, that labels the equivalence classes of $\equiv_s$ with positive integers; such as $f(\{-1, 0, 2, 3\}) = 0$ and $f(\{1, 4\}) = 1$. Construct now the string $s_h[\ell \mathinner{.\,.} r]$ by $s_h[i] = f([i]_{\equiv_s})$. For our example, $s_h[-1 \mathinner{.\,.} 4] = 001001$. We now have a necessary condition for h to be a history.

Lemma 1. *Given an array h of sets of positive integers, if h is a history, then $h \subseteq \mathsf{history}_{s_h}[0 \mathinner{.\,.} k - 1]$ (componentwise inclusion).*

The following result is useful for the proof of Lemma 1.

Lemma 2. *If h is a history, $h = \mathsf{history}_s[0 \mathinner{.\,.} k - 1]$, then $s_h[i] = s_h[j]$ implies $s[i] = s[j]$, for any $\ell \leqslant i, j \leqslant r$.*

Proof. First, both s and s_h are defined for i, j in the range $[\ell \mathinner{.\,.} r]$. This is by definition for s_h and it has to be true for s too as otherwise some runs would be in h but not in s. Then, the equality of any two letters of s_h comes from transitivity of some equalities imposed by the runs of h. Since all these runs are also in s, the statement follows. $\square$

Proof of Lemma 1. Put $\mathsf{history}_{s_h}[0 \mathinner{.\,.} k - 1] = (S_0, S_1, \ldots, S_{k-1})$ and consider a string s such that $h = \mathsf{history}_s[0 \mathinner{.\,.} k-1] = (H_0, H_1, \ldots, H_{k-1})$. By contradiction, assume there exist i and p with $p \in H_i \backslash S_i$. By construction, we have period p at i in s_h. Because $p \notin S_i$, we have that this period is either left-extendable or not primitive. In the former case we obtain $s_h[i - 1] = s_h[i - p - 1]$, which, by Lemma 2, implies $s[i - 1] = s[i - p - 1]$ contradicting the fact that there is run p at i in s. For the latter case, if $s_h[i \mathinner{.\,.} i + p - 1]$ is not primitive, then, also by Lemma 2, $s[i \mathinner{.\,.} i+p-1]$ is not primitive as well, implying the same contradiction. The lemma is proved. $\square$

Thus, for an arbitrary array of sets of positive integers h, Lemma 1 says that if $h \not\subseteq \mathsf{history}_{s_h}[0 \mathinner{.\,.} k - 1]$, then h cannot be a history. A very simple example when this happens is $h = (\{1\}, \{1\})$. The run 1 at 1 cannot exist as it simply extends the run 1 at 0. On the other hand, if $h = \mathsf{history}_{s_h}[0 \mathinner{.\,.} k - 1]$, then h is a history by definition. The case that remains to be investigated is $h \subsetneq \mathsf{history}_{s_h}[0 \mathinner{.\,.} k - 1]$, as it happens in Fig. 1.

Assume now that h is a history, $h = \mathsf{history}_s[0 \mathinner{.\,.} k - 1] = (H_0, H_1, \ldots, H_{k-1})$, and that $h \subsetneq \mathsf{history}_{s_h}[0 \mathinner{.\,.} k-1]$. Consider i and p such that $p \in S_i \backslash H_i$. Lemma 2 implies that there is period p at i in s. However, there is no run p at i in s, and therefore, the period p is either left-extendable or not primitive, or both. Precisely, there must exist a divisor d of p and a position $j \leqslant i - p + d$ such that there is run d at j in s and this run continues to the right at least until position

$i + p - 1$, that is, $s[j - d .. i + p - 1]$ has period d. There are two possibilities: $j < 0$ and $j \geqslant 0$. In the former, the period p has subperiod d that extends to the left in s at least $d + 1$ positions past position 0. In the latter, there is run d at j in both s and s_h and the period p at i in s has a subperiod d that is included in the run d at j in s. Either way, we can eliminate the run p at i in s_h by moving its center to the left until either it shifts to the left of position 0 (hence, outside the interval $[0 .. k - 1]$ we care about) or it coincides with an existing one in s_h. Denote the obtained string by t_h. The transformation done to construct t_h from s_h uses only equalities that already exist in s. This is an important observation because those equalities maintain the runs from h in s and therefore, when we apply those to s_h, the runs from h in s_h are going to stay. Only the unwanted centers in the interval $[0 .. k - 1]$ disappear. Therefore, we have $h = \mathsf{history}_{t_h}[0 .. k - 1]$.

For instance, in Fig. 1 we have the run $p = 1 \in S_0 \setminus H_0$. The reason why 1 is not in H_0 is because it extends in s to the left until position -2, which means that the run containing it has its center at -1, outside the interval $[0 .. 3]$. The string t_h includes this modification. It is one position longer than s_h and indeed $h = \mathsf{history}_{t_h}[0 .. 3]$.

However, for an arbitrary array of sets of positive integers, we do not know, a priori, for each extra run p in s_h but not in h which divisor d of p is used (in a potential string s) to eliminate this run because we are going to work with h only. We don't have s and, in fact, s may not even exist, in the case h is not a history. Therefore, we are going to try all possibilities. Denote by T_h the set of all strings t_h built as above, considering all possible combinations of divisors of integers p for which p is a run at some position $i, 0 \leqslant i \leqslant k - 1$, in s_h but not in h. Then, h is history if and only if it is $\mathsf{history}_{t_h}[0 .. k - 1]$, for some $t_h \in T_h$.

The following lemma summarizes our procedure for deciding whether a given array of sets of positive integers is a history.

Lemma 3. *Given an array h of sets of positive integers, we have:*

(i) *if $h \not\subseteq \mathsf{history}_{s_h}[0 .. k - 1]$, then h is not a history;*
(ii) *if $h = \mathsf{history}_{s_h}[0 .. k - 1]$, then h is a history;*
(iii) *if $h \subsetneq \mathsf{history}_{s_h}[0 .. k - 1]$, then h is a history iff there is $t_h \in T_h$ such that $h = \mathsf{history}_{t_h}[0 .. k - 1]$.*

We are now in position to give our algorithm for verifying that the number of microruns in any string of length n is bounded by bn. Assume our microruns have period at most $\mathsf{max\_per}$. For a history $h = (H_0, \ldots, H_{k-1})$, denote

- $\mathrm{LENGTH}(h) = k$ — the number of positions covered by h
- $\mathrm{ALLRUNS}(h) = \sum_{i=0}^{k-1} \mathrm{card}(H_i)$ — the number of all runs in h
- $\mathrm{BRUNS}(h) = \sum_{i=1}^{k-1} \mathrm{card}\{p \in H_i \mid i - p \geqslant 0\}$ — the number of the runs in h that begin within the range $[0 .. k - 1]$ (this value will be used in a technical argument in the proof when at the beginning of a string; see below).

TEST(h)
1.   **if** (LENGTH(h) = too_large) **then**
2.   PRINT("not amortized"); EXIT()   // exit the main program
3.   **if** (($h \neq$ ( )) **and** ($\frac{\text{BRUNS}(h)}{\text{LENGTH}(h)} > b$)) **then**
4.   PRINT("not amortized"); EXIT()
5.   **if** (($h \neq$ ( )) **and** ($\frac{\text{ALLRUNS}(h)}{\text{LENGTH}(h)} \leqslant b$)) **then**
6. $am\_pos \leftarrow \max(am\_pos, \text{LENGTH}(h))$
7.   **return**()
8.   **for each** $H \in 2^{\{1,\ldots,\text{max\_per}\}}$ **do**
9. $g \leftarrow (H, H_0, H_1, \ldots, H_{k-1})$ // append H in front of h
10.   **if** (ISHISTORY(g)) **then**
11.   TEST(g)
12. **return**()

Fig. 2. The TEST function

Lemma 3 will be used in the function ISHISTORY(h) to test whether h is a history or not. The function TEST(h) in Fig. 2 tries all histories that have h as a suffix and here is a brief description of it.

In case the number of microruns cannot be amortized because, say, the bound b is not true, then we stop the whole program when the histories to be tested become too long (steps 1-2). The condition in step 3 is needed for beginning of strings as follows. The TEST function attempts to amortize the number of centers of runs and the length of histories is increased until this is achieved. However, when using this procedure to prove the bound we are looking for, the beginning of an arbitrary string may appear before the number of runs is amortized. Therefore, to cover this situation, we amortize at each step the number of centers of runs which do not extend to the left past the current position. This condition turns out to be much weaker in practice than the one for all runs. That means, if we gradually decrease the bound b to the point where it cannot be amortized, then the program will exit in step 2 and not 4. However, we do not have a proof of this fact, so we need to check. The overhead imposed by checking this condition is negligible.

If the condition in step 5 is true, then the amortizing process succeeded for this branch and we update the number of positions needed to amortize, am_pos. Otherwise, we investigate all histories that add another set of periods in front of the current one.

The main program is simply calling the TEST function with the empty history, TEST(( )). If it stops normally, that is, without printing "not amortized", then it proves that the bound bn on the number of microruns with periods up to max_per holds, as we see next.

Proposition 2. *If the function* TEST(( )) *terminates normally, then, for all* $n \geqslant 1$, *we have* RUNS$_{\leqslant\text{max\_per}}(n) \leqslant bn$.

The problem with the function TEST is that there may be too many sets H of periods to be tried in step 8 for large max_per.[1] In the next section we are investigating ways to reduce drastically the number of such sets of periods.

5   Compatible Runs

The main idea in reducing the number of sets of runs to be considered in step 8 of the function TEST is that some runs are incompatible with each other, either at the same position or at different positions. We improve a result from [2] to be used for this purpose.

Lemma 4. *Consider a string s and the periods p and $p - \ell$, $0 < \ell < p$. Let h be the smallest integer such that $h\ell \geqslant p$ ($h = \lceil p/\ell \rceil$). If s has run $p - \ell$ at i and either (a) run p at $i + j$ with $j \leqslant \ell - 1$, or (b) run p at $i - j$ with $j \leqslant \ell$, then*

(i) ℓ does not divide p;
(ii) s has run $p - k\ell$ at i, for $2 \leqslant k \leqslant h - 3$ (i.e., all but the shortest two). If $p-(h-2)\ell$ is a prime or 4, then s has run $p-(h-2)\ell$ at i. If $p-(h-1)\ell = 1$, then s has run $p - (h - 1)\ell$ at i.

The following simple observation is also useful to eliminate certain periods of runs.

Lemma 5. *If s has run p at i, then it cannot have run p at j, for $i-p \leqslant j \leqslant i+p$, $j \neq i$.*

Given a run p at i, the function FORBIDDENSAMEPOS in Fig. 3 computes the runs that are forbidden by p at the same position due to Lemma 4(i) with $j = 0$. As an example, FORBIDDENSAMEPOS(6) $= \{3, 4, 5, 7, 8, 9, 12\}$.

FORBIDDENSAMEPOS(p)

1. **if** $(p = 1)$ **then return**($\{1, 2\}$)
2. $forbidden \leftarrow \{p - 1, p + 1, 2p\}$
3. **for** (each ℓ proper divisor of p) **do**
4. $forbidden \leftarrow forbidden \cup \{p - \ell, p + \ell\}$
5. **return**($forbidden$)

Fig. 3. The FORBIDDENSAMEPOS function

Next, we pre-compute all possible sets of periods of microruns at the same position. The correctness follows from Lemma 4. The function RUNSSAMEPOS in Fig. 4 uses two arguments: *current*, which contains the runs included so far and *available*, that is, those runs that are not forbidden by the existing ones. The sets of periods of runs at the same position is obtained by calling

[1] To solve the conjecture, the values of max_per that need to be tested exceed 70. That means an impossible 2^{70} H-sets any time the function reaches step 8.

RUNSSAMEPOS(*current*, *available*)

1.   **if** (*available* = ∅) **then return**({*current*})
2.   *possible_sets* ← {*current*}
3.   **for** (each *p* ∈ *available*) **do**
4.   *can_add_p* ← 1
5.   **for** (all *q* ∈ *current* with *q* < *p*) **do**
6.   **for** (each *k* ∈ GOODKS(*p*, *p* − *q*)) **do**
7.   **if** (*p* − *k*(*p* − *q*) ∉ *current*) **then** *can_add_p* ← 0
8.   **if** (*can_add_p* = 1) **then**
9.   *available2* ← (*available* ∩ [*p* + 2 .. max_per]) \ FORBIDDENSAMEPOS(*p*)
10.   *possible_sets* ← *possible_sets* ∪ RUNSSAMEPOS(*current* ∪ {*p*}, *available2*)
11. **return**(*possible_sets*)

Fig. 4. The RUNSSAMEPOS function

RUNSSAMEPOS(∅, {1, 2, . . . , max_per}). The following notation is useful in connection with Lemma 4:

$$\text{GOODKS}(p, \ell) = \{2, 3, \ldots, \lceil p/\ell \rceil - 3\}$$
$$\cup \{\lceil p/\ell \rceil - 2 \text{ if } p - (\lceil p/\ell \rceil - 2)\ell \text{ is prime or } 4\}$$
$$\cup \{\lceil p/\ell \rceil - 1 = \tfrac{p-1}{\ell} \text{ if } p - (\lceil p/\ell \rceil - 1)\ell = 1\}.$$

Each of these sets of periods has an effect on the nearby positions, which we pre-compute as well. The function FORBIDDENLEFT(*H*, *j*), $1 \leqslant j \leqslant 2\max(H)$, shown in Fig. 5, computes the periods that are forbidden at $i - j$ by the set of period H at i. Note that it does not depend on i. The steps 2-4 are due to Lemma 5, the steps 5-7 are due to Lemma 4(i), and the steps 8-11 are due to Lemma 4(ii).

Symmetrically, the effect on the positions to the right are computed by the function FORBIDDENRIGHT(*H*, *j*), $1 \leqslant j \leqslant 2\max(H)$, whose code is identical to the one of FORBIDDENLEFT except for step 10 where $j \leqslant \ell$ is replaced by

FORBIDDENLEFT(*H*, *j*)

1.   *forbidden* ← ∅
2.   **for** (each *p* ∈ *H*) **do**
3.   **if** (*j* ⩽ *p*) **then**
4.   *forbidden* ← *forbidden* ∪ {*p*}
5.   **for** (each *ℓ* divisor of *p*) **do**
6.   **if** (*j* ⩽ *ℓ*) **then**
7.   *forbidden* ← *forbidden* ∪ {*p* − *ℓ*, *p* + *ℓ*}
8.   **for** (each *q* ∉ *H* with *q* < *p*) **do**
9.   **for** (each *ℓ* divisor of *p* − *q*) **do**
10.   **if** ((*j* ⩽ *ℓ*) and ($\frac{p-q}{\ell}$ ∈ GOODKS(*p*, *ℓ*)) and (*p* + *ℓ* ⩽ max_per)) **then**
11.   *forbidden* ← *forbidden* ∪ {*p* + *ℓ*}
12. **return**(*forbidden*)

Fig. 5. The FORBIDDENLEFT function

$j \leqslant \ell - 1$. This is due to the difference between (a) and (b) in Lemma 4(ii). Note that this difference does not affect the periods in the lemma and therefore the similar condition in step 6 is unchanged.

6   The Improved Algorithm

We include now the improvements in the previous section in our TEST function. The range to be tested is $[0 .. N]$ (N is the previous *too_large*). We start at N and advance to the left by considering longer and longer histories. If we reach position 0, then the algorithm terminates without success. At each step we have some runs that are forbidden as well as some that already exist. Therefore, we use a two dimensional array, $history[- \mathsf{max\_per} .. N + \mathsf{max\_per}][1 .. \mathsf{max\_per}]$, to store this information. (The first range exceeds $[0 .. N]$ both ways by $\mathsf{max\_per}$ positions to be able to store all information required.) We put

$$
history[i][p] = \begin{cases} 1 & \text{if run } p \text{ at } i \text{ exists already,} \\ -1 & \text{if run } p \text{ at } i \text{ is forbidden,} \\ 0 & \text{if no value has been assigned.} \end{cases}
$$

We shall pass to the function TEST the *history* and current position *pos*. The current history being tested is given by the 1's in the array $history[pos + 1 .. N]$. The length of this history is $N - pos$.

The improved TEST function is given in Fig. 6. Steps 1-5 are similar to what we had before. In steps 6-17 we restrict the possible values for the runs at the current position *pos* according to the theory in the previous section. Two sets, *existing* and *forbidden*, contain periods that must and must not be, respectively, included among the runs at *pos*. They are computed from the information already in *history* (steps 8-10). We start with all possible sets of runs at the same position and eliminate all sets that do not obey the restrictions imposed by *existing* and *forbidden* (steps 11-13). In addition, we eliminate the sets whose FORBIDDENRIGHT sets conflict with the information in *history* (steps 14-17).

We try then the remaining ones in steps 18-32. For each, we update *history* in steps 19-29. We copy first the information from the current set in $history[pos]$ (steps 20-21) and then use the FORBIDDENLEFT set to impose negative restrictions in *history* (steps 22-24). Small runs included in a single period of larger runs are copied from the right period to the left to impose some positive restrictions in *history* (steps 25-29).

Finally, the function ISHISTORY uses information from the previous history in order to compute the s_h and $t_h \in T_h$ strings. This is going to be passed as an union-find data structure *graph* that is updated by the function UNIONFIND using the information from H. The function ISHISTORY has been described in detail in the previous section. All runs in a string are computed using the linear-time algorithm of Kolpakov and Kucherov [7], where the Lempel–Ziv factorization is computed by the recent algorithm of Crochemore and Ilie [3].

The main program, TESTMICRORUNSBOUND, will simply initialize all elements of *history* on 0 and then call TEST($history, N, \emptyset$).

TEST($history, pos, graph$)

1. **if** $\Big((pos = 0)$ **or** $\big((pos \neq N)$ **and** $(\frac{\text{BRUNS}(history[pos+1..N])}{N-pos} > b)\big)\Big)$ **then**
2.   PRINT("not amortized"); EXIT()   // exit the main program
3. **if** $\big((pos \neq N)$ **and** $(\frac{\text{ALLRUNS}(history[pos+1..N])}{N-pos} \leqslant b)\big)$ **then**
4. $am\_pos \leftarrow \max(am\_pos, N - pos + 1)$
5.   **return**()   // amortized: done with current history
   // steps 6-17: restrict the possible continuations $history[pos]$
6. $possible\_sets \leftarrow$ RUNSSAMEPOS($\emptyset, \{1, 2, \ldots, \mathsf{max\_per}\}$)
7. $existing \leftarrow \emptyset; forbidden \leftarrow \emptyset$
8. **for** p **from** 1 **to** max_per **do**
9.   **if** $(history[pos][p] = 1)$ **then** $existing \leftarrow existing \cup \{p\}$
10.   **if** $(history[pos][p] = -1)$ **then** $forbidden \leftarrow forbidden \cup \{p\}$
11. **for** (each $H \in possible\_sets$) **do**
12.   **if** $((existing \setminus H \neq \emptyset)$ **or** $(forbidden \cap H \neq \emptyset))$ **then**
13. $possible\_sets \leftarrow possible\_sets \setminus \{H\}$
14.   **for** i **from** 1 **to** $2\max(H)$ **do**
15.   **for** (each $p \in$ FORBIDDENRIGHT(H, i)) **do**
16.   **if** $(history[pos + i][p] = 1)$ **then**
17. $possible\_sets \leftarrow possible\_sets \setminus \{H\}$
   // steps 18-32: try the continuations that are histories
18. **for** (each $H \in possible\_sets$) **do**
   // steps 19-29: update $history$
19. $history2 \leftarrow history$
20.   **for** (each $p \in H$) **do**
21. $history2[pos][p] \leftarrow 1$
22.   **for** i **from** 1 **to** $2\max(H)$
23.   **for** (each $p \in$ FORBIDDENLEFT(H, i)) **do**
24. $history2[pos - i][p] \leftarrow -1$
25.   **for** (each $p \in H$) **do**
26.   **for** i **from** 1 **to** $p - 2$ **do**
27.   **for** q **from** 1 **to** $\lfloor \frac{p-i}{2} \rfloor$ **do**
28.   **if** $history2[pos + i + q][q] = 1$ **then**
29. $history2[pos + i + q - p][q] \leftarrow 1$
30. $graph2 \leftarrow$ UNIONFIND($graph, H$)
31.   **if** (ISHISTORY($history2[pos..N], graph2$)) **then**
32.   TEST($history2, pos - 1, graph2$)
33. **return**()

Fig. 6. The improved TEST function

7 Results

We tested the program TESTMICRORUNSBOUND on the SHARCNET high-speed clusters (www.sharcnet.ca) and obtained the results in Table 1.

Table 1. The bounds on the maximum number of runs in a string of length n, RUNS(n), obtained using TESTMICRORUNSBOUND for the given values of maximum periods of microruns, max_per, and the bound b, that is, RUNS$(n) \leqslant \frac{6}{\text{max\_per}+1}n + bn$. The columns labeled "solutions" and "amortize" give the number of histories for which amortization succeeded in step 5 and the highest number of positions needed to amortize, resp.

max_per	bound b	solutions	amortize	RUNS$(n) \leqslant$
9	0.85	630	100	$1.450n$
10	0.85	900	100	$1.396n$
15	0.89	5275	27	$1.265n$
20	0.89	34833	97	$1.176n$
25	0.91	135457	153	$1.141n$
30	0.91	471339	153	$1.104n$
35	0.93	1455422	82	$1.097n$
40	0.93	3907110	84	$1.077n$
50	0.93	22635894	139	$1.048n$

8 Conclusion

We are very close to solving the conjecture. In fact, solving the conjecture seems to be now only a matter of time. We should note however that the bound in the conjecture, RUNS$(n) \leqslant n$, may not be the optimal one. For all practical purposes, the bound n (or even the best one we obtained) is good enough, but the search for the optimal one will continue. Very likely, different tools will be needed in finding the optimal bound as approximations, no matter how good, will probably keep us even asymptotically away from it.

References

1. Crochemore, M.: An optimal algorithm for computing the repetitions in a word. Inform. Proc. Letters 12, 244–250 (1981)
2. Crochemore, M., Ilie, L.: Maximal repetitions in strings. J. Comput. Syst. Sci. (in press, 2007)
3. Crochemore, M., Ilie, L.: Computing Longest Previous Factor in linear time and applications. Inform. Process. Lett. 106, 75–80 (2008)
4. Crochemore, M., Ilie, L., Rytter, W.: Repetitions in strings: algorithms and combinatorics. Theoret. Comput. Sci. (to appear)
5. Franek, F., Simpson, R.J., Smyth, W.F.: The maximum number of runs in a string. In: Miller, M., Park, K. (eds.) Proc. 14th Australasian Workshop on Combinatorial Algorithms, pp. 26–35 (2003)
6. Giraud, M.: Not so many runs in strings. In: Martin-Vide, C. (ed.) Proc. of LATA 2008 (to appear, 2008)
7. Kolpakov, R., Kucherov, G.: Finding maximal repetitions in a word in linear time. In: Proc. of FOCS 1999, pp. 596–604. IEEE Computer Society Press, Los Alamitos (1999)

8. Main, M.G.: Detecting lefmost maximal periodicities. Discrete Applied Math. 25, 145–153 (1989)
9. Main, M.G., Lorentz, R.J.: An O(n log n) algorithm for finding all repetitions in a string. J. Algorithms 5(3), 422–432 (1984)
10. Puglisi, S.J., Simpson, J., Smyth, B.: How many runs can a string contain? (submitted, 2006)
11. Rytter, W.: The number of runs in a string: improved analysis of the linear upper bound. In: Durand, B., Thomas, W. (eds.) STACS 2006. LNCS, vol. 3884, pp. 184–195. Springer, Heidelberg (2006)
12. Rytter, W.: The number of runs in a string. Inf. Comput. 205(9), 1459–1469 (2007)

On the Longest Common Parameterized Subsequence

Orgad Keller, Tsvi Kopelowitz, and Moshe Lewenstein

Department of Computer Science, Bar-Ilan University, Ramat-Gan 52900, Israel
{kellero,kopelot,moshe}@cs.biu.ac.il

Abstract. The well-known problem of the longest common subsequence (LCS), of two strings of lengths n and m respectively, is $O(nm)$-time solvable and is a classical distance measure for strings. Another well-studied string comparison measure is that of parameterized matching, where two equal-length strings are a parameterized-match if there exists a bijection on the alphabets such that one string matches the other under the bijection. All works associated with parameterized pattern matching present polynomial time algorithms.

There have been several attempts to accommodate parameterized matching along with other distance measures, as these turn out to be natural problems, e.g., Hamming distance, and a bounded version of edit-distance. Several algorithms have been proposed for these problems.

In this paper we consider the longest common parameterized subsequence problem which combines the LCS measure with parameterized matching. We prove that the problem is NP-hard, and then show a couple of approximation algorithms for the problem.

1 Introduction

The problem of finding the *longest common subsequence*, denoted as *LCS*, of two given strings is one of the classical and well-studied problems in the area of algorithms: given two strings B and C of lengths n and m respectively (throughout this paper we will assume $n \geq m$), we wish to find the longest string that is a subsequence of both B and C.

For apparent reasons, LCS is one of the most natural measures used to test the *similarity* between two strings. While this problem and its variants are interesting theoretically, they are of fundamental practical use in the areas of molecular biology and code analysis, e.g., where one wishes to test the differences between two programming language code fragments. To name only one, the well known UNIX *diff* command applies LCS as its main tool.

The classic and well-known solution of Wagner and Fischer [19] uses dynamic programming to solve the problem in time $O(nm)$. It can be generalized to solve LCS for any *fixed* number of input strings in polynomial time. Masek and Paterson [16] improved the running time of the case where $n = m$ to $O(n^2/\log n)$, by using the *"four russians"* technique. Other solutions—e.g., [10,18,17]—in which the running time of the solutions are dependent on different parameters besides the length of the strings, have also been provided.

P. Ferragina and G. Landau (Eds.): CPM 2008, LNCS 5029, pp. 303–315, 2008.
© Springer-Verlag Berlin Heidelberg 2008

While, as mentioned, the problem for any fixed number of strings can be solved in polynomial time, Maier [15] showed that LCS on an *arbitrary* number of strings is NP-hard (by applying a reduction from *vertex cover*), and later Jiang and Li [11] showed that there exists a constant $\delta > 0$ for which there is not an n^δ-approximation algorithm for the problem, unless P = NP. Note that when the number of input strings is fixed to be 2, almost all LCS variants can be solved in polynomial time.

Another very important and interesting model for testing similarity between strings, introduced by Baker [2,3,4,5], is called *parameterized matching*, or *p-match* in short. In this model, two length-n input strings are said to p-match if (roughly, and will be detailed later) there exists a bijection on the alphabet symbols which maps the i-th symbol of the first string to the i-th symbol of the second. As the symbols of the alphabet can be, for example, programming language code tokens, this model has practical importance in testing whether two code segments are essentially the same, even when some tokens (e.g., variable names) have been globally renamed.

In parameterized pattern matching, we get a length-n text and a length-m pattern and wish to report all locations i in the text where the pattern p-matches the length-m text substring starting at location i. Extensive amount of work has been done on this problem: Amir et al. [1] showed an efficient algorithm even when the alphabet size is $O(n)$, which runs in worst-case $O(n \log \sigma)$ time, where σ is the size of the parametric alphabet. In [8] they showed how to efficiently provide an approximate solution, and in [9] they generalized the problem for the 2-dimensional case. In [13,3,4,6] it was aimed at providing *parameterized text indexing*, and was shown how to efficiently construct a *parameterized suffix tree*. Finally, Ferragina and Grossi [7] showed how to provide for efficient parameterized text indexing even in external memory.

In parameterized pattern matching, we benefit from two facts: the first, that in each match, *consecutive* symbols of the text are compared against the pattern, and the second, that in two locations where the pattern matches the text, the corresponding bijections need not be the same. It is very natural and tempting to solve the problem without using these conditions to aid us; by this, we adapt the p-match model to the LCS problem, thus defining the LCPS problem discussed in this paper. Such a setting would be very practical in the case where, for example, two code fragments—an original, and a suspected copy—are being tested for similarity after the alleged copy has been edited, besides possibly having its variable names changed. Unfortunately, we show that this problem is NP-hard. We prove this by a reduction from the problem of 3D-matching in a graph [12], and then provide a couple of approximation algorithms, which yield a $\lambda\sqrt{|\mathrm{OPT}|}$-length solution for any constant λ, where OPT is the optimal solution.

A note must be made about the similarity between LCS and *edit-distance* [14]: testing the similarity of two strings via LCS is the equivalent of doing so using edit-distance when the edit operations allowed are only insertions and deletions. Baker [5] discusses the notion of *parameterized edit-distance*, in which the operations allowed are insertions, deletions, and p-matches, where the p-match edit operation replaces a *substring* in the first string with a substring that p-matches it which appears in the second. Therefore, the aiding conditions of parameterized pattern matching still play a role there.

The rest of this paper is organized as follows: in Sect. 2 we provide the formal definitions of our problems. In Sect. 3 we provide some preliminaries. In Sect. 4 we provide a naïve algorithm for the specific case where the parametric alphabet is small. In Sect. 5 we prove that the LCPS problem is NP-hard. In Sects. 6 and 7 we provide an approximation for a specific case of the problem called *LCMS*, and for the general LCPS, respectively. In Sect. 8 we give our concluding remarks.

2   Problem Definitions

Let $S = s_1 \ldots s_n$ and $T = t_1 \ldots t_n$ be strings over alphabet set $\Sigma \cup \Pi$, such that $\Sigma \cap \Pi = \emptyset$. We say S and T are a *parameterized-match* (*p-match* for short) if there exists a bijection $f \colon \Pi \to \Pi$ for which, for each $i = 1, \ldots, n$, it holds that:

1. if $s_i \in \Sigma$, then $s_i = t_i$.
2. if $s_i \in \Pi$, then also $t_i \in \Pi$, and $f(s_i) = t_i$.

For two strings $B = b_1 \ldots b_n$ and $C = c_1 \ldots c_m$ over $\Sigma \cup \Pi$, We define their *common parameterized subsequence* (CPS for short) as a pair of two ascending sequences $I = \langle i_1, \ldots, i_k \rangle$ and $J = \langle j_1, \ldots, j_k \rangle$ of locations in B and C respectively (i.e., $i_\ell \in \{1, \ldots, n\}$ and $j_\ell \in \{1, \ldots, m\}$ for each $\ell = 1, \ldots, k$), such that B^I p-matches C^J, where $B^I = b_{i_1} b_{i_2} \ldots b_{i_k}$ and $C^J = c_{j_1} c_{j_2} \ldots c_{j_k}$.

The *longest common parameterized subsequence problem* is defined as follows:

Input: Two strings $B = b_1 \ldots b_n$ and $C = c_1 \ldots c_m$ over alphabet set $\Sigma \cup \Pi$, such that $\Sigma \cap \Pi = \emptyset$.
Output: A CPS of maximal length, denoted LCPS.

By CPS we will also denote the decision version of the problem, in which we ask whether two strings have a common parameterized subsequence of a specified length. The meaning will be clear from the context.

The specific case of the LCPS problem in which $\Sigma = \emptyset$ (i.e., the only alphabet is the parametric alphabet Π) is denoted the *longest common mapped subsequence (LCMS)* problem.

3   Preliminaries

Let $\mathcal{A}$ be an algorithm (exact or approximate) for the LCPS problem. $\mathcal{A}(B, C)$ returns a pair (I, J) of sequences of indices in B and C respectively. Denote $I = \langle i_1, \ldots, i_k \rangle$ and $J = \langle j_1, \ldots, j_k \rangle$. We define the length of the solution $|(I, J)| = |I| = |J| = k$ and denote $|\mathcal{A}(B, C)| = |(I, J)|$.

A convenient way of describing the CPS restrictions is by defining the *sequence graph*: given the input strings B and C and two sequences $\langle i_1, \ldots, i_k \rangle$ and $\langle j_1, \ldots, j_k \rangle$ of locations in B and C respectively, a *sequence graph* is a directed planar graph $G = (V, E)$ in which the vertex set V is the set of location-specific characters of B and C, set on a grid in the following manner:

1. for each $i = 1, \ldots, n$, b_i is set at grid location $(i, 1)$;
2. for each $j = 1, \ldots, m$, c_j is set at grid location $(j, 0)$;

and E is defined such that there is an edge from $(i_\ell, 1)$ to $(j_\ell, 0)$ for each $\ell = 1, \ldots, k$. Formally: $E = \{((i_\ell, 1), (j_\ell, 0)) \mid \ell = 1, \ldots, k\}$.

Remark 1. For convenience, when we refer to some edge written as "(b_i, c_j)" or described as "the edge mapping b_i to c_j", we mean the specific edge from grid-point $(i, 1)$ to grid-point $(j, 0)$ (if such exists), and not to any other edge whose endpoints are two other grid-points labeled with the symbol b_i and the symbol c_j, respectively, which might also exist in the graph.

If a sequence graph contains some edge (b_i, c_j), we say b_i is *mapped* to c_j. Two different edges $(b_i, c_j), (b_{i'}, c_{j'})$ are said to be *intersecting* if the straight line on the plane connecting grid-point $(i, 1)$ to grid-point $(j, 0)$ (which corresponds to (b_i, c_j)) crosses the straight line connecting $(i', 1)$ to $(j', 0)$ (which corresponds to $(b_{i'}, c_{j'})$). Alternatively: if $i' \geq i$, but $j' \leq j$.

Observation 1. *If the sequences $\langle i_1, \ldots, i_k \rangle$ and $\langle j_1, \ldots, j_k \rangle$ are both ascending, then the sequence graph does not contain intersecting edges.*

A sequence graph is said to be a *CPS graph* if it corresponds to some CPS, i.e., to two sequences $\langle i_1, \ldots, i_k \rangle$ and $\langle j_1, \ldots, j_k \rangle$ which comply with the conditions described in Sect. 2. Notice that there is always a one-to-one correspondence between a CPS of two strings and a CPS graph.

Let X, Y, Z be three disjoint sets such that $|X| = |Y| = |Z| = n$, and let $S \subseteq X \times Y \times Z$. In the *3D-matching* problem [12], we wish to find a subset $S' \subseteq S$ which is a *perfect matching* of X, Y, and Z, i.e., *every* element of X, Y, and Z is covered by S' exactly *once*. In the problem's decision version, denoted 3DM, when given (X, Y, Z, S), we say $(X, Y, Z, S) \in$ 3DM if there exists such a perfect matching $S' \subseteq S$. Notice that we can always assume $n < |S| < n^3$, otherwise solving the problem is trivial.

4   Solving the Problem for Asymptotically-Small Π

Theorem 1. *There exists an algorithm $\mathcal{N}$ for the LCPS problem, which solves the problem in $O(|\Pi|! \cdot nm)$ time.*

Proof. We propose the following "naïve" algorithm: for each possible bijection $f : \Pi \to \Pi$, construct a new string B_f by replacing each symbol $b_i \in \Pi$ in B with $f(b_i)$, and find $\text{LCS}(B_f, C)$ using [19]. Finally, choose the bijection f for which $\text{LCS}(B_f, C)$ gave a maximal-length result, and recover its corresponding indices in B_f (and hence, in B) and in C. Clearly, this algorithm is correct. Since there are $|\Pi|!$ possible bijections from Π to Π, and [19] runs in time $O(nm)$, the running time is $O(|\Pi|! \cdot nm)$.   □

Corollary 1. *If $|\Pi| = c$ for some constant c, then the LCPS problem can be solved in time $O(nm)$.*

Corollary 2. *Assume w.l.o.g. that $n \geq m$ and let c be a constant. If $|\Pi| \leq \frac{c \log n}{\log \log n}$, then the LCPS problem can be solved in time*

$$O((c \log n / \log \log n)! \cdot nm) = O\left(2^{\frac{c \log n}{\log \log n} \log\left(\frac{c \log n}{\log \log n}\right)} nm\right) = O(n^{c+1} m) \ . \quad (1)$$

Remark 2. Note that $\mathcal{N}$ also trivially solves the LCMS problem, and therefore will be used as such later.

5   Finding the LCPS of Two Strings Is NP-Hard

We define the decision version of the LCPS problem: for two strings B and C and an integer t, we say $(B, C, t) \in$ CPS if there exists a solution (I, J) for LCPS(B, C) such that $|(I, J)| \geq t$.

Theorem 2. *LCPS is NP-hard. Alternatively: if there exists a polynomial-time algorithm for LCPS, then* $P = NP$.

Proof. We show that LCPS is NP-hard (or rather, that CPS $\in$ NPC) using a reduction from 3DM:

The Reduction. Given the input-tuple (X, Y, Z, S) for the 3DM problem, where $|X| = |Y| = |Z| = n$ (note that in this section n denotes the size of X, Y, and Z) and $S = \{t_1, \ldots, t_s\} \subseteq X \times Y \times Z$, we choose $\Sigma = \emptyset$ and $\Pi = X \cup S \cup \{*\}$. In order to construct the reduction strings properly, we first require some notation: for a specific tuple $t_i = (x, y, z)$, we denote $x(t_i) = x$, $y(t_i) = y$ and $z(t_i) = z$. For some fixed $y_i \in Y$, we define $S(y_i) = \{(x, y, z) \in S \mid y = y_i\}$, i.e., $S(y_i)$ is the set of all tuples in S having y_i as their y-coordinate. Denote $s(y_i) = |S(y_i)|$. Furthermore, assume $S(y_i) = \{t_{r_1}, \ldots, t_{r_{s(y_i)}}\}$, where the sequence $\langle t_{r_1}, \ldots, t_{r_{s(y_i)}} \rangle$ is $S(y_i)$ sorted in ascending order of x-coordinates. We define the blocks

$$\mathcal{B}_{y_i}^B = x(t_{r_1}) x(t_{r_2}) \ldots x(t_{r_{s(y_i)-1}}) x(t_{r_{s(y_i)}}) \ , \quad (2)$$

and

$$\mathcal{B}_{y_i}^C = t_{r_{s(y_i)}} t_{r_{s(y_i)}-1} \ldots t_{r_2} t_{r_1} \ . \quad (3)$$

In other words, in $\mathcal{B}_{y_i}^B$ we list the x-coordinates of the tuples in an *ascending* order, and in $\mathcal{B}_{y_i}^C$ we list the tuples themselves (each tuple serves as a single character), only this time, in the *descending* order of their respective x-coordinates. As we shall see later, the role of $\mathcal{B}_{y_i}^B$ and $\mathcal{B}_{y_i}^C$ will be to assure that no two tuples which share the same y-coordinate value will be included in S', i.e., each y_i will be covered at most once by a tuple in S'. Finally, we define $\mathcal{B}_{z_i}^B$ and $\mathcal{B}_{z_i}^C$, using the same principle, only this time for the z-coordinates.

We now move to construct the strings, each comprised of three segments:

$$B = \overbrace{x(t_1) *^3 \ldots *^3 x(t_s) *^3}^{\text{Seg. 1}} \overbrace{\mathcal{B}_{y_1}^B *^3 \ldots *^3 \mathcal{B}_{y_n}^B *^3}^{\text{Seg. 2}} \overbrace{\mathcal{B}_{z_1}^B *^3 \ldots *^3 \mathcal{B}_{z_n}^B *^3}^{\text{Seg. 3}} \ , \quad (4)$$

and

$$C = \underbrace{t_1 *^3 \ldots *^3 t_s *^3}_{\text{Seg. 1}} \underbrace{\mathcal{B}^C_{y_1} *^3 \ldots *^3 \mathcal{B}^C_{y_n} *^3}_{\text{Seg. 2}} \underbrace{\mathcal{B}^C_{z_1} *^3 \ldots *^3 \mathcal{B}^C_{z_n} *^3}_{\text{Seg. 3}} . \tag{5}$$

Notice that each of the strings contains $s + 2n$ blocks of $*$ symbols—each block is of length 3—and $3s$ non-$*$ symbols (since each tuple appears exactly once in each segment of C, and for each such single appearance, the tuple's x-coordinate appears once in the respective segment of B). We derive that $|B| = |C| = 3(s + 2n) + 3s = 6s + 6n$. Finally, we choose $t = 3s + 9n < 6s + 6n$.

Before showing that this reduction is correct, we require some definitions: for some sequence graph of B and C, we define an $(*, *)$-*type edge* as an edge whose endpoints are both $*$ symbols, and an (x, t)-*type edge* as an edge whose endpoint in B is some x-coordinate value, and whose endpoint in C is some tuple. Likewise we define an $(x, *)$-*type edge* and an $(*, t)$-*type edge*. We continue to the following claim:

Claim. Assume a CPS of B and C is given, and is of length $3s + 9n$, and let f be its corresponding bijection. Then the following statements apply to the corresponding CPS graph and bijection f:

1. $f(*) = *$.
2. There are exactly $3n$ (x, t)-type edges, and exactly $3s + 6n$ $(*, *)$-type edges.
3. Every $*$ at some location i in B is mapped to its respective $*$ at location i in C.
4. Each segment of B contributes exactly n (x, t)-type edges. In particular, Segment 1 of B contributes n (x, t)-type edges, all of them vertical.
5. Each $\mathcal{B}^B_{y_i}$ (resp. $\mathcal{B}^B_{z_j}$) block contributes exactly one edge, to a symbol in $\mathcal{B}^C_{y_i}$ (resp. $\mathcal{B}^C_{z_j}$).

Proof. We prove each item using the previous ones:

1. Assume by contradiction that $f(*) \neq *$. In this case (as shown by a very loose analysis), there are (a) at most 3 $(*, t)$-type edges (since each unique tuple t_i appears at most 3 times in C, one in each segment), (b) at most $3s$ $(x, *)$-type edges (since a unique x-coordinate appears at most s times in each segment of B), and (c) at most $3n$ (x, t)-type edges (since there are n distinct x-coordinates, each of them may be mapped to a tuple, and each unique tuple appears 3 times in C). We derive that this scenario gives us at most $3 + 3s + 3n < 3s + 9n$ edges, which contradicts the fact that the LCPS is of length $3s + 9n$. We conclude that indeed, $f(*) = *$.
2. From the last item it follows that each x-coordinate x_i is mapped by f to some tuple t_j. Since each unique tuple appears exactly 3 times in C, and there are n distinct x-coordinates, then there are at most $3n$ (x, t)-type edges. Now, since the number of $(*, *)$-type edges is bounded by $3(s + 2n)$ (the number of $*$ symbols in each string), we conclude that in order to reach length $3s + 9n$, we require the number of $(*, *)$-type edges to be exactly $3(s + 2n) = 3s + 6n$, and the number of (x, t)-type edges to be exactly $3n$.

3. Since the number of $(*, *)$-type edges is $3s + 6n$, and no two edges can intersect each other (since it is a CPS graph), the only way to obtain this number of edges is by mapping *every* $*$ at some location i in B to the $*$ at the respective location i in C.

4. First of all, notice that an (x, t)-type edge emanating from a specific segment in B cannot go to other than its respective segment in C, otherwise it would result in the loss of $(*, *)$-type edges, which would contradict Item 3. In each segment of B, there are n distinct x-coordinates. In each segment of C, each unique tuple appears once. Therefore, each segment can contribute at most n (x, t)-type edges, and must contribute exactly n of those, otherwise we would not reach the target length. Finally, each non-vertical (x, t)-type edge emanating from Segment 1 of B would result in the loss of $(*, *)$-type edges. We conclude all (x, t)-type edges in Segment 1 of B are vertical and therefore go to symbols in Segment 1 of C.

5. First notice that an edge emanating from some block $\mathcal{B}_{y_i}^B$ cannot go to other than the block $\mathcal{B}_{y_i}^C$; the opposite would result in losing $(*, *)$-type edges. We proceed to show that there is at most a single edge from each block. Assume by contradiction that there are two edges from $\mathcal{B}_{y_i}^B$ to $\mathcal{B}_{y_i}^C$, and let them be (x_a, t_c) and (x_b, t_d). Assume w.l.o.g. that x_b appears right of x_a in $\mathcal{B}_{y_i}^B$. Since a unique tuple can appear at most once in $\mathcal{B}_{y_i}^C$, then obviously $t_c \neq t_d$. It follows that also $x_a \neq x_b$ (since f is a proper function). Notice that $x(t_c) = x_a$ and $x(t_d) = x_b$ (in words, both edges must be from an x-value to a tuple having this value as its x-coordinate), otherwise we would lose one of the n vertical (x, t)-type edges in Segment 1, which always map a value to a tuple having it as its x-coordinate. However, since x_b appears *right* of x_a and $x_a \neq x_b$, it follows that in $\mathcal{B}_{y_i}^C$, the tuples for which x_b is the x-coordinate appear *left* of the tuples for which x_a is the x-coordinate. In particular, t_d is left of t_c in $\mathcal{B}_{y_i}^C$. We conclude that the two edges intersect, which contradicts the fact that this is a CPS graph. The proof for $\mathcal{B}_{z_j}^B$ and $\mathcal{B}_{z_j}^C$ is similar. We have just proved that each $\mathcal{B}_{y_i}^B$ (resp. $\mathcal{B}_{z_j}^B$) block contributes at most a single edge, but since we require n edges from Segment 2 (resp. Segment 3) in order to obtain the target length, we conclude that each such block contributes exactly one edge. $\qquad\square$

It remains to show that the reduction described is correct:

Claim. $(X, Y, Z, S) \in 3DM$ if and only if $(B, C, 3s + 9n) \in$ CPS.

Proof. We prove both directions:

(only if) Given a subset $S' \subseteq S$, $|S'| = n$, which covers each element of X, Y, or Z exactly once (i.e., S' is a *perfect matching*), we determine the respective I, J sequences by describing a CPS graph: for each $i = 1, \ldots, 6s + 6n$:
1. If $b_i = c_i = *$, then map b_i to c_i.
2. Otherwise, c_i is some tuple in S. If it also holds that $c_i \in S'$, then:

(a) If i is a location in Segment 1, map b_i to c_i.

(b) If i is a location in Segment 2, then c_i appears as a symbol in the block $\mathcal{B}^C_{y(c_i)}$, and therefore $x(c_i)$ appears as a symbol b_j in $\mathcal{B}^B_{y(c_i)}$. Therefore, map b_j to c_i.

(c) If i is a location in Segment 3, the argument is similar, only this time with $\mathcal{B}^C_{z(c_i)}$ and $\mathcal{B}^B_{z(c_i)}$ respectively.

Claim. The above scheme yields a CPS graph and therefore a CPS of length $3s + 9n$.

Proof. First notice that the mappings of the form $(*, *)$ actually define that $f(*) = *$ and contribute $3s + 6n$ edges. Since they are all vertical, they do not intersect with each other. Since all other edges in Segment 1 are also vertical (i.e., are of the form (b_i, c_i)), they do not intersect with the above edges or each other. In addition, since S' is a matching, each unique x_i value is mapped to a unique tuple denoted $t(x_i)$ having x_i it as its x-coordinate value. Hence it defines by this that $f(x_i) = t(x_i)$ for $i = 1, \ldots, n$. Since $|S'| = n$, we conclude that this has contributed another n edges. Finally, at each $\mathcal{B}^B_{y_i}$ block, we make a *single* mapping to a value in $\mathcal{B}^C_{y_i}$ (because S' is a matching, and all tuples in $\mathcal{B}^C_{y_i}$ share the same y-coordinate, and in addition a unique tuple can appear at most once in $\mathcal{B}^C_{y_i}$). Notice that mappings in these blocks are consistent with mappings in Segment 1, and therefore agree with the definition of f made before. The argument for $\mathcal{B}^B_{z_j}$ and $\mathcal{B}^C_{z_j}$ is similar. Finally, since each $\mathcal{B}^B_{y_i}$ or $\mathcal{B}^B_{z_j}$ block contributes a single edge, we conclude that those blocks contributed $2n$ edges all together, none of them intersects with other edges. It follows that the constructed graph is a CPS graph with $3s + 9n$ edges and therefore the claim follows.   □

We thus conclude that $(B, C, 3s + 9n) \in$ CPS.

(if) Assume that $(B, C, 3s + 9n) \in$ CPS, i.e., B, C have a common parameterized subsequence of length $3s + 9n$, and consider the corresponding CPS graph and the bijection f. By Item 4 of the first claim, each (x, t)-type edge in Segment 1 is vertical and therefore agrees with the mapping of each unique x_i to a *unique* tuple t_j for which $x(t_j) = x_i$. Define $S' = \{t_j \mid \exists x_i, \quad f(x_i) = t_j\}$. Since all tuples sharing the same y-coordinate (resp. z-coordinate) appear in the same $\mathcal{B}^C_y$ (resp. $\mathcal{B}^C_z$) block, and by Item 5 such block contributes a single edge (which agrees with the mappings defined by the edges in Segment 1, since f is a bijection), we conclude each unique y-coordinate (resp. z-coordinate) is covered, and furthermore covered exactly once by S'. We conclude that S' is a perfect matching and therefore $(X, Y, Z, S) \in$ 3DM.   □

3DM $\in$ NPC, CPS is trivially in NP, and the above reduction clearly can be performed in polynomial time. We therefore conclude CPS $\in$ NPC. Therefore if LCPS admits a polynomial-time algorithm, then P $=$ NP.   □

Algorithm 1: $\mathcal{A}_{\lambda}^{\mathrm{LCMS}}(B, C)$

---

1 calculate the values $|\Pi_B|, |\Pi_C|$;

2 $\pi_{\min} \leftarrow \min\{|\Pi_B|, |\Pi_C|\}$;

3 **foreach** *possible* $\Pi' \subseteq \Pi_B$ *and* $\Pi'' \subseteq \Pi_C$, *such that* $|\Pi'| = |\Pi''| = \lambda^2$ **do**

4   construct the strings $B_{\Pi'}, C_{\Pi''}$;

5   run $\mathcal{N}(B_{\Pi'}, C_{\Pi''})$;

6 choose Π', Π'' which yielded maximal result, and let k be the length of the resulting solution;

7 **if** $k \geq \pi_{\min}$ **then**

8   construct the ascending sequences $I = \langle i_1, \ldots, i_k \rangle$ and $J = \langle j_1, \ldots, j_k \rangle$ of effective locations in $B_{\Pi'}$ and $C_{\Pi''}$, respectively, chosen by the naïve algorithm;

9 **else**

10    $k \leftarrow \pi_{\min}$;

11   choose an ascending sequence $I = \langle i_1, \ldots, i_k \rangle$ such that i_ℓ ($\ell = 1, \ldots, k$) is the first (i.e., leftmost) occurrence of the symbol b_{i_ℓ} in B;

12   choose an ascending sequence $J = \langle j_1, \ldots, j_k \rangle$ such that j_ℓ ($\ell = 1, \ldots, k$) is the first (i.e., leftmost) occurrence of the symbol c_{j_ℓ} in C;

13 **return** (I, J);

---

6 Approximating LCMS

Recall that LCMS is the specific case of the LCPS problem where $\Sigma = \emptyset$. For a given parameter $\lambda > 0$, we provide an $O(n^{2\lambda^2+1}m)$-time algorithm, $\mathcal{A}_{\lambda}^{\mathrm{LCMS}}$, for which, for two strings B and C of lengths n and m respectively, $|\mathcal{A}_{\lambda}^{\mathrm{LCMS}}(B, C)| \geq \lambda\sqrt{|\mathrm{OPT}(B, C)|}$, where $\mathrm{OPT}(B, C)$ denotes the optimal solution.

First, some notation: for a string S, let $\Pi_S = \{a \in \Pi \mid a \text{ appears in } S\}$. Given some alphabet set $\Gamma \subseteq \Sigma \cup \Pi$, we denote by S_Γ the string S with all symbols not from Γ deleted, while, for symbols not deleted, preserving their original location in S. In other words, we keep aside each symbol in S_Γ its original location in S. We will refer to this location as the symbol's *effective* location. For our two strings B and C, let $\pi_{\min} = \min\{|\Pi_B|, |\Pi_C|\}$. Finally, let $\mathrm{OPT}(B, C) = (I^*, J^*)$ be the optimal solution, and let $I^* = \langle i_1^*, \ldots, i_t^* \rangle$ and $J^* = \langle j_1^*, \ldots, j_t^* \rangle$. We define π^* to be the number of distinct symbols which appear in B^{I^*} (equivalently, in C^{J^*}; by the problem properties, it is the same).

6.1 The Algorithm

Algorithm 1 utilizes the fact that two strategies for the LCMS problem are available: for the first, notice that both $|\Pi_B| \geq \pi_{\min}$ and $|\Pi_C| \geq \pi_{\min}$ by the definition of $\pi_{\min}$. We can therefore create sequences I, J for which $|I| = |J| = \pi_{\min}$, by mapping the ℓ-th unique symbol which appears in B, to the ℓ-th unique symbol which appears in C, for $\ell = 1, \ldots, \pi_{\min}$. For the second strategy, assume we know the λ^2 symbols most frequent in B^{I^*}, and the λ^2 symbols most frequent in C^{J^*}. Running the naïve algorithm on the two strings, wherein all symbols not from the λ^2 most frequent are deleted, will yield a solution of length at least $\frac{|OPT(B,C)|}{\pi^*/\lambda^2}$ (since if we partition $\Pi_{B^{I^*}}$ to π^*/λ^2 sets, each of size λ^2, one of them must give us length of at least $\frac{|B^{I^*}|}{\pi^*/\lambda^2} = \frac{|OPT(B,C)|}{\pi^*/\lambda^2}$ when running the naïve algorithm on the strings induced by its symbols only). Since we do not know $\Pi_{B^{I^*}}$, we test every possible combination of λ^2 symbols in both strings and choose the combination yielding the maximal result. Finally, our approximation algorithm chooses the better of the two strategies.

6.2   Analysis

Theorem 3. *Given a parameter $\lambda > 0$, $\mathcal{A}_\lambda^{LCMS}$ is an $O(n^{2\lambda^2+1}m)$-time approximation algorithm for LCMS, such that $|\mathcal{A}_\lambda^{LCMS}(B,C)| \geq \lambda\sqrt{|OPT(B,C)|}$.*

Proof. We provide the approximation factor and the running-time analysis:

Approximation. From the discussion above, the algorithm returns sequences of length $\max\{\pi_{\min}, \frac{|OPT(B,C)|}{\pi^*/\lambda^2}\}$. Notice that:

$$\lambda^2|OPT(B,C)| = \pi^* \cdot \frac{|OPT(B,C)|}{\pi^*/\lambda^2} \tag{6}$$

$$\leq \min\{|\Pi_B|,|\Pi_C|\} \cdot \frac{|OPT(B,C)|}{\pi^*/\lambda^2} \tag{7}$$

$$= \pi_{\min} \cdot \frac{|OPT(B,C)|}{\pi^*/\lambda^2} , \tag{8}$$

where (7) is true because π^* is bounded by $\min\{|\Pi_B|,|\Pi_C|\}$ and (8) is true by definition. We therefore conclude that $\max\{\pi_{\min}, \frac{|OPT(B,C)|}{\pi^*/\lambda^2}\} \geq \lambda\sqrt{|OPT(B,C)|}$. Since $|\mathcal{A}_\lambda^{LCMS}(B,C)| = \max\{\pi_{\min}, \frac{|OPT(B,C)|}{\pi^*/\lambda^2}\}$, the approximation factor follows.

Running-Time. Step 1 of the algorithm can be done efficiently by sorting both strings according to the symbols of the alphabet. Step 11 and Step 12 can be efficiently executed by (a) leaving only one copy of each unique symbol in the two sorted strings, and (b) re-sort the sorted strings, this time using the indices as the keys by which the sorting is done. Since there are $\binom{|\Pi_B|}{\lambda^2} \leq n^{\lambda^2}$ options for Π', and $\binom{|\Pi_C|}{\lambda^2} \leq n^{\lambda^2}$ options for Π'', and running the naïve algorithm costs $O(nm)$, we conclude that the running-time is bounded by $O(n^{\lambda^2} \cdot n^{\lambda^2} \cdot nm) = O(n^{2\lambda^2+1}m)$. $\qquad\square$

7   Approximating LCPS

For a given parameter $\lambda > 0$, we provide an $O(n^{4\lambda^2+1}m)$-time algorithm, $\mathcal{A}_\lambda^{LCPS}$, for which, for two strings B and C of lengths n and m respectively, $|\mathcal{A}_\lambda^{LCPS}(B,C)| \geq \min\{\lambda\sqrt{|OPT(B,C)|}, \frac{1}{2}|OPT(B,C)|\}$.

Algorithm 2: $\mathcal{A}_\lambda^{\mathrm{LCPS}}(B, C)$

---

1 construct the strings B_Π, C_Π;

2 $(I', J') \leftarrow \mathcal{A}_{\sqrt{2}\lambda}^{\mathrm{LCMS}}(B_\Pi, C_\Pi)$;

3 construct the strings B_Σ, C_Σ;

4 $D \leftarrow \mathrm{LCS}(B_\Sigma, C_\Sigma)$;   `/* assume `$D = d_1 \ldots d_k$` */`

5 **if** $|(I', J')| \geq |D|$ **then return** (I', J');

6 else

7   construct the ascending sequences $I'' = \langle i_1, \ldots, i_k \rangle$ and $J'' = \langle j_1, \ldots, j_k \rangle$ of effective locations in B_Σ and C_Σ, respectively, such that $b_{i_\ell} = c_{j_\ell} = d_\ell$ $(\ell = 1, \ldots, k)$;

8   **return** (I'', J'');

7.1 The Algorithm

Note that almost all notation remains the same, except that this time, (I^*, J^*) is the solution returned by $\mathrm{OPT}(B_\Pi, C_\Pi)$ (instead of $\mathrm{OPT}(B, C)$, as before). Again, $I^* = \langle i_1^*, \ldots, i_t^* \rangle$ and $J^* = \langle j_1^*, \ldots, j_t^* \rangle$. π^* is defined as before to be the number of distinct symbols which appear in B^{I^*} (or equivalently, in C^{J^*}).

Algorithm 2 utilizes the fact that this time *three* strategies for the LCPS problem are available: while the first two remain the same as before—and thus, actually work now on B_Π and C_Π—the third corresponds to B_Σ and C_Σ: we can simply run the ordinary LCS algorithm on B_Σ and C_Σ, thus obtaining a legal CPS. As before, our approximation algorithm will choose the best of the three.

7.2 Analysis

Theorem 4. *Given a parameter $\lambda > 0$, $\mathcal{A}_\lambda^{\mathrm{LCPS}}$ is an $O(n^{4\lambda^2+1}m)$-time approximation algorithm for LCPS, such that*

$$|\mathcal{A}_\lambda^{\mathrm{LCPS}}(B, C)| \geq \min\{\lambda\sqrt{|\mathrm{OPT}(B, C)|}, \frac{1}{2}|\mathrm{OPT}(B, C)|\} \ .$$

Proof. We provide the approximation factor and the running-time analysis:

Approximation. $\sqrt{2}\lambda$ was used as the parameter when running $\mathcal{A}^{\mathrm{LCMS}}$ on B_Π and C_Π, and therefore $\mathcal{A}_{\sqrt{2}\lambda}^{\mathrm{LCMS}}$ returned a $\max\{\pi_{\min}, \frac{|\mathrm{OPT}(B_\Pi, C_\Pi)|}{\pi^*/2\lambda^2}\}$-length solution. It follows that the entire $\mathcal{A}_\lambda^{\mathrm{LCPS}}$ algorithm returned a solution of length $\max\{\pi_{\min}, \frac{|\mathrm{OPT}(B_\Pi, C_\Pi)|}{\pi^*/2\lambda^2}, |\mathrm{LCS}(B_\Sigma, C_\Sigma)|\}$. Notice that:

$$2\lambda^2|\mathrm{OPT}(B, C)| \leq 2\lambda^2|\mathrm{OPT}(B_\Pi, C_\Pi)| + 2\lambda^2|\mathrm{LCS}(B_\Sigma, C_\Sigma)| \tag{9}$$

$$= \pi^* \cdot \frac{|\mathrm{OPT}(B_\Pi, C_\Pi)|}{\pi^*/2\lambda^2} + 2\lambda^2|\mathrm{LCS}(B_\Sigma, C_\Sigma)| \tag{10}$$

$$\leq \pi_{\min} \cdot \frac{|\mathrm{OPT}(B_\Pi, C_\Pi)|}{\pi^*/2\lambda^2} + 2\lambda^2|\mathrm{LCS}(B_\Sigma, C_\Sigma)| \ , \tag{11}$$

where (9) is true because $|\mathrm{OPT}(B,C)| \leq |\mathrm{OPT}(B_\Pi, C_\Pi)| + |\mathrm{LCS}(B_\Sigma, C_\Sigma)|$ (since symbols from Π in the optimal solution cannot contribute more than $|\mathrm{OPT}(B_\Pi, C_\Pi)|$, and likewise, symbols from Σ in the optimal solution cannot contribute more than $|\mathrm{LCS}(B_\Sigma, C_\Sigma)|$), and (11) is true due to the same explanation of (7–8). We conclude that $\pi_{\min} \cdot \frac{|\mathrm{OPT}(B_\Pi, C_\Pi)|}{\pi^*/2\lambda^2} + 2\lambda^2|\mathrm{LCS}(B_\Sigma, C_\Sigma)| \geq 2\lambda^2|\mathrm{OPT}(B,C)|$ and therefore

$$\max\left\{ \pi_{\min} \cdot \frac{|\mathrm{OPT}(B_\Pi, C_\Pi)|}{\pi^*/2\lambda^2}, 2\lambda^2|\mathrm{LCS}(B_\Sigma, C_\Sigma)| \right\} \geq \lambda^2|\mathrm{OPT}(B,C)| . \quad (12)$$

We can therefore split to cases:

1. If $2\lambda^2|\mathrm{LCS}(B_\Sigma, C_\Sigma)| \geq \pi_{\min} \cdot \frac{|\mathrm{OPT}(B_\Pi, C_\Pi)|}{\pi^*/2\lambda^2}$, we get that $|\mathrm{LCS}(B_\Sigma, C_\Sigma)| \geq \frac{1}{2}|\mathrm{OPT}(B,C)|$.

2. Otherwise, $\pi_{\min} \cdot \frac{|\mathrm{OPT}(B_\Pi, C_\Pi)|}{\pi^*/2\lambda^2} > 2\lambda^2|\mathrm{LCS}(B_\Sigma, C_\Sigma)|$. Since It follows that $\pi_{\min} \cdot \frac{|\mathrm{OPT}(B_\Pi, C_\Pi)|}{\pi^*/2\lambda^2} \geq \lambda^2|\mathrm{OPT}(B,C)|$, in this case we finally conclude that $\max\{\pi_{\min}, \frac{|\mathrm{OPT}(B_\Pi, C_\Pi)|}{\pi^*/2\lambda^2}\} \geq \lambda\sqrt{|\mathrm{OPT}(B,C)|}$.

Summing up the two cases, we get:

$$\max\left\{ \pi_{\min}, \frac{|\mathrm{OPT}(B_\Pi, C_\Pi)|}{\pi^*/2\lambda^2}, |\mathrm{LCS}(B_\Sigma, C_\Sigma)| \right\}$$
$$\geq \min\left\{ \lambda\sqrt{|\mathrm{OPT}(B,C)|}, \frac{1}{2}|\mathrm{OPT}(B,C)| \right\} . \quad (13)$$

Since $|\mathcal{A}_\lambda^{\mathrm{LCPS}}(B,C)| = \max\{\pi_{\min}, \frac{|\mathrm{OPT}(B_\Pi, C_\Pi)|}{\pi^*/2\lambda^2}, |\mathrm{LCS}(B_\Sigma, C_\Sigma)|\}$, the approximation factor follows.

Running-Time. The running-time is dominated by the use of $\mathcal{A}_{\sqrt{2}\lambda}^{\mathrm{LCMS}}$ as a subprocedure. Since it is executed on B_Π and C_Π with $\sqrt{2}\lambda$ as the parameter, its running-time (and therefore the running-time of the entire algorithm) is $O(n^{2(\sqrt{2}\lambda)^2+1}m) = O(n^{4\lambda^2+1}m)$. $\qquad \square$

8   Conclusions

We have defined the very natural LCPS problem, proven its NP-hardness, and provided approximation algorithms for the general and a more specific case. The obvious problem remains to devise better approximation algorithms for the problem, or to prove their nonexistence.

References

1. Amir, A., Farach, M., Muthukrishnan, S.: Alphabet dependence in parameterized matching. Inf. Process. Lett. 49(3), 111–115 (1994)
2. Baker, B.S.: Parameterized pattern matching by boyer-moore-type algorithms. In: SODA, pp. 541–550 (1995)

3. Baker, B.S.: Parameterized pattern matching: Algorithms and applications. J. Comput. Syst. Sci. 52(1), 28–42 (1996)
4. Baker, B.S.: Parameterized duplication in strings: Algorithms and an application to software maintenance. SIAM J. Comput. 26(5), 1343–1362 (1997)
5. Baker, B.S.: Parameterized diff. In: SODA, pp. 854–855 (1999)
6. Cole, R., Hariharan, R.: Faster suffix tree construction with missing suffix links. In: STOC, pp. 407–415 (2000)
7. Ferragina, F., Grossi, R.: The string b-tree: A new data structure for string search in external memory and its applications. J. ACM 46(2), 236–280 (1999)
8. Hazay, C., Lewenstein, M., Sokol, D.: Approximate parameterized matching. ACM Transactions on Algorithms 3(3) (2007)
9. Hazay, C., Lewenstein, M., Tsur, D.: Two dimensional parameterized matching. In: Apostolico, A., Crochemore, M., Park, K. (eds.) CPM 2005. LNCS, vol. 3537, pp. 266–279. Springer, Heidelberg (2005)
10. Hunt, J.W., Szymanski, T.G.: A fast algorithm for computing longest subsequences. Commun. ACM 20(5), 350–353 (1977)
11. Jiang, T., Li, M.: On the approximation of shortest common supersequences and longest common subsequences. SIAM J. Comput. 24(5), 1122–1139 (1995)
12. Karp, R.M.: Reducibility among combinatorial problems. In: Complexity of Computer Computations, pp. 85–103. Plenum Press (1972)
13. Kosaraju, S.R.: Faster algorithms for the construction of parameterized suffix trees (preliminary version). In: FOCS, pp. 631–637 (1995)
14. Levenshtein, V.I.: Binary codes capable of correcting deletions, insertions, and reversals. Soviet Physics Doklady 10, 707–710 (1966)
15. Maier, D.: The complexity of some problems on subsequences and supersequences. J. ACM 25(2), 322–336 (1978)
16. Masek, W.J., Paterson, M.: A faster algorithm computing string edit distances. J. Comput. Syst. Sci. 20(1), 18–31 (1980)
17. Myers, E.W.: An o(nd) difference algorithm and its variations. Algorithmica 1(2), 251–266 (1986)
18. Nakatsu, N., Kambayashi, Y., Yajima, S.: A longest common subsequence algorithm suitable for similar text strings. Acta Inf. 18, 171–179 (1982)
19. Wagner, R.A., Fischer, M.J.: The string-to-string correction problem. J. ACM 21(1), 168–173 (1974)

Author Index

Lecture Notes in Computer Science

Sublibrary 1: Theoretical Computer Science and General Issues

For information about Vols. 1– 4707
please contact your bookseller or Springer

Vol. 4919: A. Gelbukh (Ed.), Computational Linguistics and Intelligent Text Processing. XVIII, 666 pages. 2008.

Vol. 4917: P. Stenström, M. Dubois, M. Katevenis, R. Gupta, T. Ungerer (Eds.), High Performance Embedded Architectures and Compilers. XIII, 400 pages. 2008.

Vol. 4915: A. King (Ed.), Logic-Based Program Synthesis and Transformation. X, 219 pages. 2008.

Vol. 4912: G. Barthe, C. Fournet (Eds.), Trustworthy Global Computing. XI, 401 pages. 2008.

Vol. 4910: V. Geffert, J. Karhumäki, A. Bertoni, B. Preneel, P. Návrat, M. Bieliková (Eds.), SOFSEM 2008: Theory and Practice of Computer Science. XV, 792 pages. 2008.

Vol. 4905: F. Logozzo, D.A. Peled, L.D. Zuck (Eds.), Verification, Model Checking, and Abstract Interpretation. X, 325 pages. 2008.

Vol. 4904: S. Rao, M. Chatterjee, P. Jayanti, C.S.R. Murthy, S.K. Saha (Eds.), Distributed Computing and Networking. XVIII, 588 pages. 2007.

Vol. 4878: E. Tovar, P. Tsigas, H. Fouchal (Eds.), Principles of Distributed Systems. XIII, 457 pages. 2007.

Vol. 4875: S.-H. Hong, T. Nishizeki, W. Quan (Eds.), Graph Drawing. XIII, 402 pages. 2008.

Vol. 4873: S. Aluru, M. Parashar, R. Badrinath, V.K. Prasanna (Eds.), High Performance Computing – HiPC 2007. XXIV, 663 pages. 2007.

Vol. 4863: A. Bonato, F.R.K. Chung (Eds.), Algorithms and Models for the Web-Graph. X, 217 pages. 2007.

Vol. 4860: G. Eleftherakis, P. Kefalas, G. Păun, G. Rozenberg, A. Salomaa (Eds.), Membrane Computing. IX, 453 pages. 2007.

Vol. 4855: V. Arvind, S. Prasad (Eds.), FSTTCS 2007: Foundations of Software Technology and Theoretical Computer Science. XIV, 558 pages. 2007.

Vol. 4854: L. Bougé, M. Forsell, J.L. Träff, A. Streit, W. Ziegler, M. Alexander, S. Childs (Eds.), Euro-Par 2007 Workshops: Parallel Processing. XVII, 236 pages. 2008.

Vol. 4851: S. Boztaş, H.-F.(F.) Lu (Eds.), Applied Algebra, Algebraic Algorithms and Error-Correcting Codes. XII, 368 pages. 2007.

Vol. 4848: M.H. Garzon, H. Yan (Eds.), DNA Computing. XI, 292 pages. 2008.

Vol. 4847: M. Xu, Y. Zhan, J. Cao, Y. Liu (Eds.), Advanced Parallel Processing Technologies. XIX, 767 pages. 2007.

Vol. 4846: I. Cervesato (Ed.), Advances in Computer Science – ASIAN 2007. XI, 313 pages. 2007.

Vol. 4838: T. Masuzawa, S. Tixeuil (Eds.), Stabilization, Safety, and Security of Distributed Systems. XIII, 409 pages. 2007.

Vol. 4835: T. Tokuyama (Ed.), Algorithms and Computation. XVII, 929 pages. 2007.

Vol. 4818: I. Lirkov, S. Margenov, J. Waśniewski (Eds.), Large-Scale Scientific Computing. XIV, 755 pages. 2008.

Vol. 4800: A. Avron, N. Dershowitz, A. Rabinovich (Eds.), Pillars of Computer Science. XXI, 683 pages. 2008.

Vol. 4783: J. Holub, J. Žďárek (Eds.), Implementation and Application of Automata. XIII, 324 pages. 2007.

Vol. 4782: R. Perrott, B.M. Chapman, J. Subhlok, R.F. de Mello, L.T. Yang (Eds.), High Performance Computing and Communications. XIX, 823 pages. 2007.

Vol. 4771: T. Bartz-Beielstein, M.J. Blesa Aguilera, C. Blum, B. Naujoks, A. Roli, G. Rudolph, M. Sampels (Eds.), Hybrid Metaheuristics. X, 202 pages. 2007.

Vol. 4770: V.G. Ganzha, E.W. Mayr, E.V. Vorozhtsov (Eds.), Computer Algebra in Scientific Computing. XIII, 460 pages. 2007.

Vol. 4769: A. Brandstädt, D. Kratsch, H. Müller (Eds.), Graph-Theoretic Concepts in Computer Science. XIII, 341 pages. 2007.

Vol. 4763: J.-F. Raskin, P.S. Thiagarajan (Eds.), Formal Modeling and Analysis of Timed Systems. X, 369 pages. 2007.

Vol. 4759: J. Labarta, K. Joe, T. Sato (Eds.), High-Performance Computing. XV, 524 pages. 2008.

Vol. 4750: M.L. Gavrilova, C.J.K. Tan (Eds.), Transactions on Computational Science I. XI, 181 pages. 2008.

Vol. 4746: A. Bondavalli, F. Brasileiro, S. Rajsbaum (Eds.), Dependable Computing. XV, 239 pages. 2007.

Vol. 4743: P. Thulasiraman, X. He, T.L. Xu, M.K. Denko, R.K. Thulasiram, L.T. Yang (Eds.), Frontiers of High Performance Computing and Networking ISPA 2007 Workshops. XXIX, 536 pages. 2007.

Vol. 4742: I. Stojmenovic, R.K. Thulasiram, L.T. Yang, W. Jia, M. Guo, R.F. de Mello (Eds.), Parallel and Distributed Processing and Applications. XX, 995 pages. 2007.

Vol. 4739: R. Moreno Díaz, F. Pichler, A. Quesada Arencibia (Eds.), Computer Aided Systems Theory – EUROCAST 2007. XIX, 1233 pages. 2007.

Vol. 4736: S. Winter, M. Duckham, L. Kulik, B. Kuipers (Eds.), Spatial Information Theory. XV, 455 pages. 2007.

Vol. 4732: K. Schneider, J. Brandt (Eds.), Theorem Proving in Higher Order Logics. IX, 401 pages. 2007.

Vol. 4731: A. Pelc (Ed.), Distributed Computing. XVI, 510 pages. 2007.

Vol. 4728: S. Bozapalidis, G. Rahonis (Eds.), Algebraic Informatics. VIII, 291 pages. 2007.

Vol. 4726: N. Ziviani, R. Baeza-Yates (Eds.), String Processing and Information Retrieval. XII, 311 pages. 2007.

Vol. 4719: R. Backhouse, J. Gibbons, R. Hinze, J. Jeuring (Eds.), Datatype-Generic Programming. XI, 369 pages. 2007.

Vol. 4711: C.B. Jones, Z. Liu, J. Woodcock (Eds.), Theoretical Aspects of Computing – ICTAC 2007. XI, 483 pages. 2007.

Vol. 4710: C.W. George, Z. Liu, J. Woodcock (Eds.), Domain Modeling and the Duration Calculus. XI, 237 pages. 2007.

Vol. 4708: L. Kučera, A. Kučera (Eds.), Mathematical Foundations of Computer Science 2007. XVIII, 764 pages. 2007.